Vehicular Ad Hoc Networks

T0341074

Vehicular Ad Hoc Networks

Futuristic Technologies for Interactive Modelling, Dimensioning, and Optimization

Edited by
Muhammad Arif
Guojun Wang
Mazin Abed Mohammed
Md Tabrez Nafis

CRC Press is an imprint of the
Taylor & Francis Group, an informa business

MATLAB® is a trademark of The MathWorks, Inc. and is used with permission. The MathWorks does not warrant the accuracy of the text or exercises in this book. This book's use or discussion of MATLAB® software or related products does not constitute endorsement or sponsorship by The MathWorks of a particular pedagogical approach or particular use of the MATLAB® software.

First Edition published 2023
by CRC Press
6000 Broken Sound Parkway NW, Suite 300, Boca Raton, FL 33487-2742

and by CRC Press
4 Park Square, Milton Park, Abingdon, Oxon, OX14 4RN

CRC Press is an imprint of Taylor & Francis Group, LLC

© 2023 selection and editorial matter, Muhammad Arif, Guojun Wang, Mazin Abed Mohammed, Md Tabrez Nafis]; individual chapters, the contributors

Reasonable efforts have been made to publish reliable data and information, but the author and publisher cannot assume responsibility for the validity of all materials or the consequences of their use. The authors and publishers have attempted to trace the copyright holders of all material reproduced in this publication and apologize to copyright holders if permission to publish in this form has not been obtained. If any copyright material has not been acknowledged please write and let us know so we may rectify in any future reprint.

Except as permitted under U.S. Copyright Law, no part of this book may be reprinted, reproduced, transmitted, or utilized in any form by any electronic, mechanical, or other means, now known or hereafter invented, including photocopying, microfilming, and recording, or in any information storage or retrieval system, without written permission from the publishers.

For permission to photocopy or use material electronically from this work, access www.copyright.com or contact the Copyright Clearance Center, Inc. (CCC), 222 Rosewood Drive, Danvers, MA 01923, 978-750-8400. For works that are not available on CCC please contact mpkbookspermissions@tandf.co.ukA

Trademark notice: Product or corporate names may be trademarks or registered trademarks and are used only for identification and explanation without intent to infringe.

ISBN: 978-0-367-74251-5 (hbk)
ISBN: 978-0-367-74253-9 (pbk)
ISBN: 978-1-003-15678-9 (ebk)

DOI: 10.1201/9781003156789

Typeset in Sabon
by codeMantra

Contents

Editors

Muhammad Arif is an Assistant Professor at University of Lahore, Pakistan. His research interests include Artificial intelligence, big data, cloud computing, and cyberspace security, data mining, image processing, medical image processing, Privacy, Security, and E-learning. Currently, he is working on the privacy and security of vehicular networks. Previously he was a lecturer at the University of Gujrat, Gujrat, Pakistan. He completed his master's and bachelor's degrees in Pakistan. He received his BS in Computer Science from the University of Sargodha, Pakistan in 2011. He obtained his MS in Computer Science from COMSATS Islamabad 2013 Pakistan. He completed his Ph.D. from Guangzhou University, China. He is the author of more than 100 SCIE journal publications, and more than 20 conference publications and a number of Best Paper Awards in the international conferences, including iSCI 2019. He has three SCI highly cited papers. He has more than 1600 citations, according to Google Scholar. His H-Index is 20 and the i10-Index is 40. He is the Editorial board member of International Journal of Advanced Intelligence Paradigms (IJAIP) and to International Journal of Computational Systems Engineering (IJCSysE). He has participated in many international conferences as Program committee member, Session Chair, Technical Program Committee, or International Program Committees. He is the program chair of The 2020 International Workshop on Smart Technologies for Intelligent Transportation and Communications (SmartITC 2020 and 2021).

Guojun Wang received a B.Sc. degree in Geophysics, an M.Sc. degree in Computer Science, and a Ph.D. degree in Computer Science, at Central South University, China, in 1992, 1996, 2002, respectively. He is a Pearl River Scholarship Distinguished Professor of Higher Education in Guangdong Province, a Doctoral Supervisor and Vice Dean of School of Computer Science and Cyber Engineering, Guangzhou University, China, a Vice Chairman of Intelligence Engineering Society of Guangzhou, and the Director of Institute of Computer Networks at Guangzhou University. He has been listed in Chinese Most Cited Researchers (Computer Science) by Elsevier in the past 6 consecutive years (2014–2019). His h-index is 49 (as of the date August 16, 2020). He had been a Professor at Central South University, China; an Adjunct Professor at Temple University, USA; a Visiting Scholar at Florida Atlantic University, USA; a Visiting Researcher at the University of Aizu, Japan; and a Research Fellow at the Hong Kong Polytechnic University, Hong Kong SAR, China. His research interests include artificial intelligence, big data, cloud computing, the Internet of Things (IoT), blockchain, trustworthy/dependable computing, network security, privacy preserving, recommendation systems, and smart cities. He has published more than 400 technical papers and books/chapters in the above areas, including top international journals like *ACM TAAS/TOSN/CSUR* and *IEEE TC/TPDS/TDSC*, and top international conferences like CCS/WWW/INFOCOM/CIKM/DSN/ESORICS. His research is supported by Key Project of the National Natural Science Foundation of China, the National High-Tech Research and Development Plan of China (863 Plan), the Ministry of Education Fund for Doctoral Disciplines in Higher Education, the Guangdong Provincial Natural Science Foundation, the Hunan Provincial Natural Science Foundation, and the Hunan Provincial Science and Technology Program. His research is also supported by talents programs including the Program for Pearl River Scholarship Distinguished Professor of Higher Education in Guangdong Province, the Hunan Provincial Natural Science Foundation of China for Distinguished Young Scholars, and the Program for New Century Excellent Talents in University. He is a recipient of several science and technology awards, including the 2018 Special Government Allowances of the State Council, the 2014 First Prize of the National Natural Science Award (name in the fifth place), the 2014 Second Prize of the Hunan Provincial Natural Science Award (name in the first place), the 2013 and 2003 First Prize of two Natural Science Awards of Ministry of Education of China (name in the fourth and fifth places, respectively), and several Best Paper Awards in the international conferences, including iSCI 2019, IEEE UIC 2018, IEEE IoP 2018, IEEE ISPA 2013, and IEEE TrustCom 2010. He is an associate editor or editorial board member of some international journals including IEEE Transactions on Parallel and Distributed Systems (TPDS), Security and Communication Networks (SCN), International Journal of Parallel, Emergent and Distributed Systems (IJPEDS), and International Journal of Computational Science and Engineering (IJCSE).

Mazin Abed Mohammed obtained his B.Sc. in Computer Science from the University of Anbar, Iraq in 2008. He obtained his M.Sc. in Information Technology from the College of Graduate Studies, Universiti Tenaga Nasional (UNITEN), Malaysia in 2011. He obtained his Ph.D. in Information and Communication Technology from the Faculty of Information & Communication Technology, Universiti Teknikal Malaysia Melaka, Melaka, Malaysia in 2018. He has produced more than 60 articles in journals, book chapters, conferences, and tutorials. His specialization and research interest include the areas of Artificial Intelligence, Biomedical Computing and Processing, Medical Image and Data Processing, Machine Learning, Deep Learning, Optimization Methods, and Software Medical IoT. ORCID (ID: 0000-0001-9030-8102), Google Scholar H-index 25 of 62 papers, Scopus (ID: 57192089894) H-index 19 of 45 papers, Publons (ID: E-3910-2018) H-index 14, total citations 435 of 45 papers.

Md Tabrez Nafis is working as an Assistant Professor in the Department of Computer Science & Engineering, Jamia Hamdard (Deemed University), New Delhi, India. He is having a rich experience of more than 12 years in the field of Computer Science and Engineering. His research areas are Big Data, Machine Learning, Health Informatics, and IoT. Dr Nafis has published several research papers in reputed International Journals and Conferences. He has chaired several sessions and is a member of TPC in International Conferences. Dr Nafis is an Associate Editor of the *International Journal of End-User Computing and Development* (IJEUCD), published by IGI Global, USA. He is also a member of the editorial board in the *International Journal of Big Data, AI & IoT* (IJBDAI), Published by Hikari, Bulgaria. He is a reviewer in the *Journal of Technology in Behavioral Science* (Springer), *Arabian Journal of Geosciences*, published by Springer (Scopus, SCI), and *International Journal of Big Data Intelligence and Applications* (IJBDIA) published by IGI Global, USA. Dr Nafis is a Senior Member IEEE and life member of several International/National professional bodies viz. CSI, IETE, ISTE. His research areas are Big Data, Machine Learning, Health Informatics, and IoT.

Contributors

M. Afshar Alam
Department of Computer Science and Engineering
School of Engineering Sciences & Technology (SEST)
Jamia Hamdard
New Delhi, India

Fakir Mashque Alamgir
Department of Electrical & Electronics Engineering
East West University
Dhaka, Bangladesh

Aitizaz Ali
School of Information Technology
Monash University Malaysia
Subang Jaya, Malaysia

Aleem Ali
Department of CSE
Glocal University Saharanpur
Mirzapur, India

Kaniz Amena
Department of Electrical & Electronics Engineering
United International University
Dhaka, Bangladesh

Rishabh Anand
Service Delivery
HCL Technologies Limited
New Delhi, India

Salma Masuda Binta
Department of EEE
Bangladesh Army University of Science and Technology
Saidpur, Bangladesh

Korhan Cengiz
Department of Telecommunication
Trakya University
Edirne, Turkey

Umme Rubaiyat Chowdhury
Department of Computer Science and Engineering
Daffodil International University
Dhaka, Bangladesh

Ummee Sabreen Daisy
Daffodil International University
Dhaka, Bangladesh

Pijush Dutta
Department of Electronics & Communication Engineering
Global Institute of Management & Technology
Krishnagar, India

Ong Hue Fang
School of Information Technology
Monash University Malaysia
Subang Jaya, Malaysia

Mohammad Farhan Ferdous
Japan-Bangladesh Robotics and Advance Technology Research Center (JBRATRC)
Fukui, Japan

Mohammed Hossam-E Haider
Military Institute of Science and Technology
Dhaka, Bangladesh

Mahmudul Hasan
Jahangirnagar University
Dhaka, Bangladesh

Md Hasanuzzaman
Department of Computer Science and Engineering
University of Dhaka
Dhaka, Bangladesh

Imran Hussain
Department of Computer Science and Engineering
School of Engineering Sciences & Technology (SEST)
Jamia Hamdard
New Delhi, India

Md. Shahinur Islam
Department of Electrical and Electronics Engineering
East West University
Dhaka, Bangladesh

Rasmeet Kaur
Department of CSE
Glocal University Saharanpur
Mirzapur, India

Aqeel Khalique
Department of Computer Science and Engineering
School of Engineering Sciences & Technology (SEST)
Jamia Hamdard
New Delhi, India

Tabrej A. Khan
Department of Computer Science and Engineering
School of Engineering Sciences & Technology (SEST)
Jamia Hamdard
New Delhi, India

Asok Kumar
Student welfare Department
Vidyasagar University
Medinipur, India

Syeda Florence Madina
Department of Electrical and Electronics Engineering
East West University
Dhaka, Bangladesh

Shakik Mahmud
Department of Computer Science & Engineering,
United International University
Dhaka, Bangladesh

Madhurima Majumder
Department of Computer Science Engineering
Global Institute of Management & Technology
Krishnagar, India

Nazmun Nessa Moon
Daffodil International University
Dhaka, Bangladesh

Arshil Noor
Department of Computer Science & Engineering
School of Engineering Sciences & Technology (SEST)
Jamia Hamdard
New Delhi, India

Muhammad Fermi Pasha
School of Information Technology
Monash University Malaysia
Subang Jaya, Malaysia

Manoj Roy
Department of Computer Science and Engineering
Daffodil International University
Dhaka, Bangladesh

Safdar Tanweer
Department of Computer Science and Engineering
School of Engineering Sciences & Technology (SEST)
Jamia Hamdard
New Delhi, India

Farjana Yeasmin Trisha
East West University
Dhaka, Bangladesh

Samar Wazir
Department of Computer Science & Engineering
School of Engineering Sciences & Technology (SEST)
Jamia Hamdard
New Delhi, India

Chapter 1

Security issues in blockchain as access control in electronic health records

Aitizaz Ali, Muhammad Fermi Pasha, and Ong Hue Fang
Monash University Malaysia

CONTENTS

1.1 INTRODUCTION

In the traditional symmetric key model, the data owner can encrypt the data using the symmetric key. Data owner data into some groups and then encrypts these groups using the symmetric key. Users who have the key can decrypt the encrypted data. In this scheme, authorized users are listed in the anterior cruciate ligament (ACL). The major drawback of this scheme is the number of keys grows linearly as the number of data groups increases. Also, if there any change occurs in the user and data owner relationship, it will lead to affect other users in the ACL. So, in summary, this scheme is not in practical use in different scenarios [2].

Standard ACL defines a set of rules on how to prevent different attacks. In this techniques network traffic is filtered out. Firewall mostly configured using standard ACL. This scheme follows the Internet Protocol only. Extended ACL can be called the gatekeeper of a network. It's always

DOI: 10.1201/9781003156789-1

implemented at the network edge. The extended ACL decides for granting or denying access based on port number, Internet Protocol, source, destination, and current time [3].

1.2 TYPES OF ACCESS CONTROL MODELS

In this section, we have discussed various access control models in a hierarchical order.

1.2.1 Discretionary access control and mandatory access control

Object-based models are considered as DAC (discretionary access control) model and MAC (mandatory access control) model. In DAC, an object is directly connected to a subject using the relationship between these two entities. MAC is the improved form of DAC in a way that it uses the security attributes of the subject and objects to grant access [6]. MAC is considered a standard and well-established approach in cryptography [8]. It was first designed for military purposes for controlling information. MAC is indeed based on a lattice-based information flow model. MAC has further two versions, such as Bell–LaPadula and Biba models. The Bell–LaPadula model provides information flow and confidentiality, whereas the Biba model is concerned with maintaining the integrity of the data.

In MAC, there is no concept of ownership. In other words, it can be described that in MAC, user rights and privileges are not resource-centric [9]. In order to understand more, the working of the MAC model, partial orders, and lattices must be understood. Partial orders use mathematical sets and set properties. Partial order modeling in MAC is used to match and order resources and users' attributes properly. A lattice consists of the greatest lower bound and least upper bound set. In lattice, every two elements have a least upper bound and greatest lower bound. Lattice is used in MAC when information flow is almost critical. That's the main reason it was purposely designed for the defense sector. MAC uses lattice to follow the information flow policy [7]. Information flow policy deals with the flow of information from one security level to another security level. Information flow can be monitored and maintained by assigning each object a security level or class. It was mainly used by the government and military departments. Access to the resources is controlled by the operating system as configured by the system admin. MAC controls labels and tags security levels on each resource or object. The security information can be classified as secrets, top secrets, and confidential. These security levels show management and access levels to specific information. When a user wants to access a particular object the operating system first checks the user classification according to the object security level. If the user satisfies the object sensitivity level then the access is granted otherwise rejected.

1.2.1.1 Drawbacks

MAC model provides most of the security but the overhead is so much as considered to other models. Another drawback of using MAC is its very expensive due to its system overhead. Another scheme that was proposed in the literature is the integration of public and symmetric keys. This approach is different from the various in a light way; it uses a public key to encrypt data.

1.2.2 Role-based access control

Role-based access control (RBAC) has been applied using cloud computing. RBAC provides flexibility to access control management at a level that is closed to the organization's policy and format. Secure RBAC was proposed by Zhou et al. using_ RBE (role-based encryption). In RBE, user revocation is supposed to be flexible. This approach provides users who have authorized roles can only decrypt data. Dynamic role-based access control was proposed by Wang et al. In this approach, trust was integrated with the RBAC using cloud computing. Many RBAC approaches have been introduced earlier, including, but it is observed that mostly these approaches can't provide flexible data access demands that depend upon trust. Another reason is that fine-grained access control can't be supported inside the role [10]. RBAC is easy to manage in the context of role assignment to user dynamically change with time but stable in permission assignments The main function of the role is it encapsulates lower-level permissions and privileges from a higher level. In other words, it can be concluded that roles are called pro-data abstraction. The context of roles depends upon the environment, organization, and time. In early studies, a role can be defined as a generalized hierarchy and agents that represent users who are assigned a role. Ting defined the application of role for application-level security. The modern definition of RBAC was defined by Ferraiolo and Kuhn.

1.2.2.1 RBAC types

RBAC can be divided into three major categories: (a) basic RBAC, (b) hierarchical RBAC, and (c) constrained RBAC. Basic RBAC consists of a set of users, resources, and a set of the universe of access permissions. The core behind this is to integrate subsets of access rights and permissions within named roles. When a user is assigned a role, it describes that this user has the authority to access a particular resource within the confinement of the role. When a user is assigned a role it describes that this user has the authority to access a particular resource within the confinement of the role. A role defines competency in a specific area. A virtual role or position in RBAC is a term used when a role is assigned without any direct user. For example, the role of a healthcare provider is considered a high-level abstraction for doctors and nurses. Role assignment represents a concrete scope of responsibility. At a lower level, access to resources is provided through a functional interface by a resource manager. Such an interface to a resource is called operation. It depends upon the semantics of the operation that one or more permission might be added to a role but disjoint. For example, a read and append operation be assigned to a role but disjoint. These two operations can be implied through write operation. Depending upon the access policy, the list of the permission to a particular operation can be further evaluated by the access control system.

In order to overcome the complexity and overhead in a worst-case scenario, a matrix model can be applied using RBAC. Assignment of user rights and permission can be best implemented inline using the least privilege principle. RBAC has been applied using cloud computing.

1.2.3 ABE and ABAC

Sahai and Waters [5] introduced ABE (attribute-based encryption). This was for the first new access control method proposed based on attributes. In this approach users' private key along ciphertext is associated with a set of attributes. Only a user who has the match of key associated with ciphertext can decrypt the encrypted message. The primary drawback of the Sahai and Waters [12] was the limitation to the general system and the threshold semantics which is not more expressive. To control data access in cloud computing the best mechanism is ABE which operates on the user's attributes.

By using ABE, it is easy for someone to share data according to a well-defined encryption policy without having any prior information of the data recipient. Different approaches are applied using access control, but they are based on a centralized mechanism which is more prone to vulnerability and risk.

The main problem with conventional encryption is that all data are visible to the cloud. The concept of a local domain arises for patients and users to register in order to send and receive digital health services. Leveraging consistently defined attributes, authentication and authorization activities can be executed and administered in the same or separate infrastructures, while maintaining appropriate levels of security. In Figure 1.1, the working of access control models is shown. The access control system consists of objects, subjects, and a set of policies which define access control.

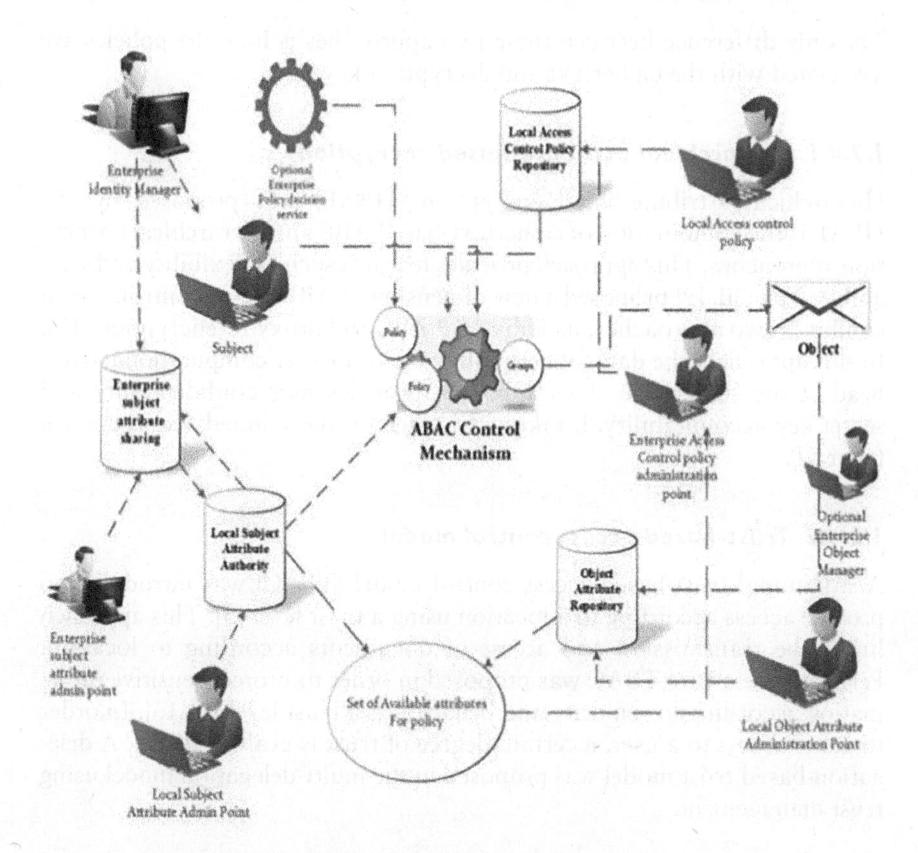

Figure 1.1 Access control system using blockchain technology.

1.2.4 ABE types

Goyal et al. proposed a more detailed and general key-policy ABE (KP ABE)-based encryption system. The construction of Goyal et al. is the extension to Sahai and Waters [7]. In this approach, a ciphertext with users' secret keys and attributes is associated with a set of monotonic structures. Pirretti et al. proposed threshold ABE. In this scheme, user revocation was introduced that how it affects the access control. Chase [11] introduced the concept of multi-authority attributes. In this construction, each authority manages a different domain of attributes. Collusion attack is the big issue while designing and implementing multi-authority attribute-based access control [4]. More generally we can define ABE as integration of security techniques with the ABAC model. Another approach is ABE to achieve flexibility and fine-grained access control. In literature, two methods have been proposed using ABE, such as KP ABE and ciphertext policy-based ABE [3].

The only difference between these two approaches is how the policies are associated with the ciphertext and decryption keys.

1.2.4.1 Hierarchical attribute-based encryption

Hierarchical attribute-based encryption (HBAE) was proposed by [5]. HBAE is the combination of ciphertext-based ABE and hierarchical encryption approaches. This approach provides features such as flexibility and scalability. Yu et al. [9] proposed a new dimension of ABE access control, which combines two approaches, including KP ABE and proxy re-encryption [13]. In this approach, the data owner is allowed to transfer computational overhead at the server-side. This approach provides user confidentiality and secret key accountability. It takes advantage of fine-grained access control from KP.

1.2.4.2 Trust-based access control model

A situational trust-based access control model (TBAC) was introduced to provide access according to a location using a trust level [3]. This approach limits the transmission and access of documents according to location. Performance aware TBAC was proposed in order to protect sensitive information according to seniority and behavior as a trust level [11,13]. In order to grant access to a user, a certain degree of trust is evaluated [12]. A delegation-based trust model was proposed in the multi-delegation model using trust management.

1.2.4.3 Bell and LaPadula

Bell and LaPadula [BELL75, MCLE88] proposed and modified the concept of MAC, which is quite similar to the Denning workflow model. It was observed that the Bell–LaPadula model is the extension to Denning and MAC based on information flow. MAC in the Bell–LaPadula model provides security classification to object and subject. Labeled assignment to object is called security classification, and those labeled which are assigned to subjects are referred to as security clearance. Bell–LaPadula approach follows two rules: the simple security policy and property policy which is also called the star policy. It's a kind of state machine used to enforce access control policy for military operations. It was developed by David Elliott Bell and Leonard J. LaPadula and widely used to preserve confidentiality only. That's the main reason that it's widely used in government organizations and defense to protect secret and confidential data.

Figure 1.2 describes the comparative analysis of ABAC and RBAC based on specific parameters. This classification is done using certain factors such as flexibility, security flaw, efficiency, authorization, and modification. It's

Issues	RBAC	ABAC
Trend in 2018	Medium	High
Global Agreement	No	Yes
Flexibility	No	Yes
Easiness	Yes	No
Dynamicity	No	Yes
Authorization Decision	Locally	Globally
Granularity	Low	High
Manageability	Simple	Complex
Conviction	Locally	Globally
Confusing deputy	No	Yes
Changing privileges	Complex	Simple
Role explosion problem	Yes	No

Figure 1.2 Comparative analysis of access control model used in blockchain.

observed from the DAC model analysis that this model is vulnerable to a Trojan horse attack. The only flexible model is RBAC. So it can be derived from the comparison table that the only rigid model is MAC whereas DAC and RBAC are considered as flexible.

1.3 CONCLUSION

The application of access control in digital healthcare systems plays an important role in the present healthcare industry, which can lead to automated data collection and verification processes, aggregated and correct data from different resources which can be immutable and tamper-resistant, and deliver secured data which helps in reduction of probability of cybercrime. Blockchain technology also supports distributed data, redundancy, and fault tolerance features for digital systems. In this chapter, current challenges and problems in the literature faced by the digital healthcare industry will be solved. In this chapter, we have compared famous models of access control with their merits and demerits. We have also explained the comparison between RBAC and ABAC based on various parameters. In conclusion, the ABAC is most trustworthy instead of RBAC because ABAC dynamicity made it more applicable and trustworthy. So, the users can facilitate themselves with the benefits of both models at a time as discussed in this chapter. In the future, we have a plan to propose a hybrid model that will overcome blockchain, RBAC, and ABAC flaws, in terms of dynamicity, flexibility, tight security, trust, and ease of access.

REFERENCES

1. A. Azaria, A. Ekblaw, T. Vieira, and A. Lippman. "Medrec: Using blockchain for medical data access and permission management", In *2016 2nd International Conference on Open and Big Data (OBD)*, pp. 25–30, 2016.
2. S. Azouvi, M. Maller, and S. Meiklejohn. "Egalitarian society or benevolent dictatorship, the state of crypto currency governance", In *International Conference on Financial Cryptography and Data Security*, pp. 127–143, 2018.
3. A. Balu and K. Kuppusamy. "Ciphertext policy attribute based encryption with anonymous access policy", arXiv preprint arXiv: 1011.0527, 2010.
4. P. Beninger and M.A. Ibara. "Pharmacovigilance and biomedical informatics: a model for future development", *Clinical Therapeutics*, 38(12):2514–2525, 2016.
5. J. Bethencourt, A. Sahai, and B. Waters. "Ciphertext-policy attribute-based encryption", In *IEEE Symposium on Security and Privacy (SP'07)*, pp. 321–334, 2007.
6. M. Blaze, S. Kannan, I. Lee, O. Sokolsky, J.M. Smith, A.D Keromytis, and W. Lee. "Dynamic trust management", *Computer*, 42(2):44–52, 2009.
7. N. Broberg and D. Sands. Para locks, "role-based information flow control and beyond", *ACM Sigplan Notices*, 45(1):431–444, 2010.
8. C. Cachin. "Architecture of the hyperledger blockchain fabric", In *Workshop on Distributed Cryptocurrencies and Consensus Ledgers*, 310, p. 4, 2016.
9. M. Chase. "Multi-authority attribute based encryption", In *Theory of Cryptography Conference*, pp 515–534, 2007.
10. R. Chinchani, A. Iyer, H.Q. Ngo, and S. Upad-hyaya. "Towards a theory of insider threat assessment", In *International Conference on Dependable Systems and Networks (DSN'05)*, pp. 108–117, 2005.
11. A. Das, M.M. Islam, and G. Sorwar. "Dynamic trust model for reliable transactions in multi-agent systems", In *13th International Conference on Advanced Communication Technology (ICACT2011)*, pp. 1101–1106, 2011.
12. N. Dimmock, J. Bacon, D. Ingram, and K. Moody. "Risk models for trust-based access control (TBAC)", In *International Conference on Trust Management*, pp. 364–371, 2005.
13. J.-M. Do, Y.-J. Song, and N. Park. "Attribute based proxy re-encryption for data confidentiality in cloud computing environments", In *2011 First ACIS/JNU International Conference on Computers, Networks, Systems and Industrial Engineering*, pp. 248–251, 2011.
14. A.A. Alan. *Donovan and Brian W Kernighan. The Go Programming Language*, Addison- Wesley Professional, 2015.

Chapter 2

Applied current software methodology of software industries in Bangladesh

Ummee Sabreen Daisy and Nazmun Nessa Moon
Daffodil International University

Mohammad Farhan Ferdous
Japan–Bangladesh Robotics and Advance Technology Research Center (JBRATRC)

Muhammad Arif
University of Lahore

CONTENTS

DOI: 10.1201/9781003156789-2

2.1 PART 1: INTRODUCTION

2.1.1 Introduction

Bangladesh has billed itself on traditions of science and technology. Today, the number of information technology (IT) companies stands more than eleven hundred according to the Association of Software and Information Service (BASIS) [1]. Besides, the government set up some software technology parks in different locations. For that, a number of IT firms stand in the future. So, it is a perfect time to know the appliance current software method which is followed internally by our software industry. However, the survey system is closely related to some other aspects. The first and foremost one is surveying data. To provide legitimate solicitations, getting surveying data and analyzing it is rudimentary. This Chapter keeps an effort to get a practical idea about the usage of a method during the development of a project.

2.1.2 Motivation

Every project has some common characteristics, in which an objective to be gained, the uniqueness of the project, the impermanent effort, and the attendance. In order to conduct a project is essential to apply some skills of project arrangement with the motive of achieving the target, especially in the case of raising project size and tangles.

The main point of project management is investigated in detail in the writing, in the theoretical and technological point of view, including the use of methodologies for the development of IT solutions.

Now there are methodologies, also called software development models, which could be chosen to manage IT projects.

2.1.3 Fundamentals of the study

In Bangladesh, there are many IT companies and increasing day by day. There are a few standards of their usage for software users in Bangladesh so as to know appliance software methods in Bangladeshi IT companies. Some key problems have been identified as follows.

Firstly, the first and foremost one is data. So effective ways of communicating with the users and getting an authentic response from them to get a thorough idea about their using methods are the main focus.

Secondly, after understanding the respondent data, the necessary criterion is done. Based on the criterion, the possibility of using Agile methodologies in Bangladesh can be understood. If it is possible, then necessary data that users are willing to share is collected.

Finally, different information filtering algorithms which can be applied to user-given data to provide personalized recommendations are studied. To come up with the best techniques of data mining and relevant research are the focuses of this phase.

2.1.4 Research questions

1. Which programming languages they are using?
2. Which development environments they are using?
3. Did they use a formal system development method?
4. Did they have any requirement analysis team?

2.1.5 Expected output

1. To know which programming languages they are using
2. To know which development environments they are using.
3. To discover which formal systems development method they use.
4. To identify which requirement analysis team they have.

2.1.6 Information design

The following description is supplied to know which Section covers which topics and their topical discussions.

1. In Section 2.1 "Introduction", the basic things of this Chapter are already described.
2. In Section 2.2 "Background", a brief discussion about methods is included.
3. In Section 2.3 "Research Methodology", a brief discussion will have made on how this study has been conducted.
4. In Section 2.4 "Experimental Results and Discussion", various aspects of surveys and how the survey has been set up to understand the scenario will be elaborately discussed.
5. In Section 2.5 "Summary of the study, conclusion and recommendations", a summary of how this Chapter can be used and continued to solve the remaining problems in the statement are discussed. The chapter "Conclusion" shows the goals of this Chapter.

2.2 PART 2: BACKGROUND

2.2.1 Introduction

This chapter will discuss related works connected to the software development methodology of software industries in Bangladesh. The first section will discuss prior studies. The second section will discuss the definition, benefits and difficulty, and conclusion.

2.2.2 Related works

As computer technology offers efficient and high-performance information processing, it has gained popularity among home and office users in the whole world. By the decade of 1990, in Bangladesh, it has also taken an important role. Since during this time personal computers become more user-friendly and attractive, the number of users had been increased. Besides the general users, in Bangladesh, the number of software developers has been increased as well. Many computer science and engineering graduates from public and private universities, as well as computer diplomas from training institutions, are getting employed at the local software companies. As time goes, the overall development of the skill of software developers has been increased with respect to Bangladesh.

Bangladesh stands out distinctly as a potential software-exporting nation, considering the analytical and technological ability of its people. Bangladesh is one of the potential countries where software development is to be grown as a software industry. According to Bangladesh BASIS, there are around 300 plus companies, which are working closely with the development of software for the local and international market for different information and communication technology services [1]. Bangladesh is a country, where the only surplus property is the human resource. Considering the earning of foreign exchanges and removing of the unemployment problem, the is software industry a very prospective field. To make this field more profitable, several plans have been done by the government and private organizations in the last several years. The Government of Bangladesh made an in-depth study on how the software sector of the country could be designed to suit the needs of the global market. To follow up on the outcome of the study and to monitor the issues associated with the sector's growth and development, a high-powered National Standing Committee on software export has been formed. This standing committee has brought together the concerned government offices organizations and leaders of the software trade to work in unison to study the problems and prospects of Sector 2. Table 2.1 shows the business application nature of software service of the software industry in Bangladesh. Notably, each software company in the software industry develops multiple categories of software services.

2.2.3 Research summary

There are many types of software development methodology. To manage a project effectively, the manager or development team must experiment with many software development methodologies to choose the one that will work best for the project at hand. Every methodology has a different capacity and bug and exists for different reasons. There is an overview of the most popular used software development methodologies and why different methodologies attend.

2.2.4 Different types of methodologies

2.2.4.1 Waterfall development methodology

The waterfall method is considered the traditional software development methodology. It's a strong linear method that stands for sequential phases (requirements, design, implementation, verification, maintenance) in which distinct goals are completed. Every phase must be 100% completed before the next phase starts, and traditionally there is no process for going back to modify the project or way. Figure 2.1 shows waterfall methodologies.

Table 2.1 Products/Service Range of Local Software Industry

Products/Services Category	*% of Companies Offering Services*
Accounting & Financial Management	69%
Inventory Management	59%
Human Resource Software	58%
Web Site/Web Application Development	57%
ERP (Enterprise Resource Planning)	48%
Software Implementation & Integration	46%
Billing	43%
Asset Management	38%
POS (Point of Sales)	37%
E-Commerce	36%
Data Entry/Data Conversion	34%
CRM (Customer Relationship Management)	32%
E-Governance Application	29%
SCM (Supply Chain Management)	27%
Data Warehousing	23%
Access Control	22%
Mobile/Wireless Application Development	18%
E-Learning	17%
Data Security	14%
Gaming Software	6%

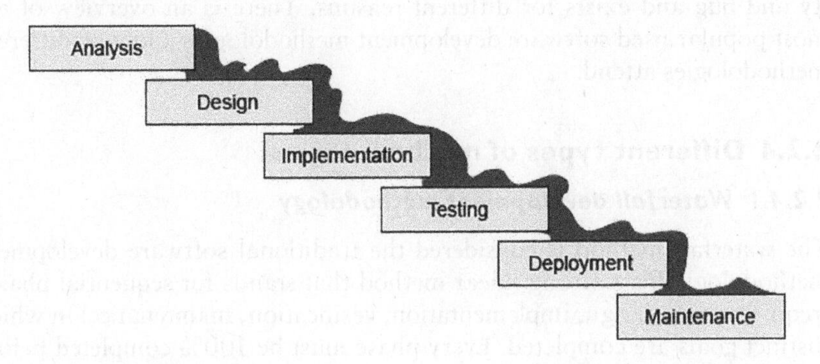

Figure 2.1 Waterfall methodology.

2.2.4.2 *Rapid application development methodology*

Rapid application development (RAD) methodology is a summarized development process that yields a high-quality system with low sending costs. In a Forbes article, Scott Stainer, CEO and president of UM Technologies, said, "This RAD process allows our developers to quickly adjust to shifting requirements in a fast-paced and constantly changing market." The power to quickly adjust is what allows such a low commission cost. The rapid application development methodology consists of four phases: requirements planning, user design, construction, and cutover. The user design and construction phases repeat until the user confirms that the product meets all demands. Figure 2.2 shows rapid application development methodologies.

2.2.4.3 *Agile development methodology (ADM)*

There are many forms of the Agile development methodology, such as Scrum, crystal, extreme programming (XP), and feature-driven development (Figure 2.3).

2.2.4.4 *DevOps development methodology*

DevOps development is centered on an organic change that improves the assistance between the departments subject for different segments of the development life cycle, such as development, quality consolation, and operations. Figure 2.4 shows DevOps development methodologies.

2.2.4.5 *Scrum*

This methodology is limber on how little performance but the Scrum philosophy would guide a team on the part of little performance as possible. Generally, a Scrum team works co-area. Yet, there have been Scrum teams that work geographically distributed whereby team members take part in

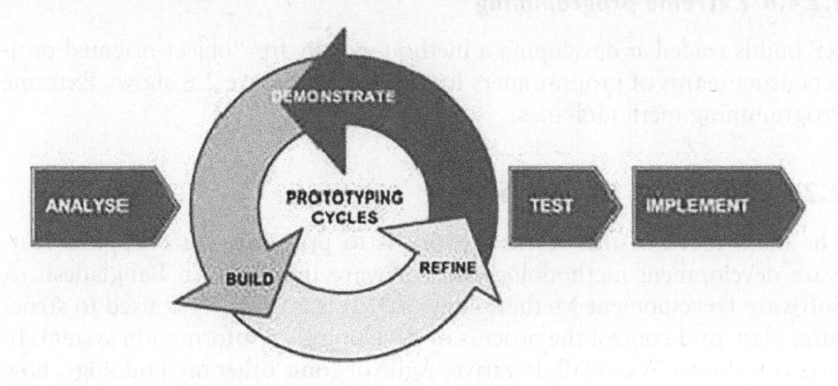

Figure 2.2 Rapid application development methodology.

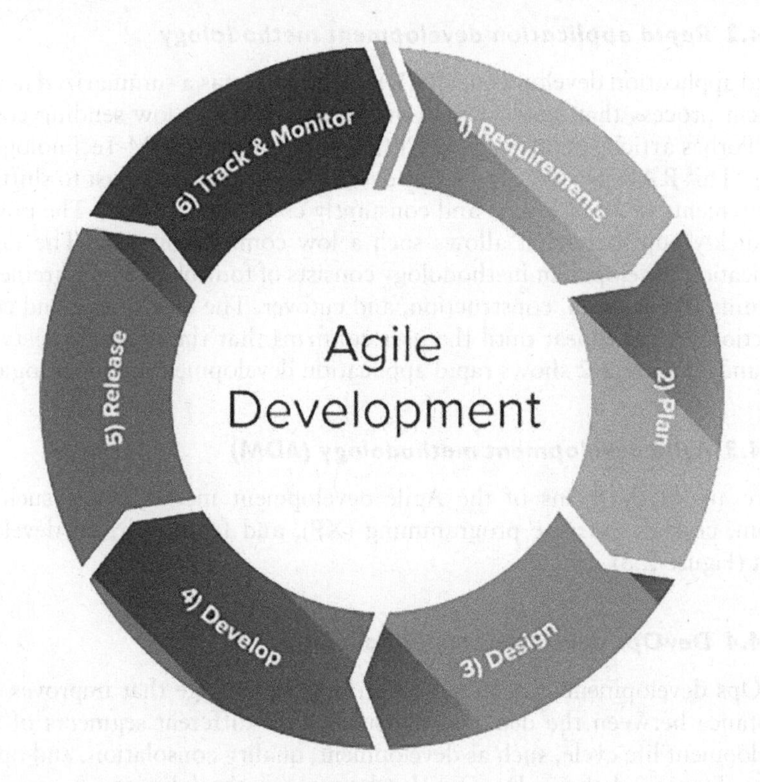

Figure 2.3 Agile development methodology.

a daily meeting via speakerphone. Scrum teams are self-directed and self-forming teams as shown in Figure 2.5.

2.2.4.6 Extreme programming

XP builds traced at developing a method worthy for "object-oriented projects using teams of programmers in one point". Figure 2.6 shows Extreme Programming methodologies.

2.2.5 Object of the problem

The main focus of this research work is to primarily study applied software development methodology on software industries in Bangladesh. A Software Development Methodology (SDM) is a framework used to structure, plan, and control the process of developing an information system. In case you choose Waterfall, Iterative, Agile or some other methodology, how well you adhere to the SDM can practically condition the success or failure of a project and/or company.

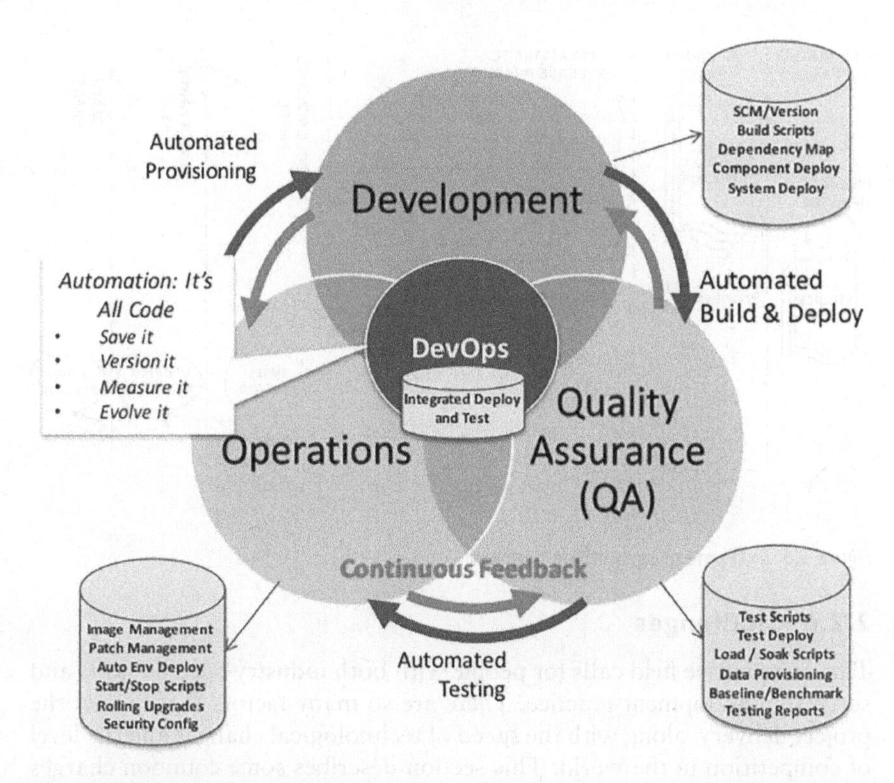

Figure 2.4 DevOps methodology.

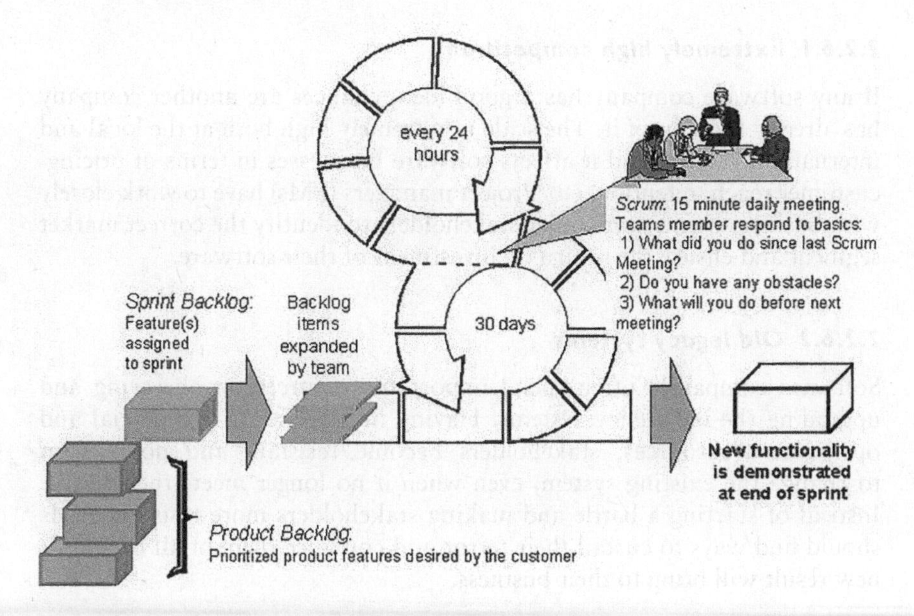

Figure 2.5 The Scrum process methodology.

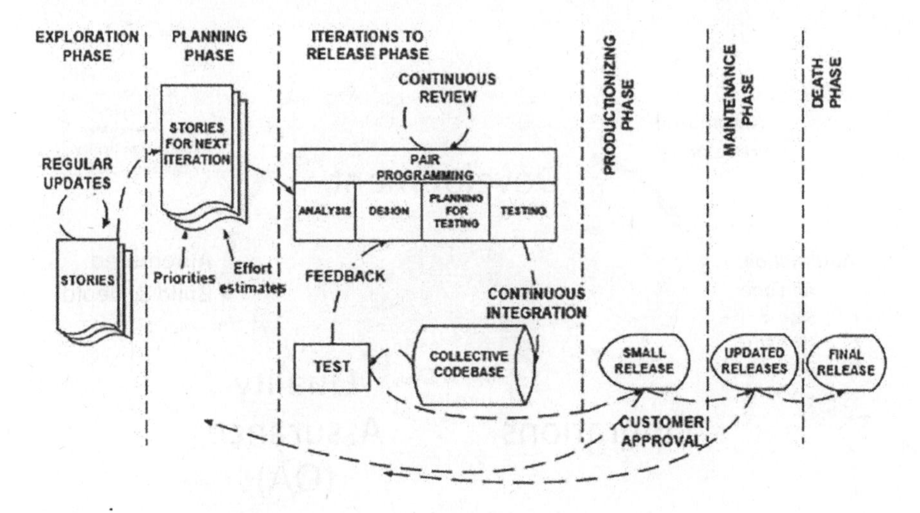

Figure 2.6 Extreme programming process.

2.2.6 Challenges

This competitive field calls for people with both industry-specific skills and software development practice. There are so many factors that impact the project delivery, along with the speed of technological changes and the level of competition in the world. This section describes some common charges that affect specific software project management.

2.2.6.1 Extremely high competition

If any software company has a good idea, chances are another company has already thought of it. The scale is extremely high both at the local and international levels, and it affects software businesses in terms of pricing, customer reach, retention, etc. Project managers (PMs) have to work closely with business owners and other stakeholders to identify the correct market segment and ensure the return on investment of their software.

2.2.6.2 Old legacy systems

Software companies often spend important resources on observing and upgrading the old legacy systems. Having invested a lot in financial and open-armed resources, stakeholders become resistant and don't want to change the existing system, even when it no longer meets their needs. Instead of starting a battle and making stakeholders more resistant, PMs should find ways to curtail their terror and convince them of all benefits a new result will bring to their business.

2.2.6.3 *High-level software expertise*

When comes to software taste and hanging, the best difference for business owners is innovation project managers with the related software worship. The more complex the software system is, the more exercise and the more specific skill set will be necessary for its simulacrum (e.g. think of large enterprise resource planning systems).

2.2.6.4 *Third-party integration*

The best companies are no longer interested in standalone solutions and are looking for third-party integration. Generally, it looks like implementing different systems in one project. For example, a PM implements a financial management system with accounting and reporting modules which interface with customer relationship management and contract management software. This puts PMs under pressure and makes them improve their practice and learn more than other software that integrates with the solution they are performing.

2.2.6.5 *Multiple-level users*

Almost all companies look for systems that allow various types of users – from basic users to strictly IT users. Project managers, who are responsible for the system execution, must be familiar with all types of users and know what user rights and permissions should be ascribed to each.

2.2.6.6 *Quality testing*

Every successful system execution requires numerous testing iterations to ensure that the outcomes align with the desired results. Project managers need to make sure all errors are discovered and all things are fixed before the system goes live. This is essential to ignore additional rework and ensure customer pleasure.

2.3 PART 3: RESEARCH METHODOLOGY

2.3.1 Introduction

This chapter gives the road map to win the research target. This research methodology was founded on a comprehensive survey directed through structured interviews and an online survey in the software development area. Because the main thing of the study and the research is to understand the usage software methodology of the software industry in the Bangladeshi context. To successfully conduct this study, the below steps were taken.

1. Various papers on software methodology based on the Agile method were studied, which were published.
2. Various papers and books on survey methods were studied; there more than 30 studies were identified, and 25 studies were selected for the final review.
3. The key factors which are needed to understand were identified are paper title, abstract, keywords list, literature review, survey, and so on.
4. The questions relevant to those facts were identified.
5. Desired answers were divided into quantitative and qualitative data. Based on the Agile Method especially Scrum, XP and DSDM, relevant questionnaires were developed.
6. The survey setup and survey conduction plan were developed.
7. Finally, a complete set of questionnaires and survey conduction plan is proposed

Which ultimately will be used to understand the usage software methodology of the software industry in the Bangladeshi context?

2.3.2 Research subject and instrumentation

Look at that a survey is only as good as a question it asks, hence the questionnaire is a critical stage in the survey research method, the all questions must be relevant and right in trying to capture the essence of the research things. To achieve these ends, a researcher will be required to make several decisions:

1. How should be asked?
2. How should each question be phrased?
3. In what sequence should the questions be arranged?
4. What questionnaire layout will best serve the researcher's objectives?
5. How should the questionnaire be pretested?
6. Does the questionnaire need to be revised?

2.3.3 Data collection procedure

The collection of data instrument is a questionnaire. It catches questions formulated based on the research questions (main and sub-questions), literature review and the theoretical positioning presented in this study. The questions were set in a way that examines the connection between software developers and development methodologies. Collected data were quantified, for instance, with observation to time spent of development methodology on the software industry in BD.

2.3.4 Statistical analysis

Statistical Analysis Software (SAS) is used all around the world in approximately 118 countries to solve complex business problems. Much of the software is driven by either menu or order. Like other programming software, SAS has its language that can control the program during its hanging.

The professionals at statistics solutions are adept in SAS and have helped thousands of doctoral candidates, master's candidates and researchers, and so on. You get proven techniques that produce accurate results, now and in the future. Figure 2.7 shows statistical analysis in software methodologies.

2.3.5 Reliability and validity

Protecting these two aspects of a study is very important. Since reliability shows the need that a study produces results that will be affirmed consistently by subsequent same as studies, validity or trustworthiness of a study requires that the instrument applied correctly obtains the type of data that it is meant to be gathered. A researcher should work objectively and carefully to ensure the fact of these two aspects of research by following applicable scientific methodology. Initially, the tool was pretested with up to 15 randomly selected software firms in Bangladesh. Doing this was viewed as a way of helping the researcher to calculate the validity of the instrument.

2.3.6 Research design

"A prerequisite to designing a good survey instrument is deciding what is to be measured". So, a research project has some survey objectives and concepts. They are given below:

1. A survey concerns a set of objects comprising various software firms.
2. The firms under work have one or more quantitative properties.
3. The main focus of the project is to describe the population by one or more parameters defined in terms of quantitative properties. This may require observing as wide of the population.
4. A sample of motive is selected from the frame in conformity with a sampling design that specifies probability machinery and sample size.
5. Watching is made on the sample in accordance with a measurement process (i.e. a measurement method and a prescription as to its use).
6. Based on the amount, a respective process is applied to compute estimates of the parameters when concluding the sample to the population.
7. The general motive of the survey, as well as the main model characteristics, i.e. the focus stated.

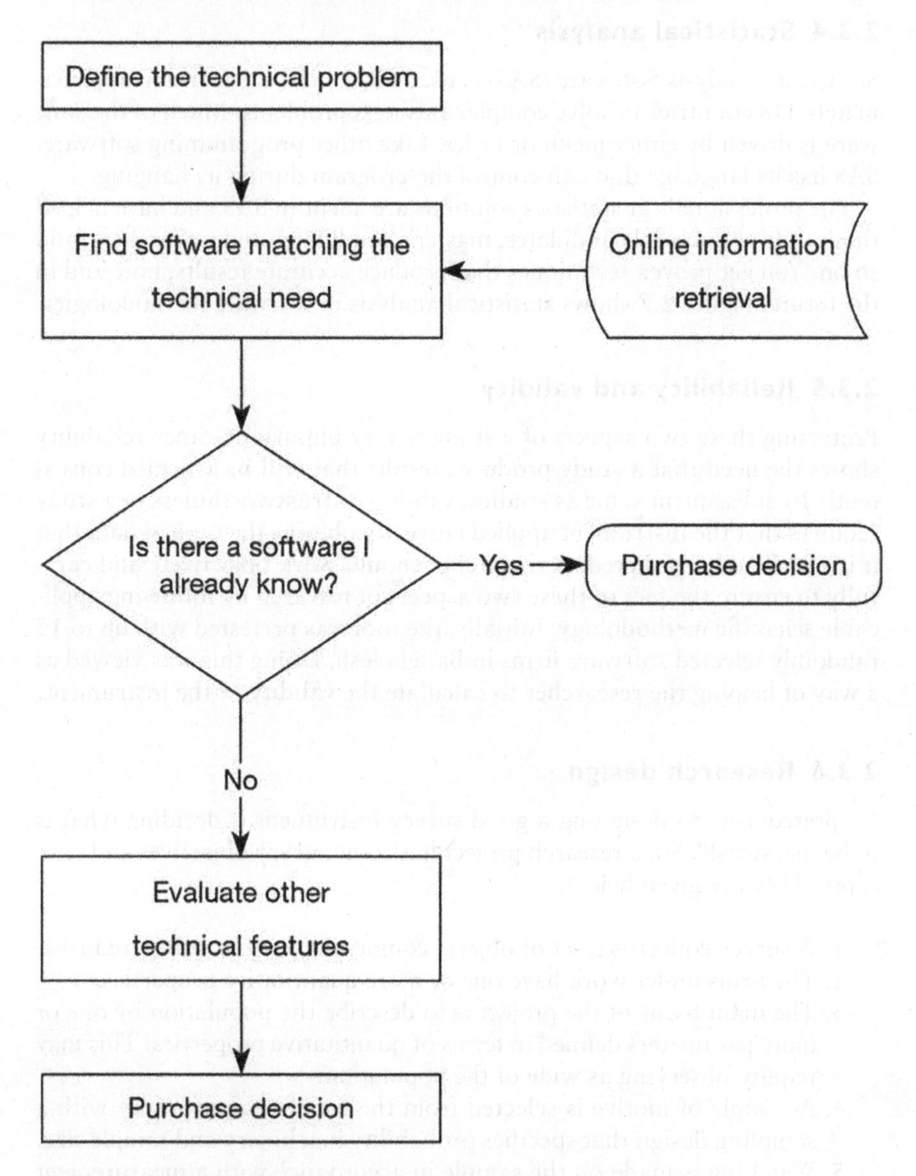

Figure 2.7 Statistical analysis in software methodology.

2.4 PART 4: EXPERIMENTAL RESULTS AND DISCUSSION

2.4.1 Introduction

This chapter will cover the presentation of data analysis and the purpose of survey results. The data exploration and purpose were based on the research objectives. The presentation and solution of the collected data were computed using rotation and share.

2.4.2 Experimental results

While showing the distribution of the respondents on many questions, tables and graphs were used in the section of data. The answer has the same questionnaire given them the sample size of the companies was 50 respondents, while the target companies are 60.

2.4.3 Descriptive analysis

2.4.3.1 The survey objective

Understanding the usage regarding the choice of specific Agile methodology.

A table is useful for identifying redundant or unnecessary questions in the questionnaire or unusual research questions. Table 2.2 shows communication between research questions and survey questions.

2.4.3.2 The target population

Profile of software industry in Bangladesh. The next step in the survey process is to define the population to be studied or the target population for the survey. The target companies or the group of persons or other units for whom the study results will apply. In Bangladesh, maximum IT companies are established after 2000. While a few IT companies are established before 2000. A little amount of IT companies is not certified from BASIS, besides all companies are citified. Not only certified BASIS but also they follow CMMI maturity level. Table 2.3 shows the "profile" of the software industry in Bangladesh.

2.4.3.3 The mode of administration

Collecting specified research motive and identifying the target population, the process is to determine the mode of administration for the survey. For developing the survey, two types of interviews were conveyed.

2.4.3.4 Face-to-face interview survey

Face to face interview technique implies the paper questionnaire and the presence of the interviewer. A total of 25 software companies were initially selected for the survey. A total of 16 companies responded. Only eight of them didn't respond to the survey. Figure 2.8 shows face to face interview survey.

Table 2.2 Communication between research questions and survey questions

Part of questionnaire	*First*	*Second*	*Third*	*Fourth*	*Fifth*
Survey question	7	4	6	3	3
Research question	5	3	6	2	3

Table 2.3 "Profile" of the software industry in Bangladesh

Serial No.	*Company Namea*	*Address*	*Established year*		*CMMI level*	*ISO certified*	*IP maintain*	*BASIS certified*
			Before 2000	*2000–2017*				
1.	RAIT Ltd.	Dhaka		✓	Level-1	✓	Copyrights, **Trade Secret Patent**	✓
2.	IPCP Services	Dhaka	✓		Level-2	✓		✓
3.	MKB Technologies	Rangpur		✓	Level-3	✓	Copyrights	✓
4.	ZSI Bd.	Rangpur		✓	Level-4	✓		✓
5.	Banglalink	Dhaka	✓		Level-5	✓	Copyrights, **Trade Secret**	✓
6.	PCN (Pvt) Ltd.	Dhaka	✓		Level-5	✓		✓

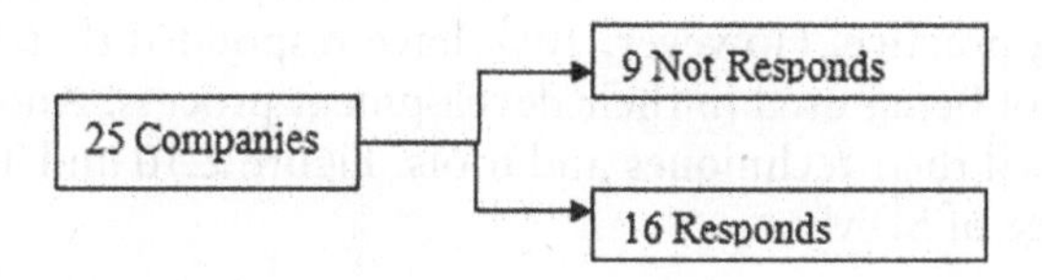

Figure 2.8 Face-to-face interview survey.

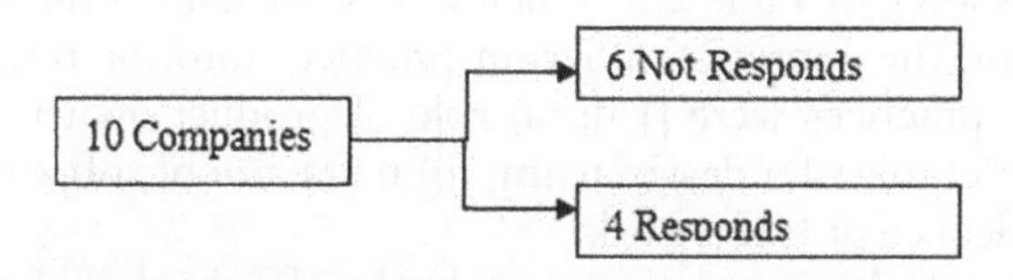

Figure 2.9 Online survey (mail & telephone).

2.4.3.5 Online survey (mail & telephone)

In this survey, paper questionnaires are sent to be answered by mail. A total of ten software companies were initially selected for the survey. There were four companies that responded. Only six of them didn't respond to the survey. Figure 2.9 shows an online survey.

2.4.4 Analysis

2.4.4.1 Adoption of SDM

The answered person was asked whether or not they were subsequent a Software Development Method.

As displayed in Figure 2.10, 65% of the respondents are currently using and involved with Agile methodologies in their developing process. This

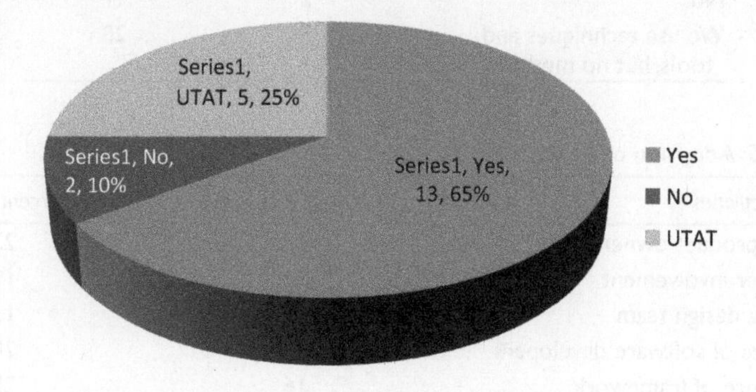

Figure 2.10 Percentage of SDM (use techniques and tools).

shows that almost 3/5 of the companies are adopting Agile methodologies as a working practice. However, 10% have responded that Agile methodologies are not being used in their development process. And 25% of companies are used their techniques and tools. Figure 2.10 and Table 2.4 show the percentage of SDM.

2.4.4.2 Adoption of Scrum

The respondents that answered "yes" were then asked the questionnaire. The result is shown in Table 2.3. When visible the data of the separate practices of Scrum, the appreciated Scrum practices into the respondents that had used the practices were [13]: (a) role of product owner, (b) customer involvement, (c) role of a design team, (d) team size of software developers and (e) significance of framework.

Most understood Scrum practice seemed to be the daily Scrum meeting (65%) as shown in Table 2.5. They also used a burndown chart for completing the project and divided it into a small group if the team size is large.

2.4.4.3 Adoption of XP

Compared with Scrum, the appreciated XP practices into the respondents that had applied the exercise were: (a) role of a tracker, (b) customer involvement, (c) role of programmer, (d) team size of software developers and (e) significance of method as shown in Table 2.6. The most used practices of XP are the 40 h week (65%) and pair programming (67%) as shown in Table 2.6, all of which can be examined in any process methodology of software development where Agile or traditional.

Table 2.4 Adoption of SDM

Answer	*Number*	*Percent (%)*
Yes	13	65
No	2	1
We use techniques and tools, but no method	5	25

Table 2.5 Adoption of Scrum

Scrum activities	*No. of responses*	*Percent (%)*
Role of product owner	15	23
Customer involvement	12	19
Role of a design team	8	13
Team size of software developers	13	20
Significance of framework	16	25
Total	64	100

Table 2.6 Adoption of XP

XP activities	*No. of responses*	*Percent (%)*
Role of tracker	1	4
Customer involvement	3	11
Role of programmer	7	2
Team size of software developers	3	11
Significance of method	13	48
Total	27	100

Table 2.7 Adoption of DSDM

DSDM activities		*No. of responses*	*Percent (%)*
Delivery time	Fixed	3	15
Resource	Fixed	3	15
Functionality	Changeable	14	70
Total		20	100

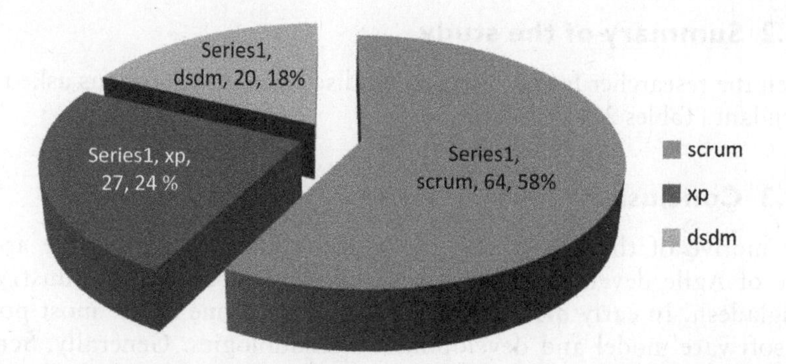

Figure 2.11 The percentages of different Agile methods used by the respondents.

2.4.4.4 Adoption of DSDM

Table 2.7 shows the results on fundamental activities of the DSDM development process. The appreciated DSDM examine into the respondents that had examined the practices were: (a) delivery time, (b) resource and (c) functionality. Negative experiences were reported of DSDM practices such as delivery time (57% negative responses) and resource (57% negative responses) as shown in Table 2.7. Because maximum respondents shared their opinion that it depends on project size.

The main cause of the survey decision is that different types of Agile methods are used in Bangladeshi IT companies. It was found that "Scrum" was leading the way at 58%, followed by "XP" at 24% and "DSDM" at 18% as shown in Figure 2.11.

2.4.5 Summary

This chapter discusses results of the survey of the results of the survey including exploration of empiric results, and also descriptive experiments, in this research questions I get huge responses that make research helpful the result.

2.5 PART 5: SUMMARY OF THE STUDY, CONCLUSION AND RECOMMENDATIONS

2.5.1 Introduction

This chapter will discuss the noticing of the final results, conclusion and recommendation of this survey: first, it will be discussed about the major decision of each study as confirmed in the research motive; second, the conclusion from the noticing of the study; and last, the researchers will give hint and recommendation of these study areas for further research.

2.5.2 Summary of the study

When the researcher found, focused on discussing the questions asked the defendant (Tables 2.8 and 2.9).

2.5.3 Conclusion

The motive of this study was to exhibit conviction about the appliance of Agile development methodologies in the software industry in Bangladesh. In early discussion, Agile is argued one of the most popular software model and development methodologies. Generally, Scrum (58%), XP (24%) and DSDM (18%) are one of the most popular software models and development methodology in Bangladesh. It could be argued based on the results of this study that studded software development firms look to be able to apply the three Agile methods, namely Scrum, XP and DSDM and their individual examine in their projects and report fairly positive results of their application.

As explained previously, the result shows that the software industry in Bangladesh is methodological and technological. This study adds evidence not only to the IT industry in Bangladesh but also to the knowledge of software engineering and software processes.

We accept that there is a limit to research. Because there are some company policies, some firms did not want to share their information. In hindsight, we recognized that the questionnaire could have been raised. More general

Table 2.8 Questionnaire with dataset

Question	*Dataset*
Do you follow the software development method?	a. Yes=65% (13) b. No=10% (2) They use techniques and tools, but no method=25% (5)
Who gathers user stories?	c. Product owner=71%(15) d. Tracker=19%(1) e. Others=10%(2)
How about your customer involvement?	a. once and again=70%(12) b. on-site customer=18%(3) c. Others=12%(2)
Who makes the planning/design part?	a. Design team=50%(8) b. Programmer=44%(7) c. Others=6%(1)
What is the number of software developers in the development department?	a. 1–10=68%(13) b. 11–20=16%(3) c. 21–50=5%(1) d. 50+=11%(2)
Which is significant according to development/coding?	a. method=6%(1) b. framework=18%(3) c. all of the above=76%(13)
Do you hold a daily Scrum meeting?	a. Yes=65%(11) b. No=35%(6)
Do you use a burndown chart for completing the project?	a. Yes=44%(7) b. No=56%(9)
If the team size is large, what do you do?	a. divided into small group=75%(12) b. unchangeable=25%(4)
Approximate working hours per week.	a. 30 hours=6%(1) b. 40"=65%(11) c. 50"=12%(2) d. others:6%(3)
Do you allow pair-programming?	a. Yes=67%(10) b. No=33%(5)
According to your project, which is changeable/fixed?	a. Time(Changeable=57%(4), Fixed=43%(3)) b. Cost (Changeable=57%(4), Fixed=43%(3)) c. Functionality/scope (Changeable=93%(14), Fixed=7%(1))

questions on the method activities could have generated more information on this issue. Yet, my survey report represents most of the Bangladeshi software development area.

Table 2.9 Questionnaires with a dataset for a company

Question	*Dataset*
Established date	a. Before 2000=15%(3) b. 2000–2017=85%(17)
What is your company/organization's main industrial section?	a. Authority b. e-commerce c. Usability, HCI d. Health and medicine sector e. IT consultancy=80% (16) f. Financial sector=5% (1) g. Telecommunication h. Game=10% (2) i. Others=5% (1)
Please select the total number of employees in your company/organization.	a. 1–20=45% (9) b. 21–50=40% (8) c. 51+=15% (3)
CMMI stages maturity level	a. Level-1 (initial) =20% (4) b. Level-2 (managed)=25% (5) c. Level-3 (defined)=30% (6) d. Level-4 (quantitatively)=20% (4) e. Level-5 (optimizing)=5% (1)
Membership of BASIS	a. Yes=60% (12) b. No=40% (8)

2.5.4 Recommendations and future work

In this research, there is a possibility to future relevant to this research, which can be hope for further extensions. Now we know which methodologies are used, and to learn how to identify and alleviate some of the more specific problems they face deploying the Agile method. The results of my study set to the understanding of how Agile methodologies are being implemented in software firms.

My finding results show that there is a positive trend to appliance software development methodology, the importance of appliance current software methodology is perceived to develop the project. It would be interesting to explore the future. My survey executes that company's access software development methodology and that's for the need to know about the usability of software development methodologies. It also would be interesting to explore the future.

The recommendations are following:

1. They need to be developed their skills in software development methodology.
2. Need to establish viable training facilities.
3. Need to improve their working environment.

There need to know about the usability of software development methodologies.

APPENDIX 2A: QUESTIONNAIRE

Company/organization name:...

1. Established date: ....

Serial	Questionnaire	Option/choice	Intimation
2.	What is your company/ organization's main industrial section?	a. Business process outsource b. Internet, e-commerce c. Software Development Firm d. Health and medicine sector e. IT consultancy f. Financial sector g. Telecommunication h. Game i. Others (please specify):	
3.	Please select the total number of employees in your company/ organization.	a. 1–10 b. 11–50 c. 51–250 d. Others (please specify):	
4.	Company size	a. Small b. Medium c. Large	

5. Which programming languages are you using?

	Not	Sometimes	Average	Extensively
C#				
Java				
C++				
C				
PHP				
ASP/AJAX				
Others (please specify):				

6. Which development environments are you using?

	Not	Sometimes	Average	Extensively
Visual studio				
Eclipse				
Net beans				
J Developer				
Dreamweaver /IT Intellij IDEA Ultimate				
Others (please specify): …………				

7. CMMI stages maturity level:

a. Level-1 (initial)
b. Level-2 (managed)
c. Level-3 (defined)
d. Level-4 (quantitatively)
e. Level-5 (optimizing)

8. Example:

a. Small project name: ……………………………………………………………
   Duration: …………………………………………………………
b. Large project name: ……………………………………………………………
   Duration: …………………………………………………………

9. Do you use a formal systems development method?

- Yes
- No
- We use techniques and tools, but no method.

Part-A: About requirements analysis

Serial.	*Questionnaire*	*Option/Choice*	*Intimation*
1.1	Do you have any requirement analysis team?	a. Yes b. No	
1.2	If Q-1.1 is "yes", how many?	a. 1–5 b. 6–10 c. Others (please specify):	
1.3	If Q-1.1 is "No", who analyzes?	a. Development team b. Others (please specify):	
1.4	Who gathers user stories?	a. Product owner (team leader) b. Tracker (programmer) c. Product manager d. Others (please specify):	
1.5	How about your customer involvement?	a. On-site customer b. Once and again c. Just once d. Others (please specify):	
1.6	Do you have fixed deadlines for requirements analysis?	a. Yes b. No	
1.7	If Q-1.6 is "yes", how long?	a. 1–5 weeks b. 6–10" c. Others (please specify):	

Part-B: About planning/design

2.1	Do you hold a daily meeting?	a. Yes b. No	
2.2	If Q-2.1 is "yes", how long?	a. 1–20 min b. 21–40" c. 41–60" d. Others (please specify):	
2.3	If Q-2.1 is "yes", what is the main purpose of this meeting?	a. Activities since the last meeting b. Obstacles faced c. Activities to perform before next meeting d. All of the above	
2.4	Who makes the planning/design part?	a. Design team b. Programmer c. Others (please specify):	

		Part-C: About development/coding	
3.1	What is the number of software developers in the development department?	a. 1–10 b. 11–20 c. 21–50 d. 50+	
3.2	If the team size is large, what do you do?	a. Divided into a small group b. Unchangeable	
3.3	Which is significant according to development/coding?	a. Method b. Framework c. all of the above	
3.4	Approximate duration for completing the development/coding part.	a. 1 month b. 2 months c. 3 months d. Others (please specify): …………	
3.5	Do you allow pair-programming?	a. Yes b. No	
3.6	Approximate working hours per week.	a. 30 h b. 40″ c. 50″ d. Others (please specify): …………	
		Part-D: Testing	
4.1	Before launching a Swede you test which is mentioned?	a. Unit test b. System test a. Integration test b. Acceptance test c. All of the above	a. Yes b. No
4.2	Do you implement unit tests before coding?	a. Yes b. No	
4.3	Can anyone change any part of the code at any time?	a. Yes b. No	
		Part-E: Miscellaneous	
5.1	Do you use a burndown chart for completing the project?	a. Yes b. No	
5.2	Do you allow excessive overtime?	a. Yes b. No	
5.3	According to your project, which is changeable/unchangeable?	a. Time and cost b. Functionality/scope c. All of the above d. None of the above	

REFERENCES

1. A. Begel, N. Nagappan, "Usage and perceptions of agile software development in an industrial context an exploratory study", *Conference Paper*, pp. 1–11, October 2007. Available at: https://www.researchgate.net/publication/4279044.
2. M. Al-Zewairi, M. Biltawi, W. Etaiwi, A. Shaout, "Agile software development methodologies: Survey of surveys", *Journal of Computer and Communications*, pp. 1–24, March 31, 2017. Available at: http://file.scirp.org/pdf/JCC_2017033115471602.pdf.
3. S. Rajagopalan, S.K. Mathew, "Choice of agile methodologies in software development: A vendor perspective", *Journal of International Technology and Information Management*, Vol. 25, pp. 1–17, 2016. Available at: http://scholarworks.lib.csusb.edu/cgi/viewcontent.cgi?article=1251&context=jitim.
4. "Software development methods and usability. Perspectives from a survey in the software industry in Norway" Available at: http://bura.brunel.ac.uk/bitstream/2438/3424/3/Software%20development%20methods%20and%20usability.pdf.
5. M.M.M. Safwan, G. Thavarajah, N. Vijayarajah, K. Senduran, C.D. Manawadu "An empirical study of agile software development methodologies: A Sri Lankan perspective", *International Journal of Computer Applications*, pp. 1–7, December 2013. Available at: http://citeseerx.ist.psu.edu/viewdoc/download?doi=10.1.1.401.9304&rep=rep1&type=pdf.
6. P.P. Biemer, L.E. Lyberg, "Introduction to survey quality". Available at: http://www.books.mec.biz/tmp/books/7YC3VBEVW3EENB22K1LW.pdf.
7. G. Brancato, S. Maccia, M. Murzia, M. Signore, G. Simeoni, K. Blanke, T. Korner, A. Nimmergut, P. Lima, R. Paolino, J. Hoffmeyer-Zlotnic, *Handbook of Recommended Practices for Questionnaire Development and Testing in the European Statistics System*, ECGA-200410300002, pp. 1–162. Available at: http://ec.europa.eu/eurostat/documents/64157/4374310/13-Handbook-recommended-practices-questionnaire-development-and-testing-methods-2005.pdf/52bd85c2-2dc5-44ad-8f5d-0c6ccb2c55a0.
8. S. Nithila, K. Priyadharshani, Y.S.G. Attanayake, T. Arani, C.D. Manawadu, "Emergence of agile methodologies: Perceptions from software practitioners in Sri Lanka", *International Journal of Scientific and Research Publications*, Vol. 3, pp. 1–6, November 2013. Available at: http://www.ijsrp.org/research-paper-1113/ijsrp-p2381.pdf.
9. H. Abimbola Soriyan, A.S. Mursu, A.D. Akinde, M.J. Korpela, "Information systems development in Nigerian software companies: Research methodology and assessment from the healthcare sector's perspective", pp. 1–18, EJISDC 2001. Available at: http://www.is.cityu.edu.hk/staff/isrobert/ejisdc/5-4.pdf.
10. R. Dave, "Patterns of success in the Indian software industry". Available at: https://web.stanford.edu/group/scip/avsgt/Dave_thesis.pdf.
11. Y. Jia, "Examining usability activities in scrum projects – a survey study", pp. 1–37, May 2012. Available at: http://www.diva-portal.org/smash/get/diva2:534627/FULLTEXT01.pdf.

12. B. Rumpe, A. Schröder, "Quantitative survey on extreme programming projects", *Third International Conference on Extreme Programming and Flexible Processes in Software Engineering*, pp. 1–6, May 26–30, 2002. Available at: https://arxiv.org/ftp/arxiv/papers/1409/1409.6599.pdf.
13. O. Salo, P. Abrahamsson, "Agile methods in European embedded software development organisations: A survey on the actual use and usefulness of extreme programming and scrum", pp. 1–8, 2008. Available at: http://digital-library.theiet.org/content/journals/10.1049/iet-sen_20070038.
14. B. Boehm, "A survey of agile development methodologies", pp. 1–19, 2007. Available at: https://pdfs.semanticscholar.org/1b68/53c52a7ed9488f9e8ca22ee9f7d62b6e4ded.pdf.
15. T. Krishna, C.P. Kanth, C.V. Krishna, T.V. Krishna, "Survey on extreme programming in software engineering", *International Journal of Computer Trends and Technology*, vol. 2, 2011. Available at: http://ijcttjournal.org/Volume2/issue-2/IJCTT-V2I2P105.pdf.

Chapter 3

Implementation of adaptive channel scheme for multiclass traffic in wireless data communication

Farjana Yeasmin Trisha
East West University

Mahmudul Hasan
Jahangirnagar University

Mohammad Farhan Ferdous
Ferdous, Japan-Bangladesh Robotics & Advanced Technology Research Center (JBRATRC)

Muhammad Arif
University of Lahore

CONTENTS

DOI: 10.1201/9781003156789-3

3.1 INTRODUCTION

Wireless communication is the type of communication in which free air is used as a communication medium. The electromagnetic waves in wireless communication transmit or receive data and voice in open space. Distance is not a barrier to wireless communication of radio waves, such as—a few meters for Bluetooth, television or as far as millions of kilometers for deep-space radio communications. It comprehends various types of fixed, portable applications and mobiles including radios, personal, cellular telephones, and wireless networking. There are several examples of applications of radio wireless technology including the Global Positioning System units, garage door openers, wireless computer mice, keyboards and headsets, headphones, receivers, broadcast, and satellite and cordless telephones. Somewhat less common methods of achieving wireless communications include the utilization of other electromagnetic wireless technologies, such as magnetic, light, or electric fields or the use of sound wireless operations give services, such as global communications, that are impractical or impossible to perform with the use of wires [1]. The expression is normally used in the telecommunications sector to indicate those telecommunications theory (e.g., radio transmitters, remote controls, and receivers) which transmits information without the use of wires by some form of energy (e.g., acoustic energy). Information is transferred in this form over both long and short distances. Wireless devices perform in a similar radio frequency as more devices. Signals from other devices can interrupt wireless transmissions, or a wireless local area network device can interfere with other devices. A wireless network is any type of computer network that uses wireless data connections for connecting network nodes. Homes, telecommunications networks, and enterprise installations avoid the costly process of introducing cables into a building by wireless networking method or use it as a connection between various equipment locations. The use of radio communication in wireless telecommunications networks is generally implemented and administered. This utilization takes place at the physical level of the Open Systems Interconnection model network structure. Examples of wireless networks include cell phone networks, wireless sensor networks, satellite communication networks, wireless local networks, and terrestrial microwave networks. The communication media executes the same as a communication channel for connecting various computing devices so that they may connect. Contemporary communication media facilitate data exchange and communication among a large number of individuals across long-winded distances via teleconferencing, email, Internet forums, etc. Traditional mass media channels such as TV, radio and magazines, newspapers, on the other hand, promote one-to-many communication. At present mobile communication is the most general and practical example of the wireless [2].

3.2 OBJECTIVES

The main objective of this thesis is to reduce the call blocking probability together with the proper use of channels by a new scheme named adaptive channel scheme. In this scheme, channel utilization is not limited. In addition, all types of communication by voice call, SMS, and internet connectivity are able to share with expanding or compressing an individual's pre-defined number of channels in each class. How we can increase the channel for large population density in a small area by using schemes.

3.3 BACKGROUND STUDY

A lot of background study is being given to prepare this paper. A brief history of background study with wireless communication is given. Wireless communication commonly works through electromagnetic signals that are published by an enabled device within the air, physical environment, or atmosphere. The sending device can be a sender or an intermediate device with the ability to propagate wireless signals. The communication between two devices occurs when the destination or receiving intermediate device captures these signals, creating a wireless communication bridge between the sender and receiver device. Wireless communication has various forms [3] of technology and delivery methods including mobile communication, wireless network communication, satellite communication, infrared communication, Bluetooth communication. Although all of these communication technologies have different underlying architecture, they all lack a physical or wired connection between their respective devices to initiate and execute communication [4]. Correspondence systems can be wired or wireless and the medium utilized for correspondence can be guided or unguided. In wired communication, the medium is a physical way, such

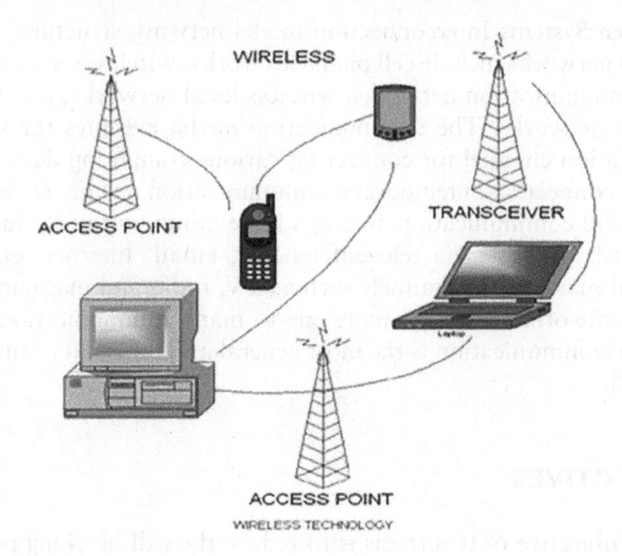

Figure 3.1 Wireless communication.

as co-pivotal cables, twisted pair cables, and optical fiber links, which passes signal one node to another. Such a kind of medium is called guided medium. However, wireless communication does not require any physical medium however proliferates the sign through space. Since space just takes into account signal transmission with no direction, the medium utilized in wireless communication is called unguided medium. On the off chance that there is no physical medium, at that point how does remote correspondence send signals? Even though there are no links utilized in remote correspondence, the transmission and gathering of signs are practiced with Antennas. Receiving wires are electrical gadgets that change the electrical signs to radio signs as electromagnetic waves and the other way around. These electromagnetic waves engender through space. Subsequently, both transmitter and beneficiary comprise a receiving wire (Figure 3.1).

3.3.1 History of wireless communication

Communication systems using electrical and electronic technology have a significant impact on modern society. As the courier speeding from Marathon to Athens in 490 BC illustrates, in early history information could be exchanged only by physical transport of messages. A few examples exist of non-electrical communication techniques for the transfer of information via other infrastructures than those for physical transport: smoke signals, signal flags in maritime operations, and the semaphore are among them. Early attempts to communicate visual signals by means of the semaphore, a pole with movable arms, were made in the 1830s in France.

A similar experimental system was used by the Dutch during the 10 days campaign against the Belgian revolt in 1831/1832. In 1837, the House of Representatives passed a resolution requesting the Secretary of the Treasury to investigate the feasibility of setting up such a system in the United States. The market interest in enhanced communication systems was also clearly illustrated by the fact that in 1860 the Pony Express started regular physical message services over land in the United States. But at the same time, electronic systems for communication had started to develop.

3.3.2 The wireless age

Telecommunication is defined by the International Telecommunication Union which is the transmission, emission, or reception of any signs, signals, or messages by using electromagnetic systems. In 1832 the verification of (electrical) telegraphy by Joseph Henry and Samuel F. B. Morse followed shortly after the discovery of electromagnetism by Hans Christian Oersted and Andre-Marie Ampere early in the 1820s. In the 1840s, the U.S. East Coast and in California built telegraph networks. In 1858 the first transatlantic cable was placed. In 1864, James Clerk Maxwell suggested wireless propagation, which was confirmed and denoted by Heinrich Hertz in 1880 and 1887, respectively. Marconi and Popov started experiments shortly about the radiotelegraph thereafter, and in 1897 Marconi permitted a complete wireless system.

We described the same thing by using radio and wireless for a long time, but the difference was that radio was the American version of the British wireless. Because no wires were connecting to the transmitting station the receiver was called wireless. The transmitting station radiated electromagnetic waves and it was called radio. The British Broadcasting Company was the first to use the term wireless around 1923, and their program guide is "The Radio Times". In 1876, Alexander Graham Bell permitted the telephone. Fleming formulated the diode in 1904 and Lee de Forest invented the triode in 1906 to make possible rapid development of long-distance (radio) telephony. Bardeen, Britain, and Shockley invented the transistor, which used integrated circuits, and covered the way for the miniaturization of electronic systems.

1864: James Clerk Maxwell proved the existence of electromagnetic waves.
1887: Heinrich Hertz sent and received wireless waves, using a spark transmitter and a resonator receiver.
1895: Guglielmo Marconi sent more radio signals over more than a mile.
1901: Marconi received the mores message "s" (...) sent across the Atlantic.
1904: J. A. Fleming patented the diode.
1906: Lee Deforest patented the triode amplifier. First speech wireless transmission, by Fessenden.

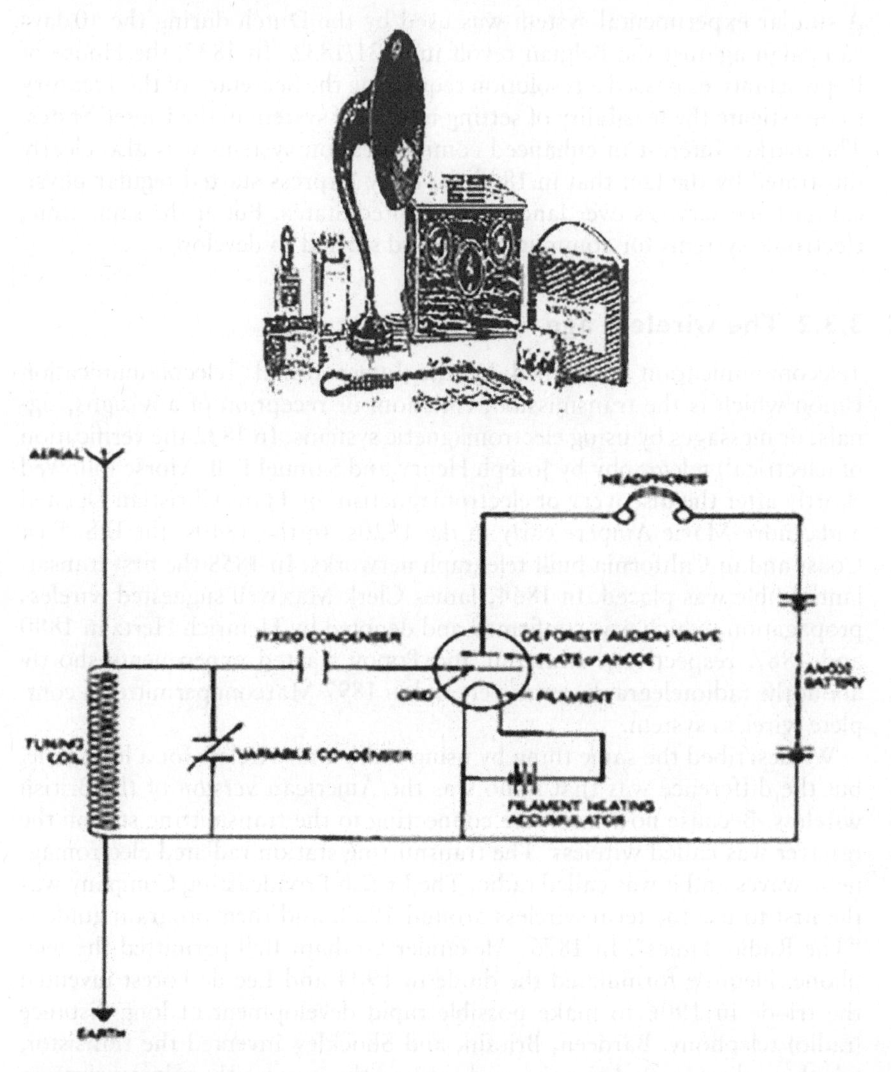

Figure 3.2 Radio channel.

- 1907: Commercial Transatlantic Wireless Service, using huge ground stations: 30×100 m antenna masts beginning of the end for cable-based telegraphy.
- 1968: Carte phone decision.
- 1974: FCC allocates 40 MHz for cellular telephony.
- 1982: European GSM and Inmarsat established.
- 1984: Breakup of AT&T.
- 1984: Initial deployment of AMPS cellular system (Figure 3.2).

3.3.3 Personal communication

Microelectronic circuits have recently made a huge rapid development of mobile and personal communication systems workable. This offers person-to-person communication; users can move freely and if desired eliminate the ineffective calls experienced with the fixed telephony service when the user is away from his or her terminal. Services on employing mobile data communication become workable, such as automatic vehicle location for fleet management, electronic mail, remote access to databases, vehicle printers, or automatic repetition of the messages even if the driver has been away from the vehicle. Additionally, data communication is doing encryption and data processing possible. In Denmark, Finland, Iceland, Norway, and Sweden, the systems (earliest) providing automatic radiotelephony, the Nordic Mobile Telephone cellular system, was based on a standard developed in close cooperation between the five different operators (Push to Talk, PTTs) and competing manufacturers. The Nordic Mobile Telephone denoted the joint drive in European countries towards international cellular networks. Initially, the United States looked less able to develop and follow a common policy for mobile networking, even though in 1970 the Bell Laboratories had played a leading role in the novel cellular technology.

GSM was the first digital cellular telephony for Europeans in 1992. But the GSM originated early in the 1980s as the French acronym for Group Special Mobile, the international working group tasked by most European PTT administrations to advance a common standard for cellular networks. The main advantages of a digital system are a larger user with the highest capacity per unit of the spectrum, ease of implementation of sophisticated encryption, authentication, and other security features, and robustness against radio channel imperfections.

3.3.4 Mobile phone

In North America, a mobile phone is known as a cell phone. It is a portable telephone. It can make and receive calls through a radio frequency link while the user is moving. The radio frequency link establishes a connection to the switching systems of a mobile phone operator, which provides access to the public switched telephone network. Around the 2000s, mobile phones provided text messaging, MMS, email, internet access, short-range wireless communications (infrared, Bluetooth), business applications, video games, and digital photography. Those capabilities are known as features. In 1973 the first handheld mobile phone was revealed by John F. Mitchell and Martin Cooper of Motorola, which weighed 2 kg (4.4 lbs). In 1983, the first commercially available handheld mobile phone was the Dynastic 8,000×. From 1983 to 2014, worldwide mobile phone subscriptions increased day by day (7 billion), penetrating 100% of the global population and reaching

Figure 3.3 Evolution of the mobile phone.

even the bottom of the economic pyramid. The top smartphone manufacturers include Samsung, Apple, and Huawei (smartphone sales represented 78% of total mobile phone sales.) [5] (Figure 3.3).

3.3.5 Radio network

Currently, two types of radio networks operate around the world: (i) one-to-many broadcast networks commonly work for public information and mass media entertainment and (ii) the two-way radio works more commonly for public safety, security, and public services such as police, fire, taxicabs, and delivery services.

The two-way type of radio network allows many of the same technologies and components as the Broadcast type radio network though it is set up with fixed broadcast points with co-located receivers and mobile receivers/transmitters. In this mobile radio units can communicate with each other over broad geographic regions, entire states/provinces, or countries. In many ways, multiple fixed transmit/receive sites can be interconnected to achieve the range of coverage which is required by the administration or authority implementing the system: conventional wireless links in numerous frequency bands, fiber-optic links, or microwave links. The signals are typically backhauled to a central switch of some type where the radio message is active and repeated to all transmitter sites where it is required to be heard. In contemporary two-way radio systems, a concept called trucking is commonly used to achieve better efficiency of radio spectrum use and provide very wide-ranging coverage with no switching of channels required by the mobile radio user as it roams throughout the system coverage. Trucking of two-way radio is identical to the concept used for cellular phone systems where each fixed and mobile radio is specifically identified to the system controller and its work is switched by the controller [6] (Figure 3.4).

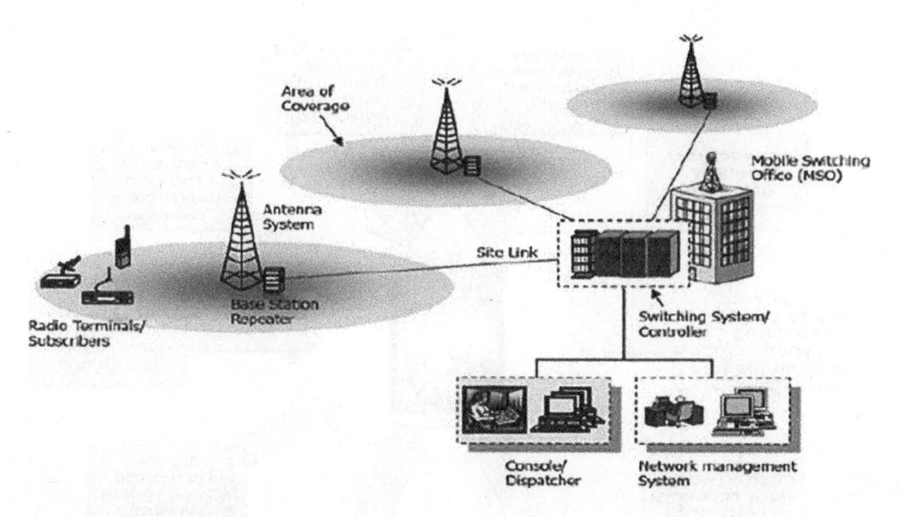

Figure 3.4 Radio network.

3.3.6 Feature of mobile phone

The features are the set of capabilities, services, and applications which offer for their users. Handsets with more approach computing ability through the use of locals so try to differentiate their own products by adding additional functions to make them more attractive to customers. Nowadays a mobile phone is our important companion that aids us in both personal and professional [7]. It is a tough decision when we go for a phone for ourselves. Some questions arise here as follows: Do we go for the latest phone? Should we give priority to battery life or the phone's design? The answer will depend on our own preference. Our personal preference will take priority. In a poll on our social media channels, we asked you what the most important mobile feature was: Screen size or Battery, Camera, Design, Display, Other [8] (Figure 3.5).

3.3.7 Voice call

The voice call can be another way of communication among two or more people to share their opinion with each other. The voice call can be performed through different operators (e.g., GrameenPhone, RobiAxiata, Banglalink) and different mobile apps (e.g., Viber, Messenger, Tango, IMO, Whatsapp).

3.3.8 Text message

When messages are sent from one cell phone to another cell phone by electromagnetically and communication is being performed by a device that is Text Message (e.g., Mobile phone messenger, Viber, Whatsapp, IMO).

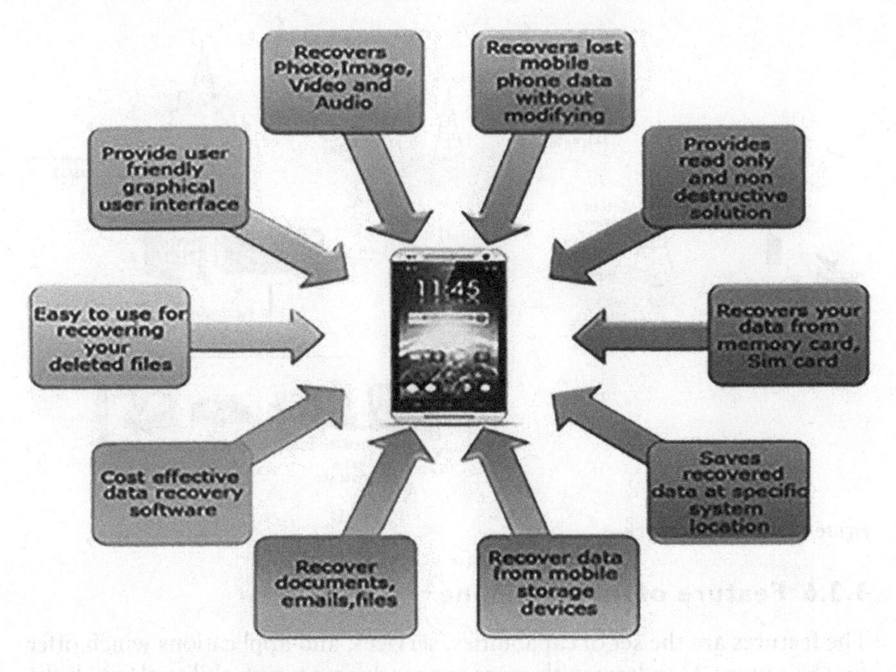

Figure 3.5 Features of mobile phone.

3.3.9 Sim card

The full form of SIM is the subscriber identity module or subscriber identification module. A SIM card contains unique information that identifies the specific mobile network, which allows the subscriber. If SIM cards cannot be inserted correctly then phones aren't working, phones cannot make calls, send messages, or connect to mobile internet services (3G, 4G, etc.). The first SIM card's size was like a credit card. But now, this time SIM cards look both mini and micro [9].

Here are the dimensions of the different types of SIM cards:

1. Full SIM: 85 mm×53mm
2. Mini SIM: 25 mm×15mm
3. Micro SIM: 15 mm×12mm
4. Nano SIM: 12.3 mm×8.8mm
5. Embedded SIM: 6 mm×5mm

3.3.10 Phone operators

A mobile phone operator or wireless provider provides wireless internet GSM services for mobile device users. The operator gives a SIM card to the customer who inserts it into the mobile device that they can access the service.

There are two types of mobile operators:

1. Mobile network operator
2. Mobile virtual network operator

In May 2016, China was the world's largest individual mobile operator, and its subscribers were around 835 million. Fifty mobile operators had around 10 million each, and 150 mobile operators had at least 1 million subscribers each by the end of 2009. By researching global penetration total mobile-cellular subscribers reached almost 6 billion by the end of 2011 [10].

R-U SIM: Removable User Identity Module (R-UIM) is a card developed for cdmaOne/CDMA2000 ("CDMA") handsets that increase the GSM SIM card to CDMA phones and networks. The R-UIM carries an early version of the CSIM application. The card also contains a SIM (GSM) application, so it can work for networks. It is physically suited with GSM SIMs and can fit into existing GSM phones. The R-UIM card has been replaced by CSIM on UICC. It allows three applications (SIM, CSIM, and USIM) to coexist on a single smartcard and allows the card to be virtually used in any phone worldwide that supports smart cards [11].

Cellular traffic: This article studies the mobile cellular network. Mobile radio networks have traffic issues but public switched telephone networks do not. Important features of cellular traffic include quality of service targets, traffic capacity and cell size, spectral efficiency, traffic capacity, and channel holding time analysis [12]. Telegraphic engineering in telecommunications network planning secures that network costs are reduced without compromising the quality of service delivered to the user of the network [13]. This field of engineering is based on probability theory, as well as other telecommunications networks. A moving mobile handset will record a signal strength that differs. Signal strength is a slow fading, fast fading, and interference from other signals, resulting in degradation of the carrier-to-interference ratio (C/I). A high C/I ratio provides quality communication. A good C/I ratio is reached in cellular systems by using optimum power levels [14]. Creating excessive interference the carrier power must be too high, and reducing the C/I ratio for other traffic and also reducing the traffic capacity of the radio subsystem. When carrier power is too low, C/I is too low [15].

3.3.11 Handover

Handover is said to have taken place. When a mobile station moves from one set to another set. It provides more reliable access continuity in the network connection. Here less chances of a call ends during moving of base stations in comparison to a hard handoff. CDMA systems use it that enables the overlapping of the repeater coverage zone and every cell phone set is always well within the range of at least one of the base stations. The

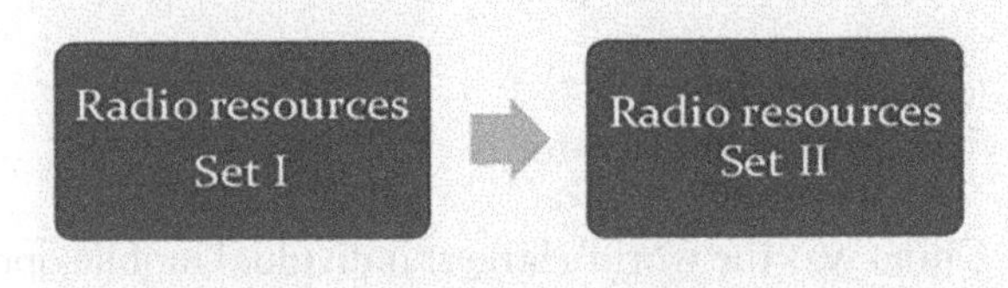

HANDOVER

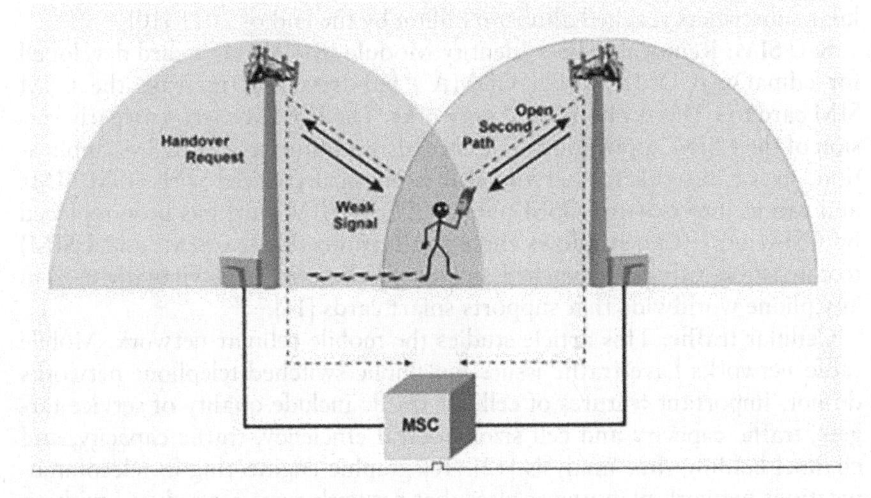

Figure 3.6 Handover.

technical execution of a Soft handoff is more expensive and complex than a hard handoff. It is used in sensitive communication services like videoconferencing [16] (Figure 3.6).

Types of handover:
There are several handover technique, such as

1. Hard handover
2. Soft handover
3. Horizontal handover
4. Vertical handover

3.3.12 Various network schemes

3.3.12.1 Call blocking

Call blocking, also known as call block, call rejection, or call screening allows a telephone subscriber to block incoming calls from specific telephone numbers. This article may require an extra payment to the sponsor's telephone agency or a third party. Call blocking is desired by individuals who wish to block unwanted phone calls. These generally include types of unsolicited calls from telemarketers and robocalls [17]. Unwanted calls to landlines may be blocked through a number of methods. Some landline phones have built-in call blocking facilities. External call blockers are sold as telephone accessories which plug into existing phones. Such devices and services enable the user to block a call as it is in progress or alternatively block the number after the call is made. These devices rely on caller ID information and thus a phone blocker requires a caller ID service active on the line for blocking to function. It may also be possible to use computer software, in conjunction with the caller ID information from the phone company and a caller ID-enabled phone modem to block an incoming call [18].

Treatment of blocked calls may include the following:

1. Sending caller to voicemail
2. Sending caller to "number no longer in service"
3. Sending caller to "keep ringing"
4. Sending caller to a busy signal

There is a multitude of third-party call blocking applications available for Smartphone's while some manufacturers provide built-in call blocking functions as standard [19].

3.3.13 Class channel

In wireless communication systems real-time voice calls, SMS, internet connectivity, etc., are the main activities. In every requested call or SMS, a cell makes a connection with a user via a different channel consisting of individual classes. These individual channels can connect with individual users based on priority. Priority plays a vital role in the selection of channels and classes [20]. The size of classes also depends on priority. For more priority size of the class becomes greater and for less priority class size becomes lower. A wireless traffic system becomes a single class for single requests like just for voice call and multiclass for voice call, SMS, internet connectivity, etc. In this paper, multiclass traffic is going to be discussed [21].

3.3.14 Multiclass traffic

Cellular traffic is the internal system of a cellular system for which the overall mobile system can communicate with different users. This traffic system is operated with the mobile operator and it deals with different predefined classes and channels. Cellular traffic is mostly used for various classes [22]. This is known as multiclass traffic. Increasing demand for multiclass traffic has become a prime concern in wireless communication systems. Quality of service is seen to be degrading to fulfill these demands. To handle multiclass traffic, already different schemes exist such as non-priority scheme, priority scheme channel reservation scheme, and queuing scheme. These schemes do not meet the requirements of minimizing the new call blocking probability and increasing the channel utilization at the same time [23]. Here, we are proposing a new adaptive channel scheme for a multiclass traffic system to minimize the new call blocking probability of higher priority calls and keep the channel utilization in a suitable range so that both the performance metrics are optimized and thus provide good quality of service [24]. In our adaptive channel scheme, guard channels are reserved adaptively based on the arrival rates of the traffic classes and the number of channels occupied. In this paper, we discuss different schemes for multiclass traffic systems and different problems in these schemes. Also, we introduce our proposed scheme, mathematical modeling, and queuing analysis of it, and we also show the improvement in our scheme through performance analysis [25].

3.3.15 Channel utilization

Channel utilization means the proper usage of channels in different classes. As channel utilization, two types of schemes are used called non-priority schemes and priority schemes. Both are going to be discussed for voice calls and SMS.

3.3.16 Queuing analysis

In this section, non-priority and priority schemes have been discussed for a single traffic system such as either a voice call or an SMS system. Here, it is assumed that a system has many cells, with each having S channels [26]. The channel holding time has an exponential distribution with a mean rate μ. Both originating and handover calls are generated in a cell according to Poisson processes with mean rates [27]. The system is assumed to be with a homogeneous cell. We focus on a single cell (called a marked cell). Newly generated calls in the marked cell are labeled originating calls (or new calls). A handover request is generated in the marked cell when a channel holding MS approaches the marked cell from a neighboring cell with a single strength below the handover threshold [28].

3.3.17 Non-priority scheme

Consider the number of channels is "S". In this scheme, all "S" channels are shared by both real-time voice calls and SMS requests. The SMS request is handled exactly in the same way or shares the overall channel as like a voice call. A system model for the non-priority scheme is shown in Figure 3.7.

Here, λ_1, λ_2, and λ_3 denote the arrival rate (probability of arrival per time unit) of class 1, class 2, and class 3, respectively. Both classes can utilize

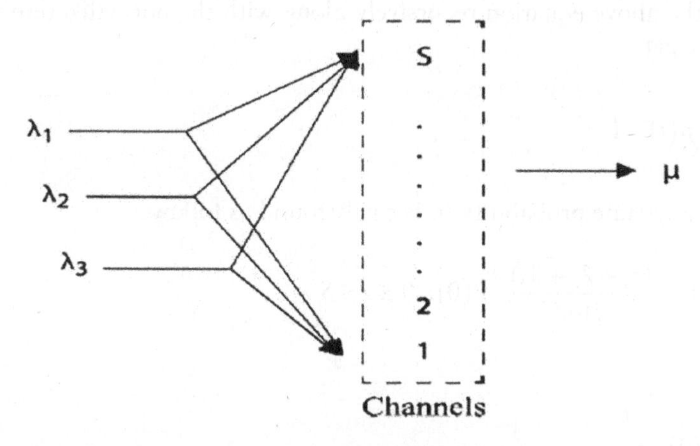

Figure 3.7 System model for the non-priority scheme.

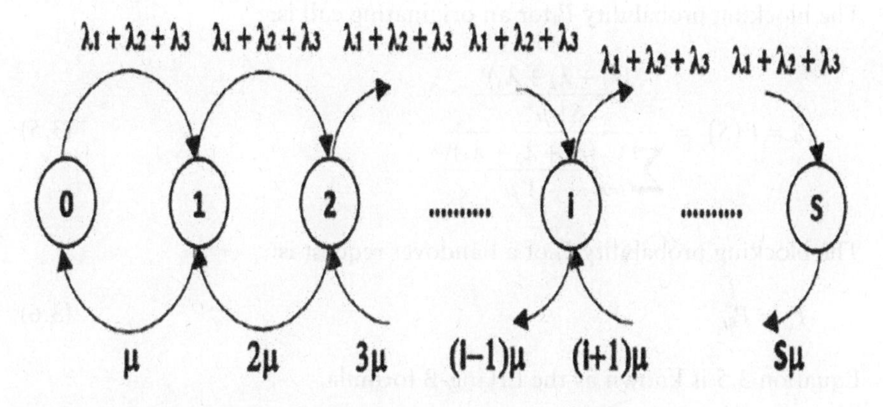

Figure 3.8 State transition diagram for the non-priority scheme.

total channels. The other factor here is the mean rateμ. If there is no empty channel available, the call becomes blocked. In this scheme, priority in different classes is absent. As a result forced termination probability is relatively higher (Figure 3.8).

Let (i) be the probability and the system be in state i. The probabilities (i) can be determined in the usual way for the birth–death process. The state equilibrium equation is:

$$P\ (i)\ = \frac{\lambda_1 + \lambda_2 + \lambda_3}{i\mu} p(i-1),\ 0 \le i \le S \tag{3.1}$$

Using the above equation recursively along with the normalization condition, we get

$$\sum_{i=0}^{s} p(i) = 1 \tag{3.2}$$

The steady-state probability (i) is easily found as follows:

$$P(i) = \frac{(\lambda_1 + \lambda_2 + \lambda_3)^i}{i!\,\mu^i} P(0),\ 0 \le i \le S \tag{3.3}$$

where,

$$P(0) = \frac{1}{\sum_{i=0}^{s} \frac{(\lambda_1 + \lambda_2 + \lambda_3)^i}{i!\,\mu^i}} \tag{3.4}$$

The blocking probability $P_B$for an originating call is:

$$P_B = P(S) = \frac{\frac{(\lambda_1 + \lambda_2 + \lambda_3)^s}{S!\,\mu^s}}{\sum_{i=0}^{s} \frac{(\lambda_1 + \lambda_2 + \lambda_3)^i}{i!\,\mu^i}} \tag{3.5}$$

The blocking probability $P_D$of a handover request is:

$$P_D = P_B \tag{3.6}$$

Equation 3.5 is known as the Erlang-B formula.

3.3.18 Priority scheme

In this scheme, priority for calls or data is divided into individuals among the "S" channels. Let λ_1 be the arrival rate for class 1, λ_2 for class 2, and λ_3 for class 3, and the mean rate be μ. Class 1,class 2 and class 3 can be used 1 *to* S_c channels. But the $S_b - S_c$channels are only to be used for class 2 and class 3. Finally $S - S_b$ channels are used for only class 3. Here, the priority for class 1 is less than class 2 as well as class 3. So, class 3 has the higher priority and class 1 has the lower priority. A system model for the priority scheme is shown in Figure 3.9.

If the number of available channels in the cell is less or equal to $S - S_b$, the call will be blocked. But here, the call dropping probability is less than the non-priority scheme for specific prioritized channels.

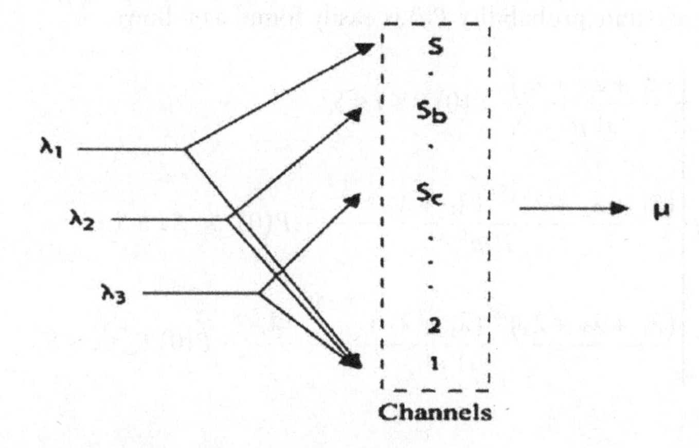

Figure 3.9 System model for priority scheme.

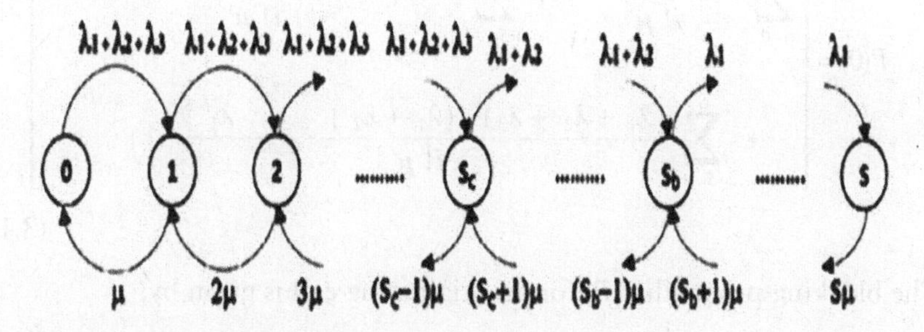

Figure 3.10 State transition diagram for priority scheme.

But here, having higher priority in a specific class, channel utilization becomes decreased for lower prioritized class. As a result, the call blocking rate of the lower-class increases (Figure 3.10).

The state balance equations are

$$\begin{cases} i\mu P(i) = (\lambda_1 + \lambda_2 + \lambda_3)P(i-1) & 0 \le i \le S_C \\ i\mu P(i) = (\lambda_1 + \lambda_2)P(i-1) & S_C \le i \le S_b \\ i\mu P(i) = \lambda_1 P(i-1) & S_b \le i \le S \end{cases} \tag{3.7}$$

Using this equation recursively along with the normalization condition, we get

$$\sum_{i=0}^{s} P(i) = 1 \tag{3.8}$$

The steady-state probability $P(i)$ is easily found as follows:

$$P(i)\begin{cases} \dfrac{(\lambda_1+\lambda_2+\lambda_3)^i}{i!\,\mu^i}P(0) & 0\le i\le S_c \\[2ex] \dfrac{(\lambda_1+\lambda_2+\lambda_3)^{S_c}(\lambda_1+\lambda_2)^{i-S_c}}{i!\,\mu^i}P(0) & S_C\le i\le S_b \\[2ex] \dfrac{(\lambda_1+\lambda_2+\lambda_3)^{S_c}(\lambda_1+\lambda_2)^{S_b-S_c}\lambda_1^{i-S_b}}{i!\,\mu^i}P(0) & S_b\le i\le S \end{cases} \tag{3.9}$$

where,

$$P(0)=\left[\sum_{i=0}^{S_c}\frac{(\lambda_1+\lambda_2+\lambda_3)^i}{i!\,\mu^i}+\sum_{i=S_c+1}^{S_b}\frac{(\lambda_1+\lambda_2+\lambda_3)^{S_c}(\lambda_1+\lambda_2)^{i-S_c}}{i!\,\mu^i}+\sum_{i=S_b+1}^{S}\frac{(\lambda_1+\lambda_2+\lambda_3)^{S_c}(\lambda_1+\lambda_2)^{S_b-S_c}\lambda_1^{i-S_b}}{i!\,\mu^i}\right]^{-1} \tag{3.10}$$

The blocking probability P_B for an originating call is given by:

$$P_B=\sum_{i=S_c}^{S}P(i) \tag{3.11}$$

Here again, a blocked handover request call can still maintain the communication via current Base Station (BS) until the received signal goes below the receiver threshold or the conversation is completed before the received signal strength goes below the receiver threshold. The equations of channel utilization for non-priority and priority schemes are respectively:

$$\frac{(1-P_B)\lambda_1+(1-P_B)\lambda_2+(1-P_B)\lambda_3}{\mu\times N}\times 100 \tag{3.12}$$

and

$$\frac{(1-P_{B1})\lambda_1+(1-P_{B2})\lambda_2+(1-P_{B3})\lambda_3}{\mu\times N}\times 100 \tag{3.13}$$

Equations 3.3 and 3.9 are simulated in MATLAB® (see Figure 3.11).

Equations 3.12 and 3.13 are simulated in MATLAB (see Figure 3.12).

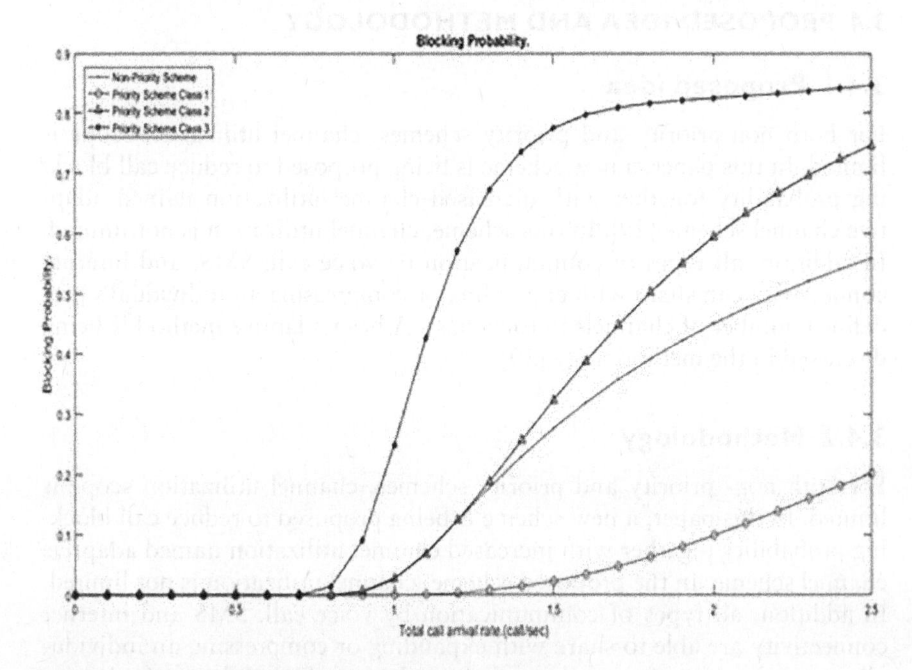

Figure 3.11 Comparison of new call blocking probability.

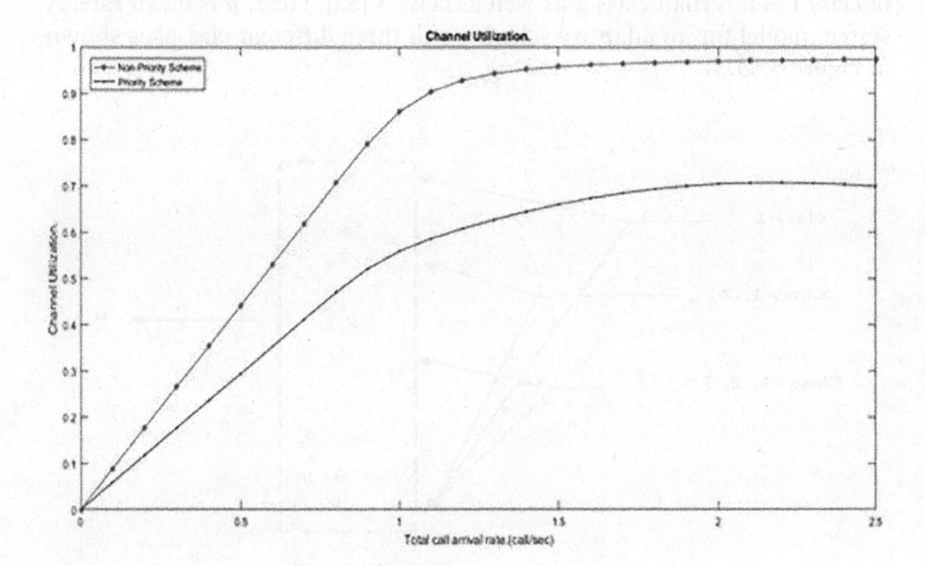

Figure 3.12 Comparison of channel utilization.

3.4 PROPOSED IDEA AND METHODOLOGY

3.4.1 Proposed idea

For both non-priority and priority schemes, channel utilization scope is limited. In this paper, a new scheme is being proposed to reduce call blocking probability together with increased channel utilization named adaptive channel scheme [29]. In this scheme, channel utilization is not limited. In addition, all types of communication by voice call, SMS, and internet connectivity can share with expanding or compressing an individual's predefined number of channels in each class. A brief adaptive method is being discussed in the methodology [30].

3.4.2 Methodology

For both non- priority and priority schemes, channel utilization scope is limited. In this paper, a new scheme is being proposed to reduce call blocking probability together with increased channel utilization named adaptive channel scheme. In the proposed scheme, channel utilization is not limited. In addition, all types of communication by voice call, SMS and internet connectivity are able to share with expanding or compressing an individual's pre-defined number of channels in each class .[31]. A brief of adaptive method is being discussed in methodology. To describe the adaptive channel scheme, consider the total number of channels is "*S*". Total "*S*" channel consists of three classes: class 1, class 2, and class 3 [32]. Here, the priority of class 1 is less than class 2 as well as class 3 [33]. Here, μ is mean rate. A system model for an adaptive scheme with three different classes is shown in Figure 3.13.

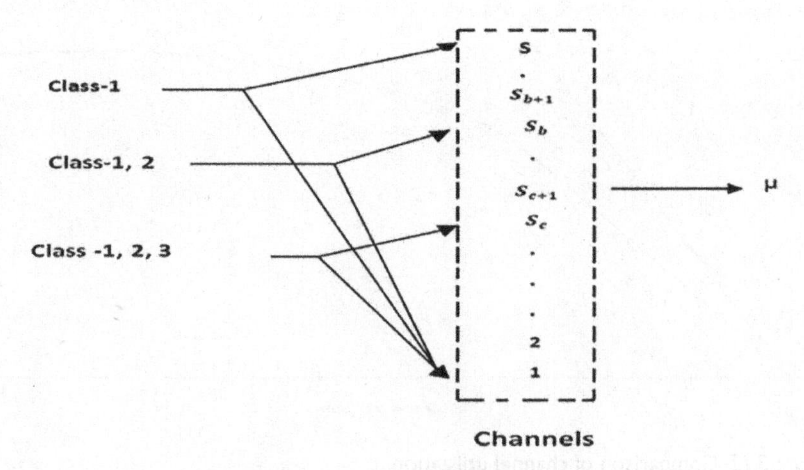

Figure 3.13 System model for adaptive channel scheme.

This scheme is similar to the priority scheme. But, little difference changes the overall concept. In priority scheme lengths of classes are fixed: class 1 from 1 to S_c, length of class 2 from 1 to S_b and so on. But, in adaptive schemes the length of classes is variable. That means for class 1, S_c can be expanded to S_c+1 or more and compressed to S_c-1 or less and so on. This method becomes a proper utilization for the overall channel.

3.4.3 Adaptive channel scheme

There are many types of schemes to handle multiclass traffic in wireless networks. We propose an adaptive channel reservation scheme in which we deal with three classes: class 1, class 2, and class 3. Here, class 3 has the lowest priority and class 1 has the highest priority. Here no priority channel is reserved for priority class1 and priority class 2. We use some factors based on which the number of reserved channels for class 1 and class 2 traffics have been made adaptive. Thus the channel utilization is high and the blocking probability of the classes reduces with respect to the other schemes [34].

3.4.4 Mathematical modeling

The mathematical modeling of our proposed adaptive scheme has been done in the almost same way as the non-priority and priority schemes. But here we introduce some factors based on which the reservation of channels for the different classes will be determined.

Let us suppose,
The total number of channels $= N_1$
Number of classes $= 3$
Call arrival rate for class $1 = \lambda_1$
Call arrival rate for class $2 = \lambda_2$
Call arrival rate for class $3 = \lambda_3$
Fixed number of minimum common channels for class 1 and class $2 = M_1$
Fixed number of minimum common channels for all classes $= M_2$
Blocking probability for class $1 = B_{P_1}$
Blocking probability for class $2 = B_{P_2}$
Blocking probability for class $3 = B_{P_3}$
Average call life time $= \frac{1}{\mu_c}$

Dwell time $= \frac{1}{\eta}$
Channel holding time becomes

$$\frac{1}{\mu} = \frac{1}{\eta + \mu_c} \tag{3.14}$$

$$\text{Proposed factor for priority } 1: f_1 = \frac{\left[(1 - B_{P_1})\lambda_1\right]\frac{1}{\mu}}{N_1}; \alpha_1 = \frac{\lambda_1}{\lambda_1 + \lambda_2 + \lambda_3} \tag{3.15}$$

$$\text{Proposed factor for priority } 2: f_2 = \frac{\left[(1 - B_{P_2})\lambda_2\right]\frac{1}{\mu}}{N_2}; \alpha_2 = \frac{\lambda_2}{\lambda_1 + \lambda_2 + \lambda_3} \tag{3.16}$$

$$\text{Number of channels reserved only for class 1 traffic, } X_1 = f_1\alpha_1(N_1 - M_1) \tag{3.17}$$

$$\text{Number of channels reserved only for class} - 1 \text{ and class 2 traffic, } X_2 = f_2\alpha_2(N_1 - M_2) \tag{3.18}$$

$$\text{The number of channels available for the traffic class 2, } N_2 = N_1 - X_1 \tag{3.19}$$

$$\text{The number of channels available for the traffic class 3, } N_3 = N_2 - (X_1 + X_2) \tag{3.20}$$

A generic model of the proposed scheme has been shown in Figure 3.14.

We define the state i (i=0, 1...N_1) of a cell as the number of calls in progress for the BS of that cell.

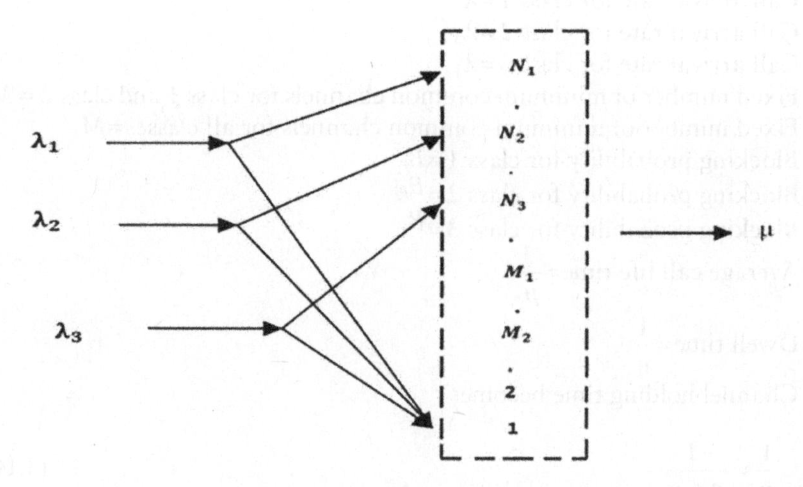

Figure 3.14 A generic system model for proposed adaptive channel reservation scheme.

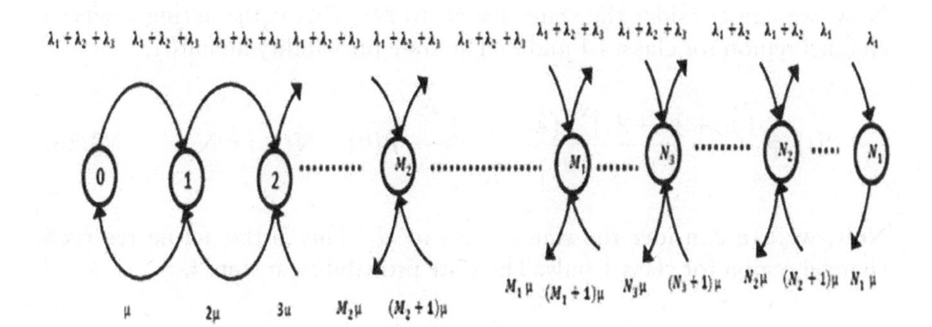

Figure 3.15 State transition diagrams for the proposed adaptive scheme.

Let (*i*) represent the steady-state probability and BS in the state *i*. The probabilities (*i*) can be determined in the usual way for birth–death processes. The relevant state transition diagram is shown in Figure 3.15.

A system in statistical equilibrium will possess the average rates of entering a state equally. So, at state 0, we find

$$\mu P(i) = (\lambda_1 + \lambda_2 + \lambda_3) P(0) \tag{3.21}$$

So, solving for (*i*), we have

$$P\ (i)\ = \frac{(\lambda_1 + \lambda_2 + \lambda_3)}{\mu} P(0) \tag{3.22}$$

Now consider state 1. The average rate of entering state 1, 2(2)+$(\lambda_1 + \lambda_2 + \lambda_3)$ *P*(0). Proceeding in the same manner, the average rate of leaving state 1 is found to be given by $(\lambda_1 + \lambda_2 + \lambda_3 + \mu)(1)$. Equating these two transition rates to provide a balance of arrivals and departures from state 1, we get

$$(\lambda_1 + \lambda_2 + \lambda_3 + \mu)(1) = (\lambda_1 + \lambda_2 + \lambda_3) P(0) + 2\mu P(2) \tag{3.23}$$

This given equation is

$$(2)\ = \frac{(\lambda_1 + \lambda_2 + \lambda_3)^2}{2\mu \cdot \mu} P(0) \tag{3.24}$$

Continuing in this manner by setting up a balance equation at each state $i \le (N_3 - 1)$, equating average arrivals and departures at a state to maintain statistical equilibrium, we find, for the state probability at state *i*, $1 \le i \le N_3$

$$(i)\ = \frac{(\lambda_1 + \lambda_2 + \lambda_3)^i}{i!\ \mu^i} P(0)\ 1 \le i \le N_3 \tag{3.25}$$

Now, we can consider the states $i = N_3$ to N_2. This is the acting reserved channel region for classes 1 and 2. The state probability at state i,

$$P(i) = \frac{(\lambda_1 + \lambda_2 + \lambda_3)^{N_3} (\lambda_1 + \lambda_2)^{i-N_3}}{i!\,\mu^i} P(0) \quad N_3 < i < N_2 \tag{3.26}$$

Now, we can consider the states $i=N_2$ to N_1. This is the acting reserved channel region for class 1 only. The state probability at state i,

$$P(i) = \frac{(\lambda_1 + \lambda_2 + \lambda_3)^{N_3} (\lambda_1 + \lambda_2)^{N_2-N_3} \lambda_1^{i-N_2}}{i!\,\mu^i} P(0) \quad N_3 < i < N_2 \tag{3.27}$$

Using this equation recursively along with the normalization condition, we get

$$\sum_{i=0}^{S} P(i) = 1 \tag{3.28}$$

The steady-state probability (i) is easily found as follows:

$$P(i)\begin{cases} \dfrac{(\lambda_1 + \lambda_2 + \lambda_3)^i}{i!\,\mu^i} P(0) & 1 \le i \le N_3 \\ \dfrac{(\lambda_1 + \lambda_2 + \lambda_3)^{N_3} (\lambda_1 + \lambda_2)^{i-N_3}}{i!\,\mu^i} P(0) & N_3 \le i \le N_2 \\ \dfrac{(\lambda_1 + \lambda_2 + \lambda_3)^{N_3} (\lambda_1 + \lambda_2)^{N_2-N_3} \lambda_1^{i-N_2}}{i!\,\mu^i} P(0) & N_2 \le i \le N_1 \end{cases} \tag{3.29}$$

where,

$$P(0) = \left[\sum_{i=0}^{N_3} \frac{(\lambda_1 + \lambda_2 + \lambda_3)^i}{i!\,\mu^i} + \sum_{i=N_3+1}^{N_2} \frac{(\lambda_1 + \lambda_2 + \lambda_3)^{N_3} (\lambda_1 + \lambda_2)^{i-N_3}}{i!\,\mu^i} + \sum_{i=N_2+1}^{N_1} \frac{(\lambda_1 + \lambda_2 + \lambda_3)^{N_3} (\lambda_1 + \lambda_2)^{N_2-N_3} \lambda_1^{i-N_2}}{i!\,\mu^i} \right]^{-1} \tag{3.30}$$

The blocking probability B_{P_1} for class 1 is given by

$$B_{P_1} = \frac{(\lambda_1 + \lambda_2 + \lambda_3)^{N_3} (\lambda_1 + \lambda_2)^{N_2 - N_3} \lambda_1^{N_1 - N_2}}{N_1!\, \mu^{N_1}} P(0) \tag{3.31}$$

The blocking probability B_{P_2} for class 2 is given by

$$B_{P_2} = \sum_{i=N_2}^{N_1} \frac{(\lambda_1 + \lambda_2 + \lambda_3)^{N_3} (\lambda_1 + \lambda_2)^{N_2 - N_3} \lambda_1^{i - N_2}}{i!\, \mu^{i}} P(0) \tag{3.32}$$

The blocking probability B_{P_3} for class 3 is given by

$$B_{P_3} = \sum_{i=N_3}^{N_1} P(i) = B_{P_2} + \sum_{i=N_3}^{N_2 - 1} \frac{(\lambda_1 + \lambda_2 + \lambda_3)^{N_3} (\lambda_1 + \lambda_2)^{i - N_3}}{i!\, \mu^{i}} P(0) \tag{3.33}$$

The equation for channel utilization is

$$\%\ \text{Channel utilization} = \frac{\left[(1 - B_{P_1})\lambda_1\right] + \left[(1 - B_{P_2})\lambda_2\right] + \left[(1 - B_{P_3})\lambda_3\right]}{\mu N_1} \times 100 \quad \ldots \tag{3.34}$$

The flow diagram for the proposed adaptive channel scheme is shown in Figure 3.16.

The guard channel reservation strategy of the proposed adaptive scheme is shown in Figure 3.16. For any traffic that arrived in the cell at first, it searches whether the minimum number of fixed channels for all the classes are empty or not, and if it finds free channels in the minimum fixed channels range M_3 the call will be accepted [35]. If no channel is empty in this range, then by calculating the factors α_1, α_2, f_1 and f_2 based on the channels occupied and the arrival rate of the traffics, the range N_3 and N_2 are determined is the variable channels range available for the priority 3 and priority 2 traffic classes. If the call arrived is of traffic class 3 and no channel is empty in the range then the call will be rejected. When a class 2 call arrives then if in the range N_2 any channel is empty, the call will be assigned and if no channel is empty then the call is rejected. Now when the call is of class 1 then in the range N_1if no channel is empty, the call will be rejected but if any channel is empty then the call will be assigned [36].

3.4.5 Analysis of proposed scheme

In this state, we have established the act of the proposed scheme. We compared the act of the proposed scheme with priority scheme and non-priority

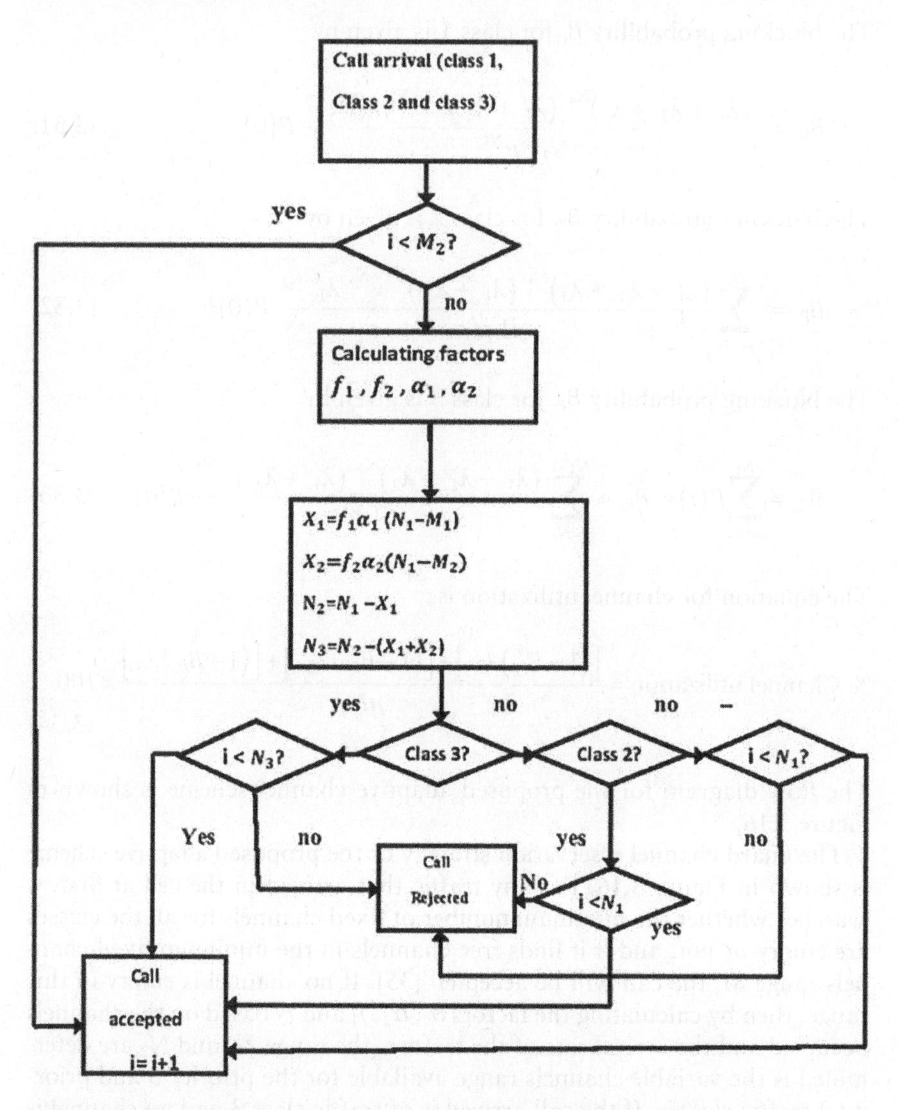

Figure 3.16 The flow diagram of the proposed adaptive channel scheme.

channel reservation scheme for multiclass traffic system. The major performance metrics considered here are new call blocking probability and channel utilization. Here we are assigned class 1, class 2, and class 3 for voice call, SMS, and internet connectivity, respectively. Voice call has been given more priority than SMS and internet connectivity, while SMS has been given more priority than internet connectivity. So voice call, SMS, and internet connectivity are priority 1, priority 2, and priority 3 traffic classes for our analysis [37].

Table 3.1 The basic assumptions of the parameters for performance analysis

Average call lifetime $\left(\frac{1}{\mu_C}\right)$	120 s
Average dwell time $\left(\frac{1}{\eta}\right)$	360 s
Total number of channels in a cell (N)	100
Call arrival rate:λ_1:λ_2:λ_3	1:2:3

In a non-priority scheme, no priority is given to any class of traffic. So the overall new call blocking probability of all classes remains the same. But in this case because of giving no priority to the upper-class traffic so that there will be more blocking of these traffics which is totally unexpected and not desirable [38].

In priority channel schemes there are some priority channels for the upper class traffics and they have been given more priority. So that the blocking probability of the highest priority traffic is lowered by giving some increase in the blocking of the lowest priority traffic classes. But due to the reservation of the channel for the highest priority traffic gives poor channel utilization.

In the proposed adaptive channel scheme the guard channels have been made adaptive so that they can be used most efficiently by all the classes based on their priority, their arrival rate, and the number of channels used. So in this case the channel utilization can be found very high than the priority channel scheme and slightly lower than the non-priority scheme [39]. This scheme also minimizes the blocking probability of the highest priority traffics which is the main advantage of this scheme [40].

The input parameters for the analysis are a total number of channels (N_1), a number of guard channels, average call lifetime $\left(\frac{1}{\mu_c}\right)$, average dwell time $\left(\frac{1}{\eta}\right)$, and new call arrival rate (calls/sec). Table 3.1 shows the assumptions of the parameters, which were used to perform the analysis. The parameters were the same for all the schemes which have been stated here

Figure 3.17 presents the performance of the multiclass traffic system in the case of a new call blocking probability. Here we can see that our proposed schemes show a better performance than the priority and non-priority channel schemes. For non-priority schemes, the blocking probability of all three classes is the same and so the blocking probability of class 1 and class 2 traffic is very high with respect to our proposed scheme. In the case of the priority channel scheme, the blocking probability of class 1 and class 2 is lower than the non-priority scheme but for class 3 it is higher than the non-priority

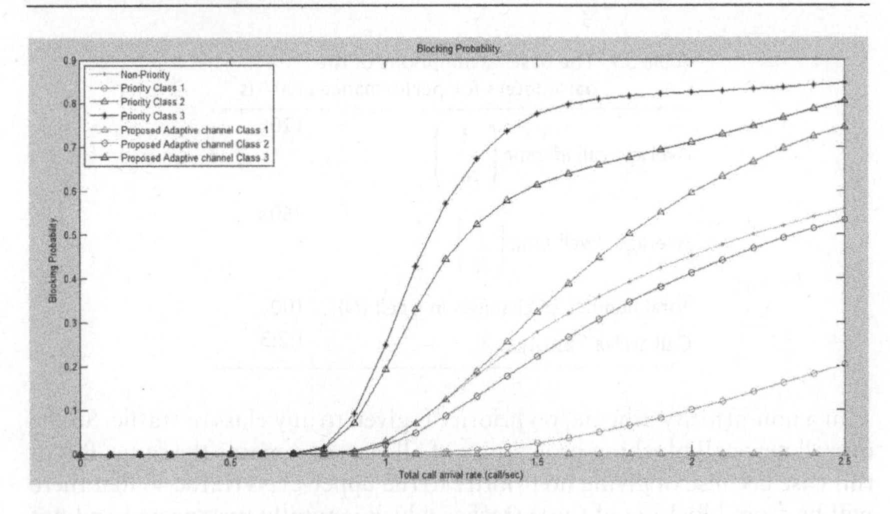

Figure 3.17 Comparison of new call blocking probability.

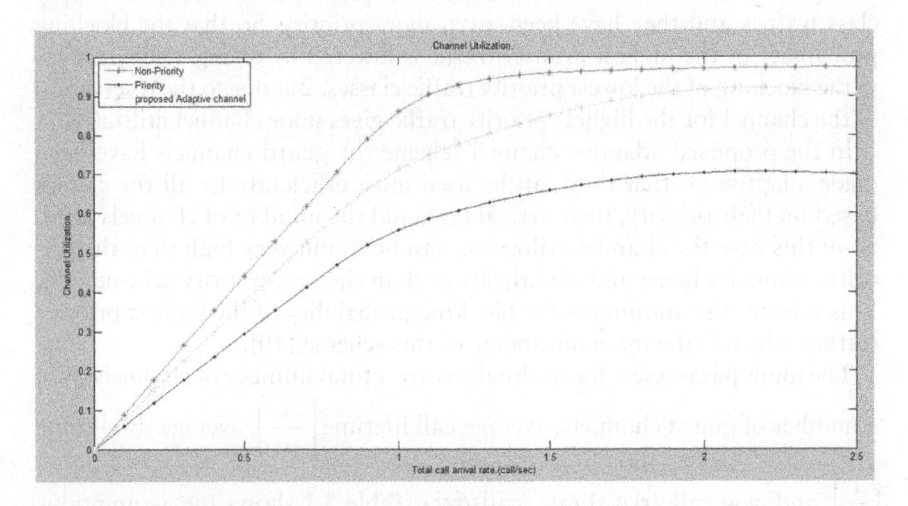

Figure 3.18 Comparison of channel utilization.

scheme because class 1 and class 2 have been given more priority than class 3 traffic and the blocking probability of class 1 and class 2 are improved by sacrificing for class 3. But in our proposed scheme the blocking probability for class 2 and class 3 have been more improved than the priority channel scheme. The blocking probability of the class 1 traffic is slightly greater than the priority channel scheme but the amount is negligible and we can take these two quantities equal. So our proposed scheme shows an overall better performance than the non-priority and the priority channel schemes [41].

Figure 3.18 shows the comparison of the channel utilization of the proposed adaptive scheme with the non-priority and the priority channel schemes. In this figure, we can see that our proposed adaptive scheme has more utilization than the priority channel scheme and slightly less than the non-priority scheme. Non-priority scheme always shows a better utilization because here no reserved channels for any classes and so that channels are always more utilized. But in the case of priority channel scheme because of reservation of the channels for the higher priority traffic sometimes all channels are not utilized and remain empty if the number of traffic arriving for the higher priorities is less. But by making the guard channels adaptive it has been made possible to use all the channels more perfectly and the maximum channel utilization for our proposed scheme has been found 91.5%.

3.5 CONCLUSION

In this paper, we proposed an adaptive channel scheme for multiclass services in wireless network systems. The proposed scheme can also be successfully applied to other communication systems, where multiple traffic classes are provided and resources are allocated based on the priority of the traffic classes. The idea behind the proposed scheme is to reserve an adaptive number of channels for the higher priority users. Reserving channels is equivalent to the guard channels; however, the number of reserved channels is not fixed in our proposed scheme to maintain higher channel utilization and to provide always lower call blocking probability for the higher priority users. More channels are reserved for the higher priority traffic class when the call arrival rate of the higher priority traffic class is higher compared to the lower priority users to support a large number of higher priority users. Thus, the scheme gives higher priority for higher priority calls over the lower priority calls without sacrificing channel utilization.

It has been shown that the proposed adaptive scheme is quite effective in reducing the call blocking probability of higher priority users without sacrificing channel utilization. While the proposed scheme blocks some more calls of lower priority calls instead of blocking higher priority calls during heavy traffic conditions [42].

3.6 FUTURE WORK

The cellular system is the most common wireless network communication system in the present. In this cellular system, non-priority and priority schemes are common methods for cellular priority. But, the major limitation of both the non-priority and priority schemes method channel utilization rate is limited. The adaptive channel scheme that is established in this paper is more advanced than non-priority and priority schemes. The new method can improve the rate of proper channel utilization. Here, three classes are discussed. But, on the rate of progress of technology is that it does it make sense, it's likely possible to use more than three classes that depend on the number of the type of call requests [43].

REFERENCES

1. M. Sanabani, S. Shamala, M. Othman, and Z. Zukarnain, "An enhanced bandwidth reservation scheme based on road topology information for QoS sensitive multimedia wireless cellular networks," in *Computational Science and Its Applications – ICCSA 2007*, Berlin, Heidelberg, 2007, pp. 261–274, doi: 10.1007/978-3-540-74477-1_26.
2. "User (computing)," *Wikipedia*. Aug. 03, 2020, [Online]. Available: https://en.wikipedia.org/w/index.php?title=User_(computing)&oldid=970929387 (accessed Aug. 05, 2020).
3. "What is wireless communications? - Definition from Techopedia," Available: https://www.techopedia.com/definition/10062/wireless-communications (accessed Aug. 05, 2020).
4. "History of wireless communication," Available: http://www.wirelesscommunication.nl/reference/chaptr07/history.htm (accessed Aug. 05, 2020).
5. "Mobile phone," *Wikipedia*. Jul. 28, 2020, [Online]. Available: https://en.wikipedia.org/w/index.php?title=Mobile_phone&oldid=969942891 (accessed Aug. 05, 2020).
6. "Radio network," *Wikipedia*. [Online]. Available: https://en.wikipedia.org/wiki/Radio_network (accessed Aug. 05, 2020).
7. "IEEE standard for local and metropolitan area networks–Part 15.7: Short-range wireless optical communication using visible light," *IEEE Std 802157-2011*, pp. 1–309, Sep. 2011, doi: 10.1109/IEEESTD.2011.6016195.
8. "Mobile phone features," *Wikipedia*. [Online]. Available: https://en.wikipedia.org/wiki/Mobile_phone_features (accessed Aug. 05, 2020).
9. R. W. W. F. L. writer R. W. has written hundreds of cell phone, S. Reviews, G. in D. of Books, and magazines our editorial process R. Ware, "What is a SIM card and why do you need one?" *Lifewire*. https://www.lifewire.com/what-are-sim-cards-577532 (accessed Aug. 05, 2020).
10. "Mobile phone operator," *Wikipedia*. [Online]. Available: https://en.wikipedia.org/wiki/Mobile_phone_operator (accessed Aug. 05, 2020).
11. "Removable user identity module," *Wikipedia*. Mar. 26, 2020. [Online]. Available: https://en.wikipedia.org/w/index.php?title=Removable_User_Identity_Module&oldid=947486524 (accessed Aug. 05, 2020).
12. M. P. Mishra and P. C. Saxena, "Survey of channel allocation algorithms research for cellular systems," *Int. J. Netw. Commun.*, vol. 2, no. 5, Art. no. 5, 2012.
13. S. Choi and K. G. Shin, "Adaptive bandwidth reservation and admission control in QoS-sensitive cellular networks," *IEEE Trans. Parallel Distrib. Syst.*, vol. 13, no. 9, Art. no. 9, Sep. 2002, doi: 10.1109/TPDS.2002.1036063.

14. "Cellular traffic," *Wikipedia*. Dec. 14, 2018. [Online]. Available: https://en.wikipedia.org/w/index.php?title=Cellular_traffic&oldid=873673202 (accessed Aug. 05, 2020).
15. C. Lin, Y. Wei, and Z. Shan, "Adaptive resource management for mobile multimedia communications in asymmetric wireless network," in *Second International Conference on Quality of Service in Heterogeneous Wired/Wireless Networks (QSHINE'05)*, Aug. 2005, pp. 6–45, doi: 10.1109/QSHINE.2005.8.
16. M. Srivastava, "Handover," 23:33:12 UTC. [Online]. Available: https://www.slideshare.net/mansri123/handover-16638241 (accessed Aug. 05, 2020).
17. W.-C. Kim, C.-S. Bae, S.-Y. Jeon, S.-Y. Pyun, and D.-H. Cho, "Efficient resource allocation for rapid link recovery and visibility in visible-light local area networks," *IEEE Trans. Consum. Electron.*, vol. 56, no. 2, Art. no. 2, May 2010, doi: 10.1109/TCE.2010.5505965.
18. M. Z. Chowdhury, Y. M. Jang, and Z. J. Haas, "Call admission control based on adaptive bandwidth allocation for wireless networks," *J. Commun. Netw.*, vol. 15, no. 1, Art. no. 1, Feb. 2013, doi: 10.1109/JCN.2013.000005.
19. "Call blocking," *Wikipedia*. May 29, 2019. [Online]. Available: https://en.wikipedia.org/w/index.php?title=Call_blocking&oldid=899289499 (accessed Aug. 05, 2020).
20. A. Sgora and D. D. Vergados, "Handoff prioritization and decision schemes in wireless cellular networks: a survey," *IEEE Commun. Surv. Tutor.*, vol. 11, no. 4, Art. no. 4, 2009, doi: 10.1109/SURV.2009.090405.
21. M. K. Luka, A. A. Atayero, and O. I. Oshin, "Call admission control techniques for 3GPP LTE: A survey," in *2016 SAI Computing Conference (SAI)*, Jul. 2016, pp. 691–700, doi: 10.1109/SAI.2016.7556057.
22. *Handbook of Wireless Networks and Mobile Computing | Wiley, Wiley.com*, Aug. 04, 2020. Available: https://www.wiley.com/en-bd/Handbook+of+Wireless+Networks+and+Mobile+Computing-p-9780471419020 (accessed Aug. 04, 2020).
23. O. T. W. Yu and V. C. M. Leung, "Adaptive resource allocation for prioritized call admission over an ATM-based wireless PCN," *IEEE J. Sel. Areas Commun.*, vol. 15, no. 7, Art. no. 7, Sep. 1997, doi: 10.1109/49.622906.
24. N. Nasser and H. Hassanein, "Prioritized multi-class adaptive framework for multimedia wireless networks," in *2004 IEEE International Conference on Communications (IEEE Cat. No.04CH37577)*, Jun. 2004, vol. 7, pp. 4295–4300, doi: 10.1109/ICC.2004.1313358.
25. M. Salamah, "An adaptive multi-guard channel scheme for multi-class traffic in cellular networks," in *IEEE International Conference on Computer Systems and Applications, 2006*, Mar. 2006, pp. 716–723, doi: 10.1109/AICCSA.2006.205169.
26. S. Rezvy, S. Rahman, A. Lasebae, and J. Loo, "System capacity improvement by on request channel allocation in LTE cellular network," in *2014 48th Annual Conference on Information Sciences and Systems (CISS)*, Mar. 2014, pp. 1–5, doi: 10.1109/CISS.2014.6814105.
27. M. Z. Chowdhury, Y. M. Jang, and Z. J. Haas, "Call Admission Control based on adaptive bandwidth allocation for multi-class services in wireless networks," in *2010 International Conference on Information and Communication Technology Convergence (ICTC)*, Nov. 2010, pp. 358–361, doi: 10.1109/ICTC.2010.5674699.

28. M. Mamman, Z. M. Hanapi, A. Abdullah, and A. Muhammed, "An adaptive call admission control with bandwidth reservation for downlink LTE networks," *IEEE Access*, vol. 5, pp. 10986–10994, 2017, doi: 10.1109/ACCESS.2017.2713451.
29. A. L. Enlil Corral-Ruiz, F. A. Cruz-Pérez, and G. Hernandez-Valdez, "Channel holding time in mobile cellular networks with generalized Coxian distributed cell dwell time," in *21st Annual IEEE International Symposium on Personal, Indoor and Mobile Radio Communications*, Sep. 2010, pp. 2348–2353, doi: 10.1109/PIMRC.2010.5671709.
30. E. Cianca, A. D. Luise, M. Ruggieri, and R. Prasad, "Channel-adaptive techniques in wireless communications: an overview," *Wirel. Commun. Mob. Comput.*, vol. 2, no. 8, Art. no. 8, 2002, doi: 10.1002/wcm.98.
31. N. Lu, J. Bigham, and N. Nasser, "An intra-class and inter-class utility-fair bandwidth adaptation algorithm for multi-class traffic in wireless networks," in *2006 Asia-Pacific Conference on Communications*, Aug. 2006, pp. 1–5, doi: 10.1109/APCC.2006.255822.
32. M. J. Hossain, P. K. Litthaladevuni, M.-S. Alouini, V. K. Bhargava, and A. J. Goldsmith, "Adaptive hierarchical modulation for simultaneous voice and multi-class data transmission over fading channels," in *2003 4th IEEE Workshop on Signal Processing Advances in Wireless Communications - SPAWC 2003 (IEEE Cat. No.03EX689)*, Jun. 2003, pp. 105–109, doi: 10.1109/SPAWC.2003.1318931.
33. J. Jobin, S. K. Tripathi, M. Faloutsos, and S. Gokhale, "Using statistical data for reliable mobile communications," *Wirel. Commun. Mob. Comput.*, vol. 2, no. 1, Art. no. 1, 2002, doi: 10.1002/wcm.36.
34. M. Jain and R. Mittal, "Adaptive call admission control and resource allocation in multi server wireless/cellular network," *J. Ind. Eng. Int.*, vol. 12, no. 1, Art. no. 1, Mar. 2016, doi: 10.1007/s40092-015-0129-3.
35. R. Bhattacharjee, T. A. Chowdhury, and M. Z. Chowdhury, "Priority based adaptive guard channel for multi-class traffic in wireless networks," in *2013 International Conference on Electrical Information and Communication Technology (EICT)*, Feb. 2014, pp. 1–4, doi: 10.1109/EICT.2014.6777832.
36. L. Peng, L. Jiandong, L. Hongyan, W. Kan, and M. Yun, "Efficient resource allocation scheme to maximise number of users with quality of service demands in small cells," *China Commun.*, vol. 11, no. 1, Art. no. 1, Jan. 2014, doi: 10.1109/CC.2014.6821306.
37. W. Zhuang, K. C. Chua, and S. M. Jiang, "Measurement-based dynamic bandwidth reservation scheme for handoff in mobile multimedia networks," in *ICUPC '98. IEEE 1998 International Conference on Universal Personal Communications. Conference Proceedings (Cat. No.98TH8384)*, Oct. 1998, vol. 1, pp. 311–315, doi: 10.1109/ICUPC.1998.732845.
38. W. Castellanos, P. Acelas, P. Arce, and J. C. Guerri, "Evaluation of a QoS-aware protocol with adaptive feedback scheme for mobile Ad hoc networks (short paper)," in *Quality of Service in Heterogeneous Networks*, Berlin, Heidelberg, 2009, pp. 120–127, doi: 10.1007/978-3-642-10625-5_8.
39. M. Z. Chowdhury, Y. M. Jang, and Z. J. Haas, "Priority based bandwidth adaptation for multi-class traffic in wireless networks," . arXiv preprint arXiv:1412.4322, p. 6.

40. Y. C. Kim, D. E. Lee, B. J. Lee, Y. S. Kim, and B. Mukherjee, "Dynamic channel reservation based on mobility in wireless ATM networks," *IEEE Commun. Mag.*, vol. 37, no. 11, Art. no. 11, Nov. 1999, doi: 10.1109/35.803649.
41. C. Oliveira, J. B. Kim, and T. Suda, "An adaptive bandwidth reservation scheme for high-speed multimedia wireless networks," *IEEE J. Sel. Areas Commun.*, vol. 16, no. 6, Art. no. 6, Sep. 2006, doi: 10.1109/49.709449.
42. M. Z. Chowdhury, M. S. Uddin, and Y. M. Jang, "Dynamic channel allocation for class-based QoS provisioning and call admission in visible light communication," *Arab. J. Sci. Eng.*, vol. 39, no. 2, pp. 1007–1016, Feb. 2014, doi: 10.1007/s13369-013-0680-4.
43. "The challenges of M2M massive access in wireless cellular networks - ScienceDirect." Available: https://www.sciencedirect.com/science/article/pii/S235286481500005X (accessed Aug. 06, 2020).

Chapter 4

Quantum blockchain

A state-of-the-art study in the intelligent transportation system

Aleem Ali and Rasmeet Kaur
Glocal University Saharanpur

CONTENTS

4.1 INTRODUCTION

In recent years, quantum blockchain (QB) technology has advanced countless services in digital data to deliver clarity, redundancy, and liability. Blockchain is a technique of IT innovation. As the fundamental innovation of bitcoin, it is a decentralized information base. Blockchain has the qualities of decentralization, receptiveness, and the data put away in it can't be messed with, so it has a wide scope of uses in digital money, in the data security industry, and smart contracts [1]. A blockchain can be considered as a distributed database containing records, or public records that have been implemented and distributed amidst partaking parties.

DOI: 10.1201/9781003156789-4

Table 4.1 Structure of block header

Field	*Size*	*Description*
Version	4 bytes	It is a number that tracks software or protocols upgrade
Merkle tree root hash	32 bytes	The hash approximation of the relative multitude of block's transactions
Timestamp	4 bytes	It is the estimated creation time of a block
Difficulty target	4 bytes	It is the POW algorithm difficulty target of a block
Nonce	4 bytes	A field generally begins with 0, later step up for individual hash computation
Parent block hash	32 bytes	The 256-digit value which paid attention to the preceding block

Every transaction inside the public record is confirmed by the agreement of the plethora of members in the framework. When entered, it's not feasible to delete the data. The blockchain holds a definite and unquestionable record of the entire exchange constructed at any time. Bitcoin, the decentralized distributed digital cash, is a well-known model that utilizes blockchain innovation [2]. The computerized money bitcoin is profoundly questionable yet the hidden blockchain innovation has worked immaculately and discovered a vast scope of utilization in monetary as well as non-monetary sectors.

Blockchain, commonly regarded as one of the revolutionary technologies developed in recent years, is rapidly evolving and has the potential to revolutionize applications with increasingly centralized intelligent transport systems (ITS). To create a secure, trusted, and decentralized autonomous ITS ecosystem, Blockchain can be used to create better use of the legacy ITS infrastructure and resources, particularly effective for crowdsourcing technology.

Blockchain is a succession of blocks, alongside a preceding block hash carried inside the header part. There are two parts inside a block: header and body [3] and a block has only one parent block. The components of the Header part are described in Table 4.1.

The body of a block is formed from an exchange counter and transactions. The highest count of exchanges retain by a block presumes the size of the block and also the size of every exchange. To approve the authenticity of exchanges blockchain makes use of an asymmetric cryptography system. In the distributed network of blockchain, the communication and trust between nodes depend on the digital signature, which understands the identifiable proof and the validness of data.

4.1.1 Digital signature

Individual client claims a pair of private keys and a public key. To sign the transactions, the private key will be used [3]. The transactions that are digitally signed are conveyed across the entire organization. The regular digital signature is engaged with two stages: the signing stage and the verification stage. For instance, one user named A needs to transmit a message to another user named B.

i. During the signing stage, user A encrypts the information using its private key and circulates the encoded outcome and the actual data to B.
ii. At the time of the verification stage, B approves the value through A's public key. So, B could undoubtedly examine if the information has been altered or not. The normal digital signature calculation utilized in the blockchain is ECDSA.

4.1.2 Major attributes of blockchain

Some vital attributes of blockchain are undermentioned.

i. **Decentralization:** In a traditional centralized exchange system, the individual exchange must be confirmed through the central trusted organization surely coming about to the outlay and the performance issues at the main servers. Compared with the centralized model, during this time, no outsider is required in the blockchain.
ii. **Persistency:** Exchanges might be confirmed swiftly and illegal exchanges wouldn't be acknowledged by authorized miners. It's almost hard to undo or roll back transactions whenever they are placed for the blockchain.
iii. **Anonymity:** An individual client can agree with the blockchain with a generated address that does not reveal the actual character of the client. Blockchain never ensures the ideal protection conservation due to the characteristic requirement.

4.2 PRELIMINARIES

Some of the crucial preliminaries used in the development and to withstand QB are discussed below.

4.2.1 Key technologies used in blockchain

i. **Hash algorithm:** The hash calculation can change an info estimation of self-assertive length into a binary estimation of fixed length. This binary value is called a hash value, which can be utilized to confirm the integrity of the information. The famous proof-of-work (POW) algorithm is the utilization of the hash algorithm. The hash estimation of the data is kept in the block of the blockchain. Moreover, the signature normally utilized in blockchains is likewise produced by hashing the private key, and the information that should be agreed upon.

ii. **Proof of work:** POW can be perceived as proof that you have done a specific measure of work. In a blockchain system, any node that needs to produce another block and compose it to the blockchain must purpose the POW puzzle in the blockchain network. POW puzzle is an NP-hard issue. Nodes that compute and understand the POW puzzle can regularly get cryptocurrency as remunerations.

iii. **Difficulty value:** The difficulty value in the POW is a significant reference for miners in mining, and it decides the number of hash tasks miners need to run to create a legitimate block. During the mining cycle, the difficulty can be progressively changed by the computing power in the entire blockchain network. In the bitcoin framework, the difficulty value is set to decide that the new block generation rate is kept up at 10 min paying little heed to the mining capacity.

iv. **Timestamp:** The blockchain framework utilizes the timestamp to demonstrate that the exchange happened right now. In this way, the responsibility for money in the exchange has been moved, and the past proprietor can't utilize the currency once more. Moreover, each block is additionally stepped with the right timestamp to frame a right linked list in sequential order.

4.2.2 Structure of blockchain

The blockchain framework comprises various information blocks, which have perfectly organized records. Each block contains a timestamp, a hash estimation of its content, and the hash estimation of the preceding block. The blockchain is shaped by connecting each block with the hash value. Each block is created after the previous block in sequential requests. When the block is affirmed to be substantial, it can barely be altered [4]. The schematic of the traditional blockchain has appeared in Figure 4.1.

Organization of blockchain: Nakamoto has portrayed the means to run the network of blockchain in Nakamoto. Following are the steps:

i. Fresh exchanges are communicated to each node of the network.
ii. Every single node gathers new exchanges into a block.
iii. Every node runs the POW calculation for its block.
iv. When a node illuminates the POW puzzle, it circulates the block to all nodes.
v. Other nodes acknowledge the block just if the entireties of the exchanges are substantial and unused.
vi. Nodes express the concede of the block by dealing with making the following blocks in the chain, utilizing the hash of the conceded block as the past hash.

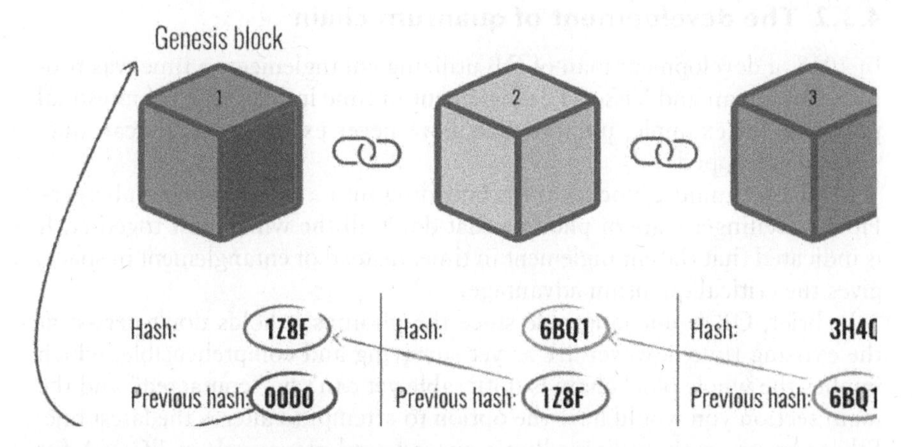

Figure 4.1 Illustration of blockchain.

4.3 BUILDING QUANTUM BLOCKCHAIN

In QB, the records in a block are encoded into a progression of photons that are trapped with one another. These blocks are connected in a sequential request through entanglement as expected.

4.3.1 Construction of quantum blockchain

As the blocks making up QB are moved inside a network of quantum computers, photons encoding each block get made and afterward consumed by the nodes making up the network. Notwithstanding, QB depends on entanglement. At the point when at least two particles, for example, photons get entangled, they can impact each other all the while regardless of how far separated they are [4].

Regular blockchains gather records into blocks of data, which cryptography joins in sequential order. If a hacker endeavors to alter a specific block, the cryptography is intended to negate all future blocks following the altered block. In this situation, a programmer can't mess with any photon encoding records of the past, since those photons presently don't exist in the current time they previously got assimilated. In the best-case scenario, a hacker can endeavor to alter the latest photon, the most current block, and effectively doing so would discredit that block, advising others it got hacked.

The analysts state that with entanglement in time, estimating the last photon in a block impacts the primary photon of that block in the past before it got estimated. Existing records inside QB are not just connected to an earlier record but instead a record previously, one that doesn't exist any longer. "This work can be seen as a quantum time machine," says senior creator Matt Visser, a hypothetical physicist at the Victoria University of Wellington.

4.3.2 The development of quantum chain

In 2018, a development plan of QB utilizing entanglement in time was proposed by Rajan and Visser. Entanglement in time implies that infinitesimal particles, for example, photons that have never existed together can likewise be entrapped.

In this technique, blockchain is encoding into a temporal Greenberger–Horne–Zeilinger state of photons that don't all the while exist together. It is indicated that the entanglement in time, instead of entanglement in space, gives the critical quantum advantage.

In brief, QB is unchangeable since the photons it holds don't persist at the existing time however are as yet surviving and comprehensible, which implies the whole blockchain is noticeable yet can't be "contacted" and the main section you would have the option to attempt to alter is the latest one. Ethereum presently utilizes elliptic curve-based plans such as ECDSA for signing exchanges and BLS for signature collection; in any case, as mentioned, the elliptic curve cryptography wherein security depends on the trouble of comprehending the discrete logarithm is powerless against quantum processing and should be supplanted with a quantum-safe plan.

The SHA-256 hash function is quantum-safe, which implies that there is no productive familiar calculation, traditional or quantum that can alter it. While Grover's algorithm executes quantum search over a black-box function, SHA-256 has demonstrated to be a safe resistance to collision and reimage threats. Grover's algorithm can just decrease N inquiries of the black-box function, SHA for this situation, to $\sqrt{N}$, so as opposed to looking through 2^256 potential outcomes, we just need to look through 2^128, which is even slower as compared to other algorithms.

In the Ethereum 2.0 Serenity overhaul, records will have the option to determine their plan for approving transactions, involving the alternative to change to a quantum-safe signature plan. Hash-based signature plans like the Lamport mark are accepted to be quantum-safe, quicker, and less perplexing than ECDSA. Tragically, this plan experiences a size problem. The size of the Lamport public key and signature together is over and above the ECDSA public key and signature. Thus, the utilization of the Lamport Signature plan will require 231× more space than ECDSA, which is lamentably too huge to be in any way functional right now.

To comprehend blockchain with regards to quantum computing and quantum-enhanced attacks we should comprehend two essential calculations: Grover's Algorithm and Shor's Algorithm. The previous is an input search algorithm to locate an exceptional input to a black-box function which works fundamentally quicker contrasted to brute force search, along these lines seriously bargaining hash functions of lacking length.

The last gives an outstanding rate increment in factoring integers when contrasted with the overall number field sieve and can be applied to the hidden subgroup and discrete logarithm issues. These issues are at the core of breaking many familiar asymmetric ciphers and accordingly apply to breaking things like public-key cryptography and digital signatures. On the whole, both quantum algorithms present a critical threat to frameworks executing blockchain [8].

i. **Grover's algorithm:** Blockchain depends on the calculation of hashes to give protection from the adjustment of the earlier blocks. The chain is protected in contempt of expanded correction by both its dispersed nature and the computational exertion needed to reprocess a chain of blocks. Alteration of a solitary block is made sure about by the trouble of resulting in a hash collision with the current hash, which adds up to the issues of rearranging the hash function.

   Grover's calculation is explicitly an answer for the issue of generating a pre-picture of an estimation of a function that is hard to rearrange. If a signature is given that is the hash estimation of some data $s=H(d)$, and the function $H(d)$ can be executed on a quantum computer, at that point, Grover's algorithm permits us to discover d for guaranteed s in time of order $O(\sqrt{n})$ where n is the size of the space of substantial hashes.

   As such, it permits us to create a hash collision more effectively than a brute force search, which would be (n). For a hash of length k bits, this implies that we have a critical speedup by a factor of $2\ k/2$. This can be extremely huge in any event, for little estimations of k.

ii. **Shor's algorithm:** Shor's Algorithm gives a sensational advancement in the effectiveness of factoring large numbers [5]. Accordingly, Shor's algorithm can be utilized to attack RSA encryption and associated issues. The multifaceted nature of the overall number field sieve is super-polynomial yet sub-exponential, while Shor's calculation is polynomial in the input length, making the increase in speed generally exponential.

   The outcome is that any part of a blockchain usage that depends on RSA or comparable calculations would be helpless against quantum computational attacks. The primary objective of Shor's calculation was the factoring of large composite whole numbers comprising a result of two large primes [6].

As expressed over, Grover's algorithm looks for the pre-picture to a function value and can do as remarkably quicker as the old brute force search of creating each output and contrasting it with secluding the producing input.

Grover's calculation can be utilized in two different ways to assault the blockchain. The one and generally self-evident is that it may be utilized to look for hash impacts that can be utilized to supplant impedes in a situation without upsetting the respectability of the blockchain. The other is that it can accelerate the age of nonces, possibly to the point that whole chains of records can be reproduced with predictably changed hashes adequately rapidly to sabotage the trustworthiness of the chain. In the two cases, the calculation is utilized to locate the pre-picture of a given incentive under a hard to reverse capacity.

As an auxiliary danger, in any part of a blockchain usage that utilizes public/private key cryptography, regardless of whether it is in data trade among parties or advanced marks, a quantum PC might have the option to break the security of the encryption.

4.3.3 Quantum-resistant blockchain

A blockchain is a record that records data identified with exchanges. The exchanges are persistently added by block size. Toward the finish of a given period, the block is encrypted utilizing a hashing function [7]. Public permission-less blockchain networks permit disturbance of incorporated players.

Public blockchains guarantee unchanging records and the security of exchanges. Quantum computers can break the hash signature utilizing Shor's algorithm. There is a requirement for a post-quantum secure signature conspires for post-QB security [8,9]. The quantum-resistant record is cryptographic money that endeavors to stay at the forefront of security and usefulness. It highlights quantum-safe cryptographic conventions and a custom verification of stake framework. The digital money record is impervious to both classic and quantum computing threats. It utilizes a hash-based digital signature that is quantum-safe [10].

The record gives a super secure reinforcement store of significant worth in case of an unexpected development in quantum computing. The underlying point of the chain is to offer a low volume of super-secure exchanges in the primary emphasis with ensured life span. Quantum safe hash tag-based signature tree-like Extended Merkel signature plot and a low force confirmation of stake algorithm are utilized for quantum-safe records [11].

Extended Merkel signature utilizes a one-timing signature (OTS) plan. This plan signs one message with one key. A signature OTS key is utilized to sign two distinct messages so an attacker could produce a legitimate signature for a third message you had never signed. An attacker can produce a legitimate exchange that is rarely affirmed. One can utilize an alternate OTS key for each message.

4.4 QUANTUM BLOCKCHAIN SECURING FUTURE OF BLOCKCHAIN

As recommended by the scientists QB can be utilized to prevent future blockchains from hackers utilizing quantum computers [12].

As per the studies, QB would take the benefit of entanglement, which mostly applies to circumstances concerning space [13]. Yet, it could likewise be helpful for circumstances including time, for example, blockchains. In such a blockchain exchange records could be presented by sets of entangled photons connected in sequential requests. At the point when exchanges occur, photons would be generated and consumed by the hubs that involve a network. Since entangled photons are connected across time, they can be caused to have never existed simultaneously.

Accordingly, any estimations of the later photon in a record would be impacted by the photon that came first, previously, before it was estimated which means if an attacker attempted to access such a record, they would find it impossible, as the entangled photon that represented it no longer existed in the current time. This implies that an attacker would just actually have the option to get to the most recent blocks, and if they did it would negate the others, which would make clients aware of the network for the intruder.

Timelines for the rise of quantum computers may be fuzzy, yet the danger they pose to the vaunted security of blockchain technology is significantly genuine.

Initially, blockchain is making advances across various enterprises, most eminently as a track-and-follow instrument demonstrating the provenance of merchandise across immense supply chains. Notwithstanding, as blockchain administrations develop and quantum computers start to arise, right now is an ideal opportunity to begin contemplating quantum-safe blockchain.

At present, it is hard to go in reverse on a blockchain's permanent record and change unique data in each block of the chain [14]. This is particularly the situation as blocks are added with more information. Individuals can only with significant effort modify history on its unchanging record because different hubs on the chain would consequently dismiss any changes. Additionally, customary blockchains depend on asymmetric cryptography, which forestalls fake signing. Lamentably, quantum computers could hypothetically break the changelessness of any block in the chain and misrepresent verifiable exchanges.

4.4.1 Prepare now for post-quantum security

Regardless of whether a small amount of the forecasts about blockchain come true, the security a lot is on the line for buyers and organizations.

Blockchain made Gartner's rundown of top ten key innovation patterns for 2020 and was anticipated to penetrate everything from handling protection claims, credits, and reviews to personality the executives for understudies, patients, and residents. By 2022, IDC experts said 10% of the world's grown-up populace will enlist for a blockchain-based self-sovereign ID, making an extending business sector of 485 million individuals who need to claim and control their digital IDs. Regardless of whether it is checking exchanges for bitcoin mining or the following food from ranch to the table, blockchain's security skyline relies upon the interesting circumstance.

4.4.2 Resecuring cryptocurrency

Cryptographic money isn't only for buyers exchanging bitcoins. IDC investigators anticipated that more than 12 nations, generally arising economies, will start giving advanced currency utilizing blockchain innovations to advance financial dependability and energize electronic trade by 2023. As some administrations start utilizing cryptographic forms of money, organizations should start taking a gander at post-QB innovation for business-to-business exchanges, for example, acquisition including cooperation among purchasers and providers.

4.4.3 Embracing cryptography agility

It is difficult to excuse the security suggestions around blockchain and quantum computers. Prominent blockchain models will in general highlight following the credibility of remarkable exchanges like uncommon fine art or jewels. Blockchain could support numerous regular activities, accelerating possession recordkeeping, settlement installments, and even devotion and prizes following for clients in numerous businesses.

Smart cities that depend on the Internet of Things innovation can utilize blockchain as a component of the foundation to exchange energy, charge electrical vehicles, and oversee smart lattices. By 2023, Gartner experts think blockchain will be adaptable actually and will uphold confided in private exchanges with vital information secrecy.

Another novel and hypothetical framework for blockchain-based information storage could guarantee that hackers won't have the option to break cryptographic forms of money once the quantum period begins [15]. The thought, proposed by specialists at the Victoria University of Wellington in New Zealand, would make sure about digital currency prospects for quite a long time utilizing a blockchain innovation that resembles a time machine.

As indicated by the scientists, QB is immutable due to the photons that it contains that don't exist at the current time yet are as yet surviving and coherent. This implies the whole blockchain is obvious yet can't be "contacted" and the lone passage you would have the option to attempt to alter is the latest one. Indeed, the analysts express, "In this spatial entrapment case, if a hacker attempts to alter any photon, the full blockchain would be nullified right away."

The system encodes a blockchain into these transiently entangled states, which would then be able to be coordinated into a quantum network for additional helpful tasks. The author writes, the entanglement in time, assumes the significant part for the quantum advantage over an old-style blockchain.

4.4.4 Quantum computers and cryptography

A private–public key pair is created in asymmetric cryptography in such a way that the two keys have a mathematical connection between them. The private key is kept secret, as the name implies, while the public key is made publicly accessible. This enables a person to create a digital signature that can be checked by someone who has the corresponding public key.

The protection of asymmetric cryptography is based on a one-way mechanism called a mathematical principle. The theory dictates that it is easy to extract the public key from the private key, but not the other way around. All known (classical) algorithms need an astronomical amount of time to perform such a calculation to extract the private key from the public key and are therefore not realistic. However, the mathematician Peter Shor presented a quantum algorithm in 1994 that could crack the privacy hypothesis of the most popular asymmetric cryptography algorithms. This means the algorithm could be used by everyone with a significantly strong quantum computer to extract a private key from its corresponding public key and therefore to invalidate any digital signature.

"Cryptography is the field concerning semantic and numerical methods for making sure about data, especially in interchanges" and the safety attributes for Distributed Ledger Technology. Quantum computers some time named as the Achilles' impact point of bitcoin and the entire digital money industry [16]. Because of their staggering benefits in figuring speed, quantum computers could hypothetically be utilized to upset the action not just of a decentralized framework or a blockchain however of any product utilizing any sort of encryption.

Quite possibly the most notable use of quantum PCs is breaking the numerical trouble basic the greater part of as of now utilized cryptography. Quantum computing utilizes subatomic particles. These particles will be available in more than one state at one specific time. On account of this special attribute, the computing cycle is a lot quicker than the ones that we are utilizing today. Aside from making the computation quick, quantum processing additionally utilizes less energy when contrasted with the current computing machines.

The quantum computing innovation attempts to make some sorts of computing issues much easier to solve as compared to present classical computers. These problems include breaking specific kinds of encryption, specifically the methodology utilized in public key infrastructure (PKI), which underlies practically all of the present online communications.

4.5 WHAT IS QUANTUM CRYPTOGRAPHY

Cryptography is a scheme to encode information or transform plain text into coded content and anyone who has the accurate "key" can understand it. Quantum cryptography (QC), likewise, basically employs the principles of quantum mechanics to encode the data and communicate it in such a way that it is not possible to hack it.

QC, also termed quantum encryption, implements postulates of quantum mechanics to encrypt information in such a manner that it can never be accessed by an unauthorized person [17]. To perform this task, quantum computers are required, which have the strong computing power to encrypt and decrypt data. The attribute of entanglement in quantum computation has a suggestion for QC.

Numerous endeavors have been made to investigate post-QC methods to plan for the rise of quantum machines and quantum processing. Post-QC alludes to cryptographic computations that are subjected to be safe against a threat by a quantum PC. These difficult numerical conditions take traditional PCs months or even a long time to break. But, quantum computers executing Shor's algorithm will have the option to break these mathematical systems in minutes.

QC possesses secure transmission of messages by using the standards of quantum mechanics. QC is dissimilar to conventional cryptographic frameworks [18]. QC depends on physics; on the other hand, traditional cryptography depends on mathematics.

The key aspects of QC encompass quantum entanglement, quantum measurement, and quantum teleportation [19].

Quantum entanglement: It alludes to the phenomena of the entangled relationship among two microscopic particles from a single source in a microscopic environment. Anyways, as far as the state of one particle alters, it is also possible to alter the state of other particles instantly.

Quantum measurement: This can be employed to retrieve the data encrypted in the quantum state. This procedure is a dynamic procedure that determines distinct quantum states and quantum state evolution. Quantum measurement and evolution are two foundational types of quantum state evolution; quantum measurement is directly related to the modification of the quantum state.

Quantum teleportation: It is a process to transfer the unknown quantum data to the distant light quantum, with which exists the entanglement performance using quantum entanglement features. However, the physical carrier that originally carried the quantum data is left in place without being transferred [20].

4.5.1 How quantum cryptography works

QC sends the information using a series of photons, starting from one area over the next through a fiber-optic channel. These two endpoints can estimate the key and if it is secure to utilize by contrasting approximations of the properties of a little portion of photons.

Here is the process:

I. The photons are sent by the sender via a channel which gives them one of four polarizations and bit assignments: vertical (1 bit), horizontal (0 bit), 45° right (1 bit), or 45° left (0 bit).
II. The photons transmit to a destination, which utilizes two beam splitters to "read" the polarization of each photon. The receiver doesn't realize which beam splitter to use for every photon and needs to figure out which one to use.
III. After the stream of photons is transmitted, the receiver tells the sender which beam splitter was used for each photon, and the sender contrasts that data and the grouping of polarizers utilized to transmit the key. The photons read utilizing some unacceptable beam splitter are rejected, and the subsequent bits turn into the key.

Consider an instance for working on QC. Suppose there are two users, A and B, who wish to pass a secret message to each other that cannot be accessed by anyone else. Using the quantum key distribution (QKD), A passes a sequence of polarized photons towards B via a fiber-optic channel (see Figure 4.2). This channel should not be made secure as the photons have a randomized quantum state. QKD strategy involves conveying the

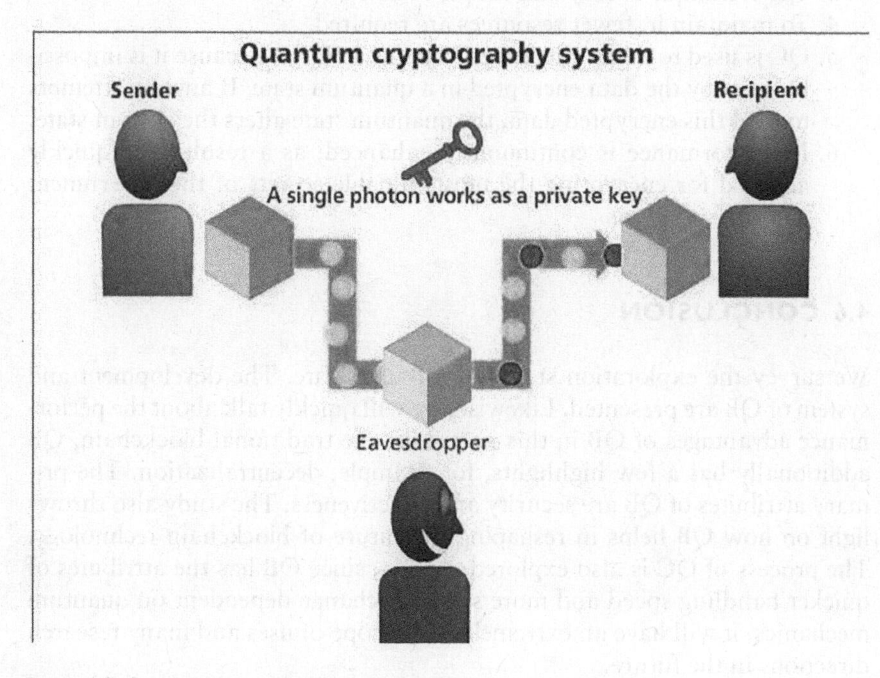

Figure 4.2 Quantum cryptography system.

encryption keys utilizing some very distinct nature of subatomic particles that is, un-hackable.

If an eavesdropper, attempts to listen to the communication, he needs to read every photon to access the secret information. Here, he should give that photon to B. He alters the quantum state of the photon by reading it, which causes faults in the quantum key. It alerts both A and B about the interference of somebody. They are aware of the attack on the key hence they discard the key. A need to send a different key (that is not attacked) to B, so that B can utilize that key to read the information [21].

The growth of QC is favorable, yet there are also some limitations. QKD, being the most significant element of QC, is also a key component of secure quantum communication. The appearance of QC comes from the risk that quantum PC will utilize Shor's secure algorithm to repudiate the current public-key cryptography. Although, if it is proven that the traditional algorithm is capable to sustain the attack of the quantum PC, then the persistence of QKD is subjected to doubt [22].

Here are some benefits of QC:

1. QC provides secure communication by employing fundamental principles of physics rather than mathematical algorithms or computing technologies used these days.
2. QC is virtually un-hackable.
3. The technique is easy and simple to use.
4. To maintain it, fewer resources are required.
5. QC is used to determine eavesdropping in QKD because it is impossible to copy the data encrypted in a quantum state. If anyone attempts to read this encrypted data, the quantum state alters the current state.
6. Its performance is continuously enhanced; as a result, it is quickly adopted for encrypting the most valuable secrets of the government and organizations.

4.6 CONCLUSION

We survey the exploration status of QB as of late. The development and system of QB are presented. Likewise, we will quickly talk about the performance advantages of QB in this part. Like the traditional blockchain, QB additionally has a few highlights, for example, decentralization. The primary attributes of QB are security and effectiveness. The study also throws light on how QB helps in reshaping the future of blockchain technology. The process of QC is also explored. At last, since QB has the attributes of quicker handling speed and more secure exchange dependent on quantum mechanics, it will have an extremely wide scope of uses and many research directions in the future.

REFERENCES

1. Z. Zheng, S. Xie, H. Dai, X. Chen, and H. Wang, "An overview of blockchain technology: architecture, consensus, and future trends", *IEEE 6th International Congress on Big Data*, pp. 557–564, 2017.
2. N. Kshetri, "Can blockchain strengthen the internet of things?" *IEEE IT Professional*, Vol. 19 no. 4, pp. 68–72, 2017.
3. C. Li, Y. Xu, J. Tang, and W. Liu, "Quantum blockchain: A decentralized, encrypted and distributed database based on quantum mechanics", *Journal of Quantum Computing (JQC)*, Vol. 1, no. 2, pp. 49–63, 2019.
4. N. Raychev, "Quantum blockchain, quantum review letters", *Web of Open Science*, pp. 15–47, 2020.
5. A. Ekert and R. Jozsa, "Quantum computation and Shor's factoring algorithm", *Reviews of Modern Physics*, Vol. 68, no. 3, pp. 733–753, 1996.
6. P. Shor, "Algorithms for quantum computation: Discrete logarithms and factoring", *Proceedings of the 35th Annual IEEE Symposium on Foundations of Computer Science*, pp. 124–134, 1994.
7. K. Ikeda, "Security and privacy of blockchain and quantum computation", *Advances in Computers*, Elsevier, Vol. 111, pp. 199–228, 2018.
8. D. Deutsch and R. Jozsa, "Rapid solution of problems by quantum computation", *Article published by The Royal Society*, Vol. 439, no. 1907 pp. 553–558, 1992.
9. L. K. Grover, "A fast quantum mechanical algorithm for database search", *28th Annual ACM Symposium on the Theory of Computing*, pp. 212–219, 1996.
10. T. M. Fernández-Caramés and P. Fraga-Lamas, "Towards post-quantum blockchain: A review on blockchain cryptography resistant to quantum computing attacks", *IEEE Access*, Vol. 01, pp. 21091–21116, 2020.
11. F. Mazzorana-Kremer, "Blockchain-based equity and STOs: Towards a liquid market for SME financing?" *Theoretical Economics Letters*, Vol. 9 no. 5, pp. 1534–1552, 2019.
12. B. Rodenburg and P. Pappas, *Blockchain and Quantum Computing*, MITRE Technical Report, June 2017.
13. D. Rajan and M. Visser, *Quantum Blockchain Using Entanglement in Time*, Article in Quantum Reports MDPI, Published: 17 April 2019.
14. M. Swan, *Blockchain: Blueprint for a New Economy*, O'Reilly Media, Inc. ISBN: 978-1-4919-2049-7, February 2015.
15. T. Zhou et al., "Quantum cryptography for the future internet and the security analysis", *Hindawi Security and Communication Networks*, Vol. 2018, Article ID 8214619, 7 p, 2018.
16. L. Wang and C. A. Alexander, "Quantum science and quantum technology: Progress and challenges", *American Journal of Electrical and Electronic Engineering*, Vol. 8, no. 2, pp. 43–50, 2020.
17. N. Lütkenhaus, T. Mor, and B.C. Sanders, "Limitations on practical quantum cryptography gilles brassard", *Physical Review Letters*, Vol. 85, no. 6, pp. 1330–1333, 2000.
18. C. H. Bennett, "Quantum cryptography using any two nonorthogonal states", *Physical Review Letters*, Vol. 68, no. 21 pp. 3121–3124, 1992.

19. C.H. Bennett, "Experimental quantum cryptography", *Journal of Cryptography*, pp. 3–28, 1992.
20. L. Jian et. al: "A survey on quantum cryptography", *Chinese Journal of Electronics*, Vol. 27, no. 2, pp. 223–228, 2018.
21. A. Goyal, S. Aggarwal, and A. Jain, "Quantum cryptography & its comparison with classical cryptography: A review paper", *5th IEEE International Conference on Advanced Computing & Communication Technologies [ICACCT-2011]* ISBN 81-87885-03-3, pp. 428–432, 2011.
22. V. Kaur and A. Singh, "Review of various algorithms used in hybrid cryptography", *International Journal of Computer Science and Network*, Vol. 2, no. 6, pp. 157–173, December 2013.

Chapter 5

Computer vision-based street-width measurement for urban aesthetics identification

Umme Rubaiyat Chowdhury and Manoj Roy
Daffodil International University

Md Hasanuzzaman
University of Dhaka

Shakik Mahmud
United International University

Mohammad Farhan Ferdous
Japan–Bangladesh Robotics and Advance Technology Research Center (JBRATRC)

CONTENTS

DOI: 10.1201/9781003156789-5

5.1 PART 1: INTRODUCTION AND OVERVIEW

"Computer vision-based street-width measurement for urban aesthetics identification" is a thesis work with procedural image-processing techniques. In this work, we tried to define urban aesthetics based on the street-width criterion. Though the study of the aesthetic characteristics of cities must go beyond concern only for the design of some of their parts, such as boulevards, parks, and civic centers, it is necessary to maintain some standard. According to the "interstate highway standards" of the United States of America, a single-lane road should use a 12-foot (3.7 m) [1] standard lane width.

Our paper mainly focuses on the streets of Dhaka, the capital of Bangladesh. Dhaka is the most populated city, with a population of 18.237 million [2]. Being the capital city, the daily crowd in the street of Dhaka is superfluous. So, the necessity of standard maintenance is a must need. The aesthetics of any urban area can also be defined through the condition of streets, buildings, and Shops, etc. Nowadays, modern algorithms and processors enable us to extract confidential data from an image using various specialized techniques. So, we have decided to measure the roads from the image and analyze them with the standard to identify the street aesthetics of Dhaka.

5.2 PART 2: MOTIVATION

Most of the streets and flyovers in Dhaka are in poor condition. The condition has become much serious as accidents are going on every day. It additionally causes serious harm to the running vehicles. It is a right of the citizens to have better streets for their cars. The entire transport framework is badly influenced because of the poor states of the roads. The concerned experts need to make an essential move in such a manner. The current circumstance has genuinely turned out to be difficult and troubling for the citizens. To check and verify road condition and status, only the authority can take the necessary steps. But to take any action, they have to go through the existing manual procedure. From here, the motivation of our project comes. A digital image is straightforward and reliable data nowadays. High-resolution photos can capture and store very subtle information to use for further processing. We want to measure the street width of various locations of Dhaka and match it with standard road width to find out the disproportion with standard roads. Through this, we will provide a statistical analysis of the urban aesthetics of the streets of Dhaka.

5.3 PART 3: BACKGROUND STUDY

As most of the techniques have already been developed, we had to study all the existing methodologies. In this section, we have added some literature reviews to support the proper methods of our work.

Nan Jiang and Zhongding Jiang, in their research paper entitled, "Distance Measurement from Single Image Based on Circles," described processing separation. Processing separation from captured images is a typical undertaking in image investigation and scene understanding. Separation fills in as a building obstructs for computing other geometry data, such as territory and volume. Besides straightforwardly estimating separation on the spot, it can be processed utilizing projective geometry from scene images. Existing work on separate estimation considers the point or line limitations. Since circles are basic in day-to-day life, this paper proposes two techniques for estimating separation utilizing hovers in one caught image. The primary technique manages two separate processes, which can be coplanar or parallel planes. The second one handles two concentric circles on a similar aircraft, which offers higher precision than straightforwardly fathoming conditions of concentric circles. The two strategies are checked by exploring different avenues regarding mimicked information and genuine images.

The immediate result was two strategies for estimating separation in light of circles from a solitary image. The principal technique handles two separate parallel circles, while the second one manages two concentric circles. Reproduced and actual information tests confirm that the proposed methods offer high precision and strength. They are helpful for straightforwardly estimating the separation [3] between two focuses on the reference plane from the single uncalibrated image.

Khandaker Abir Rahman, Md. Shafaeat Hossain, Md. Al-Amin Bhuiyan, Tao Zhang, Md. Hasanuzzaman, and H. Ueno, in their research paper entitled "Person to Camera Distance Measurement Based on Eye-Distance," described removing estimation framework in light of eye-separation. The separation between the focuses of two eyes is utilized for estimating the individual to camera separate. The variety in eye-separation (in pixels) with the adjustments in camera to individual separation (in inches) is utilized to define the separation estimating framework. The framework begins with figuring the separation between two eyes of a man, and at that point, individual to camera separate is estimated. The proposed different estimation framework is moderately straightforward and modest to actualize as it doesn't require some other outer separation evaluating devices. The trial comes about to show the viability of the framework with a standard exactness of 94.11%.

The proposed strategy has considerable significance due to its lower cost and less complex calculation for ongoing execution. Because of the straightforwardness of the proposed approach, equipment concentrated methods, for example, echo detection, extra charged-coupled device (CCD) cameras, laser projector, spotlights, and so on, are never again required for getting a palatable individual to separate camera estimation. Conversely, the precision of the deliberate face to different camera techniques diminishes as the person moves far from the camera [4].

Limeng Pu, Rui Tian Hsiao-Chun Wu, and Kun Yan, in their research paper entitled "Novel object-size measurement using the digital camera," described a novel way to deal with the measurement utilizing a customary advanced camera. These days, remote protest estimation is exceptionally critical to numerous interactive media applications. The proposed method depends on another profundity data extraction (extend discovering) to utilize a general advanced camera. The customary rangefinders are frequently completed using the detached strategy, such as stereo cameras, or the dynamic technique, such as ultrasonic and infrared hardware. The proposed approach requires just an advanced camera with specific image-handling procedures and depends on the essential standards of noticeable light. The average blunder level of this technique is under 2% [5].

This research tends to the development of urban lifestyle. We want to characterize the road standard of Dhaka by measuring the width of roads from the image. We found several methodologies to perform object measurement. But in this proposed methodology, we have focused on camera to object distance ratio to measure the road width correctly. Finally, this paper tries to define the road aesthetics of Dhaka, whether it is standard or not.

5.3.1 Related works

We found that lots of methodologies exist to measure the object size. The approach we proposed is quite similar to Nan Jiang and Zhongding Jiang [3]. They have measured the distance based on circle volume on the uncalibrated image. The other methodology proposed by Limeng Pu, Rui Tian Hsiao-Chun Wu, and Kun Yan doesn't require any external hardware to measure the actual distance. But it requires multiple images of the same object to finally obtain the measurement of the object [5]. In this section, we have also tried to cover all the existing commercial software that exists with the capability with object measurement features.

5.3.2 Related software

5.3.2.1 ImageJ

ImageJ [6] is a public domain Java picture handling program enlivened by NIH Image for the Macintosh. It runs, either as an online applet or as a downloadable application, on any PC with a Java 1.4 or later virtual machine. Downloadable disseminations are accessible for Windows, Mac OS, Mac OS X, and Linux. It has the feature of selection from a particular image and then measure the object (Figure 5.1).

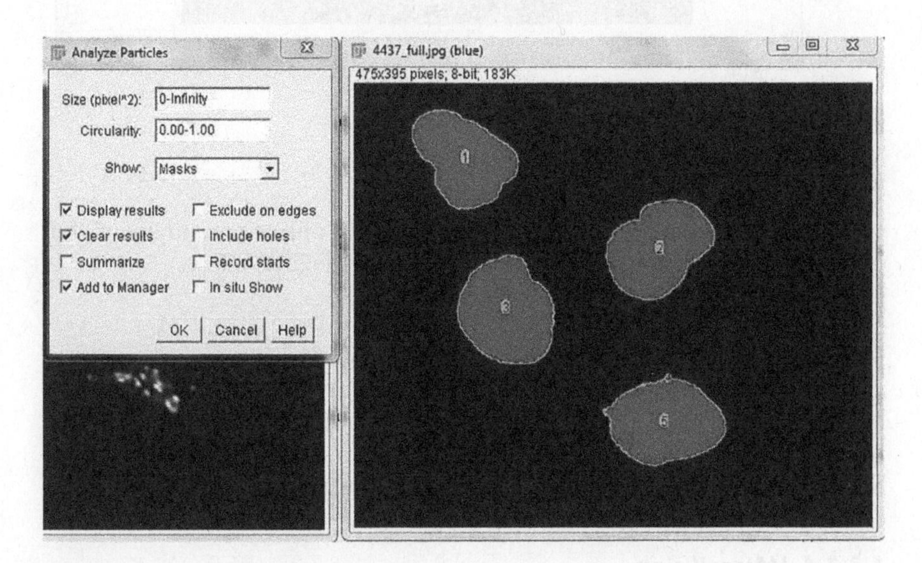

Figure 5.1 ImageJ object measure example.

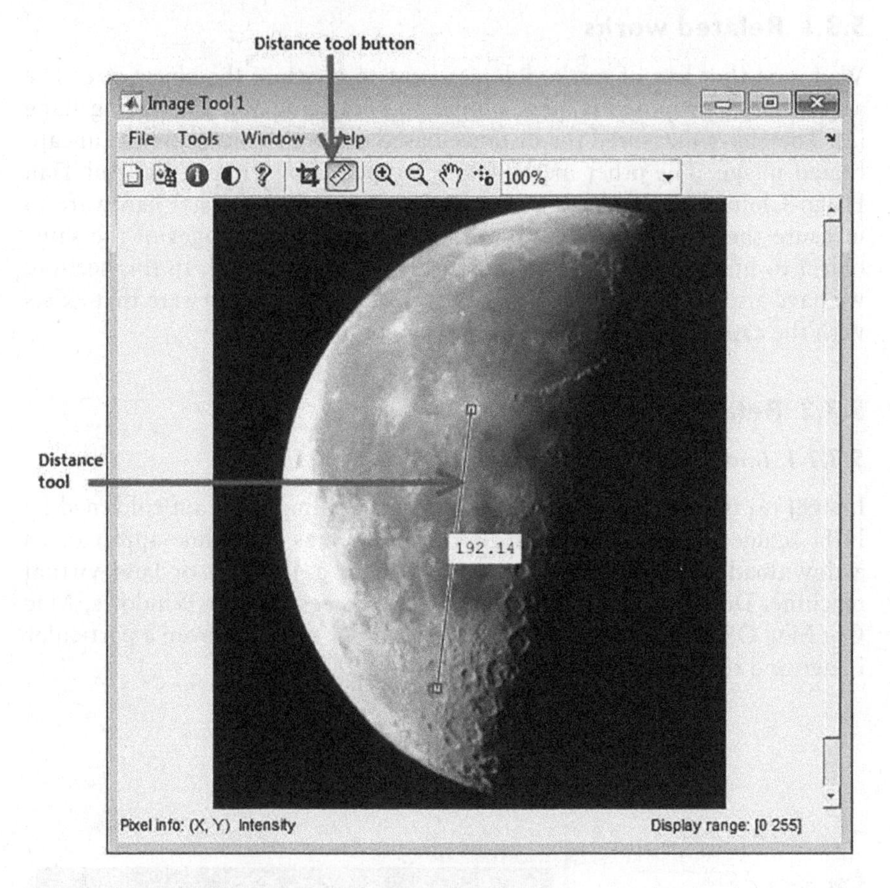

Figure 5.2 MATLAB® object measure example.

5.3.2.2 MATLAB

In MATLAB®, there is also the feature entitled object measurement. But this measurement required calibrated [7] image (Figure 5.2).

5.3.2.3 Fiji

Fiji is an image-preparing bundle. It can be portrayed as a dispersion of ImageJ together with Java, Java 3D and a ton of modules sorted out into a lucid menu structure. Fiji looks at ImageJ as Ubuntu analyzes Linux. The fundamental focal point of Fiji is to help inquire about life sciences.

5.3.2.4 JMicroVision

JMicroVision is an image examination tool [8] compartment for estimating and measuring parts of top-quality images. The program contains

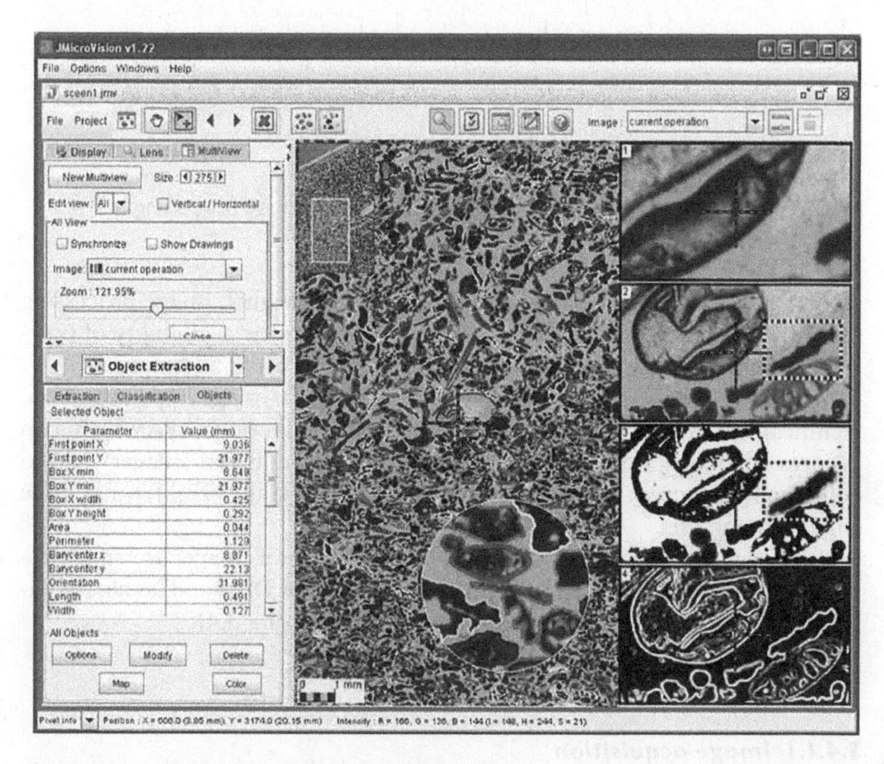

Figure 5.3 JMicroVision object measure example.

the majority of the fundamental image-handling tasks, has a straightforward and instinctive UI, a proficient representation framework, and inventive highlights. It includes devices to evaluate either physically or naturally. But this tool is currently not available on the official domain (Figure 5.3).

5.3.3 Scope of the problem

In this work, we have shown a methodology to determine object width from a road image. The expected outcome can resolve lots of issues like urban planning, event management, interior design, and of course medical processing.

5.4 PART 4: PROPOSED SYSTEM DESCRIPTION

Lots of techniques have been developed in the field of image processing in the last four to five decades. Among them, most of the methods are developed for enhancing images that have been acquired from unpopulated space

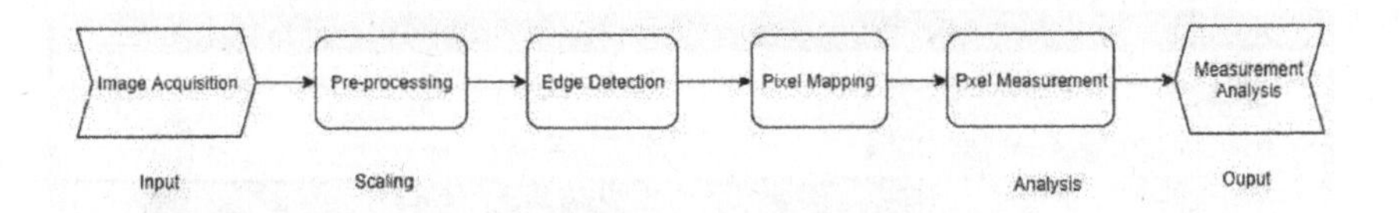

Figure 5.4 Block diagram of "object-width detection using image processing."

pierces, space shuttle, and military reconnaissance flights. Image-processing programs have become increasingly common due to the availability of powerful computers, memory equipment, technical software, and more.

Image processing involves problems related to image representation, technical pressure, and complex complicated functionality, which can be presented on photo data. Functions performed under imagery photo enhancement activities such as sharpening, blurring, brightening, edge enhancement, etc.

Using the features mentioned above to process an image, we suggest a procedure that can be used to measure the roads of Dhaka. The block diagram of the proposed algorithm is given below (Figure 5.4).

5.4.1 Research subject and instrumentation

5.4.1.1 Image acquisition

Typically, the image is a two-dimensional plane (x, y). If the amplitude of the image anyhow is f, it is known as the image intensity. It is likewise called the gray level of an appearance at that point. We must change these x and y values to complete the correct value for the discrete digital image. The input image is a fundus taken from the stare database. We need to change the simple image (analog) over to a computerized image (digital) to process it through a digital computer. Each digital image is composed of finite elements, and each finite element is called a pixel [9].

In this project, we have used images of the main road of Dhaka. The project needed a vast amount of image database. As a result, we decided to use Google street-view image.

5.4.1.2 Image preprocessing

Image scaling in the digital image occurs at specific points, whether in Bayer demosaicing or photo enlargement.

It happens at any time when we need to resize our image from the one-pixel grid to another grid. Image conversion is required if we want to increase or decrease the total number of transmissions in a snap. Although the same dimension is made, the result may differ depending on the algorithm.

In this project, photos have been resized for several reasons, but one is very important to our project. All cameras have their solution. So, when the system is created for some camera details, it will not go correctly with any other camera depending on the same features. So, it is necessary to make a regular solution to the application and then make an image of conversion.

The original image we have captured in a mobile device is with resolution 5,520*4,140. If the original resolution is processed to detect the object width, it outdoes the screen resolution as the testing device screen resolution is 1,366*768. For this, the image is resized using a python code to a resolution of 600*450. This one is the only one and major preprocessing used in our project.

5.4.1.3 BGR to grayscale conversion

People analyze colors using wavelength-sensitive cells known as cones. There are three different types of cones, each with a different sensitivity of magnetic radiation of a wide range of waves. Different cones are sensitive to different lights: the first one is sensitive to green light, the second one is to red, and the third one is to blue light. When it flashes a combination of three colors (red, green, and blue) and promotes three types of cones, it is also possible to make almost any of the colors to be seen by us. This is why the features of colors are always kept as three separate image matrices: the first one keeps as red (R) in each pixel, the second one in green pixel (G), and the third one as blue (B). This model is known as the RGB color model. Although we do not consider the amount of effuse in the grayscale image, we emit the same amount in each channel of a grayscale image. Here little light gives dark pixels and much light is perceived as bright pixels. During the conversion of an RGB image into a grayscale image, we need to consider the RGB parameters for each of the pixels and make a single value output reflecting the brightness of the respective pixel $(R+B+G)/3$. Since the light is understood to be governed by the green section, a different, human-oriented method, the way to make it is to consider a weighted average of the notes, for example, $0.2R+0.39G+0.47B$.

5.4.1.4 Gaussian blur

In image preparing, a Gaussian [10] obscure (otherwise called Gaussian smoothing) is the consequence of obscuring an image by a Gaussian capacity (named after mathematician and researcher Carl Friedrich Gauss). It is a broadly utilized impact in designs programming, commonly to lessen image clamor and decrease detail. The visual impact of this obscuring procedure is a smooth obscure looking like that of a survey of the image through a translucent screen, particularly not the same as the Bokeh impact delivered by an out-of-center focal point or the shadow of a question under normal enlightenment. Gaussian smoothing is likewise utilized as a pre-preparing stage in PC vision calculations keeping in mind the end goal to upgrade image structures at various scales like scale-space portrayal and scale-space execution.

Numerically, if we apply a Gaussian dim to an image it becomes the same as involving the image with a Gaussian capacity. This is otherwise called a two-dimensional Weierstrass change.

5.4.1.5 Edge detection

Edge detection incorporates an assortment of scientific techniques that go for distinguishing focuses in an advanced image at which the image splendor changes forcefully or, all the more formally, has discontinuities. The guides at which image brilliance changes firmly are commonly sorted into an arrangement of bent line portions named edges. A similar issue of discovering discontinuities in one-dimensional signs is known as step identification, and the issue of learning signal discontinuities after some time is known as change location. An edge location is a principal device in image handling, machine vision, and PC vision, especially in highlight discovery and highlight extraction regions [11].

In the perfect case, the consequence of applying an edge finder to an image may prompt an arrangement of associated bends that demonstrate the limits of items, the limits of surface markings, and additionally bends that relate to discontinuities in a surface introduction. In this manner, applying an edge identification calculation to an image may essentially decrease the measure of information to be handled and may, along these lines, sift through data that might be viewed as less applicable while saving the basic imperative properties of an image. On the off chance that the edge identification step is effective, the next assignment of translating the data substance in the first image may, along these lines, be significantly improved. In any case, it isn't generally conceivable to get such perfect edges from genuine images of direct multifaceted nature [12,13].

5.4.1.6 Edge detection techniques

In our task, we utilize the "Canny edge recognition system" due to its different points of features over other edge discovery procedures.

5.4.1.7 Canny edge detection

The Canny edge detection [14] is a standout amongst the most ordinarily utilized image-handling devices identifying edges in an extremely vigorous way. It is a multi-step process, which can be actualized on the GPU as an arrangement of channels. Canny edge recognition strategy depends on the following three fundamental targets:

- Low mistake rate
- Edge point ought to be very much restricted
- Single edge point reaction

5.4.1.8 Dilation and erosion

Morphology is an expansive arrangement of image-handling activities that procedure images in light of shapes. Morphological tasks apply an organizing component to an information image, making a yield image of a similar size. In a morphological study, the estimation of every pixel in the yield image depends on a correlation of the relating pixel in the information image with its neighbors [15]. By picking the size and state of the area, you can build a morphological task that is delicate [16] to particular shapes in the info image.

The most fundamental morphological tasks are widening and disintegration. Enlargement adds pixels to the limits of articles in an image, while disintegration expels pixels on question limits. The number of pixels included or removed from the pieces in an image relies upon the size and state of the sorting-out component used to process the image. In the morphological expansion and disintegration tasks, the condition of any given pixel in the yield image is controlled by applying an administer to the relating pixel and its neighbors in the information image. The government used to process the pixels characterizes the activity as an enlargement or a disintegration. Table 5.1 records the principles for both widening and disintegration.

Widening and disintegration are two major morphological tasks. Widening adds pixels to the limits of articles in an image, while disintegration expels pixels on question limits. The number of pixels included or removed from the pieces in an image relies upon the size and state of the sorting-out component used to process the image.

5.4.1.9 Contour tracing

Also called border following, boundary following, or limit following, contour tracing is a system that is connected to advanced images [17] to extricate their limit.

Table 5.1 Principles for dilation and erosion

Operation	*Rule*
Dilation	The estimation of the yield pixel is the most extreme estimation of the considerable number of pixels in the information pixel's neighborhood. In a paired image, if any of the pixels is set to the esteem 1, the yield pixel is set to 1.
Erosion	The estimation of the yield pixel is the base estimation of the considerable number of pixels in the information pixel's neighborhood. In a paired image, if any of the pixels is set to 0, the yield pixel is set to 0.

An advanced image is a gathering of pixels on a square decoration each having specific esteem. We will consider this with bi-level images, i.e., every pixel can have one of two conceivable qualities to be specific:

1, in which case we'll think of it as a "dark" pixel and it will be a piece of the example.
0, in which case we'll think of it as a "white" pixel and it will be a piece of the foundation.

5.4.1.10 Image calibration

A CCD cluster is mechanically very steady; the pixels have a settled geometric relationship. Every pixel inside the exhibit, notwithstanding, has special light affectability attributes. Since these qualities influence camera execution [5], they should be expelled through adjustment. The procedure by which a CCD image is aligned is known as fat fielding or shading rectification.

Geometric camera alignment, additionally alluded to as camera re-sectioning [18], gauges the parameters of a focal point and image sensor of an image or camcorder. You can utilize these parameters to revise for focal point twisting, measure the extent of a protest in world units, or decide the area of the camera in the scene. These errands are utilized as a part of uses, for example, machine vision to distinguish and measure objects. They are likewise utilized as a part of apply autonomy, for route frameworks, and 3D scene remaking.

5.4.2 Data collection procedure

We have performed the width detection in 2D images. We tried to collect 2D images by doing a simple python code. To find out the desired location, we first searched for the longitude and latitude of those locations from Google Map. But we faced some difficulty in doing so. The main motto of our project was to determine the road width from a street-view image. But the longitude and latitude value functionality were not so much of ease to use. The code needed a generated API key to perform the search operation. After providing all the longitude and latitude value the code returned all the images. But the output images were not in the same orientation we were expecting. After lots of attempts finally we decided to capture image manually.

All the images have been captured using a mobile device. The specification of the device we have used to capture images is given below:

Model: Sony Xperia X
Specification: 23 MP (f/2.0, 24 mm, 1/2.3″)

The images of the roads were captured from the foot-over bridges of the main roads of Dhaka. A reference object used in all the images is known as a sized object. The whole procedure is analyzed in the "statistical analysis" section.

5.4.3 Statistical analysis

Estimating the extent of items in an image is like registering the separation from our camera to a protest—in the two cases, we have to characterize a proportion that measures the number of pixels as per a given metric.

We call this the "pixels per metric" proportion, which we have all the more formally characterized in the accompanying segment.

5.4.4 Formula

To decide the extent of a question in an image, we first need to play out a "calibration" utilizing a reference object. Our reference object ought to have the following two critical properties:

- We should know the measurements of this object (regarding width or tallness) in a quantifiable unit (for example, millimeters, inches, and so on.).
- We ought to have the capacity to effortlessly discover this reference object in an image either through the situation of the object (for example, the reference object continually being set in the upper left corner of an image) or appearances (like being an unmistakable shading or shape, special and not quite the same as every single other object in the image). In either case, our reference ought to be interestingly identifiable in some way.

In this example, we'll use the Bangladeshi coin (1 Taka Coin) as our reference object, and in all examples, make sure it's always the leftmost object in our image.

We'll utilize the coin as our reference object and ensure that it is constantly set at the furthest left in the image, making it simple for us to extricate it by arranging forms keeping in view their area.

By placing the coin on the bottom left, we can edit the line of objects from left to right (which will be the first contour in the scheduled list) and use it to define our per_pixel_ratio, which defines:

per_pixel_ratio = object_width_per_pixel/the_known_width

The coin has a known width of 1.35 inch. Now, think that our object_width (measured by pixels) is calculated to be 150 pixels wide (based on the joint box).

per_pixel_ratio = 150 px/1.35 in = 111.11 px

That is why there are approximately 111.11 pixels per 1.35 inches in our image. Using this ratio, we can estimate the number of objects in the image.

5.4.5 Implementation requirements

The capacity and execution apportioned to programming as a major aspect of framework building are defined by setting up an entire data depiction, the nitty-gritty of utilitarian and social portrayal, a sign of execution prerequisites and configuration compels, suitable approval criteria, and other information correlated to necessities. The tools and dependencies used to implement the project are provided below:

Tools requirement

- OpenCV
- Python 2.7

Image processing package list

- Scipy
- Numpy
- Imutils

5.5 PART 5: EXPERIMENT RESULT AND ANALYSIS

5.5.1 Image acquisition

We have used a mobile camera for image acquisition. The road images are captured from various main roads of Dhaka. For road images, we have chosen the flag of Bangladesh on a board as our reference object (Figure 5.5).

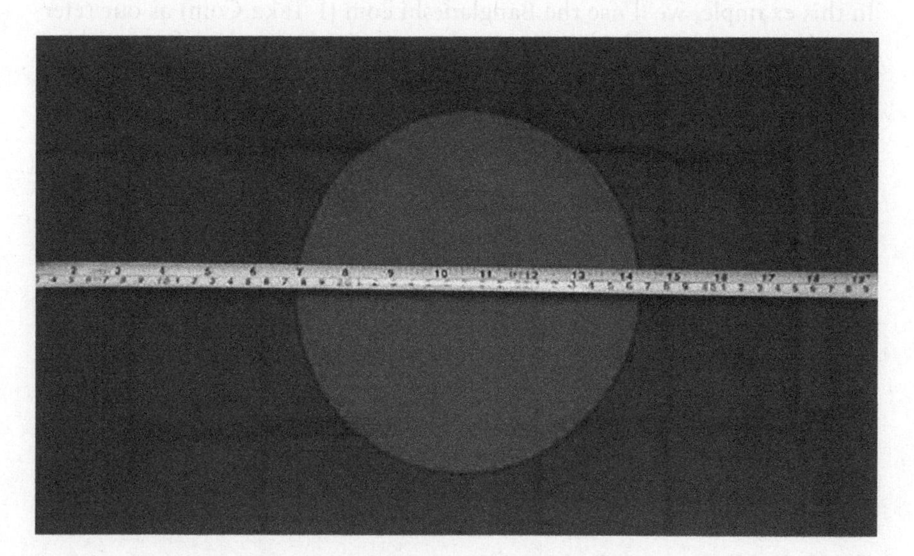

Figure 5.5 Reference object used in the road.

5.5.2 Preprocessing

The image is then scaled to a lower-resolution image due to the screen size of the testing device as follows:

- Original image resolution: 5,520*4,140
- Image resolution after preprocessing: 600*450

We also apply grayscale filtering in this section to make further processing easier.

5.5.3 Edge detection

Then we perform edge detection along with erosion+dilation to remove any gaps in between edges in the edge lines. We find outlines that correspond to the objects in our edge lines (Figure 5.6).

5.5.4 Pixel mapping

These contours are then sorted from left to right. Through this procedure, we can find out the reference object from the input image. We also initialize our per-pixel metric (Figure 5.7).

5.5.5 Analysis

This section provides the calculated value in table for being analyzed the accuracy of our approach.

See Figure 5.8 and Table 5.2.

5.5.6 Descriptive analysis

The result we have got capturing a single image worked with some error rate. The error percentage rate is discussed in the next section. After

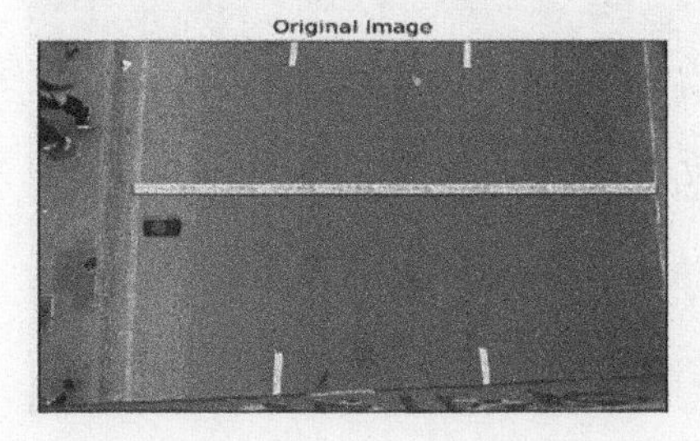

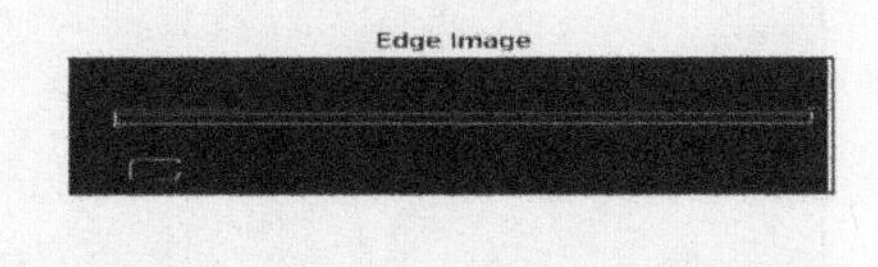

Figure 5.6 Image after edge detection.

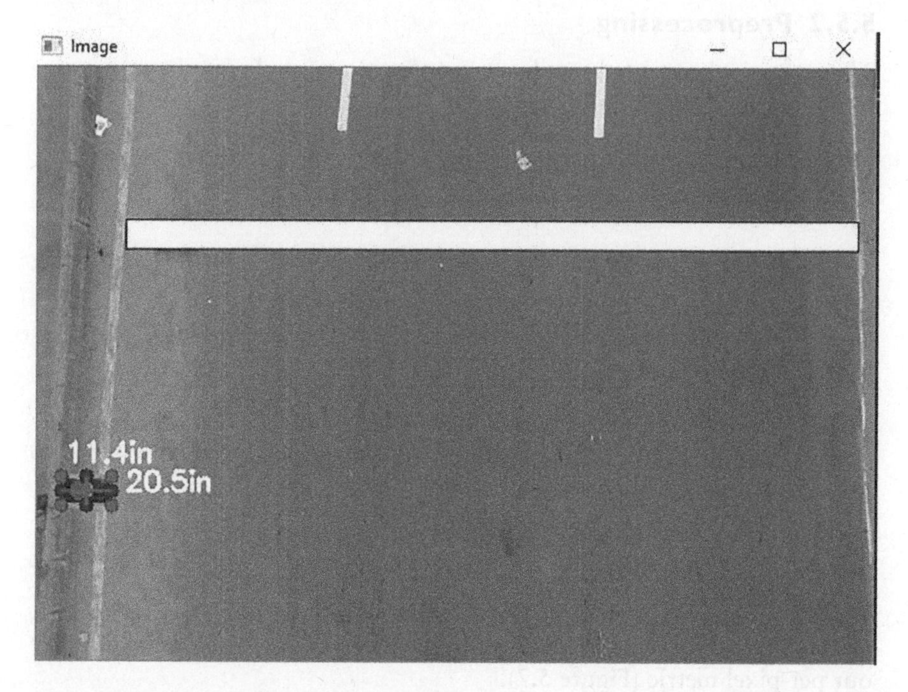

Figure 5.7 Reference object size detection.

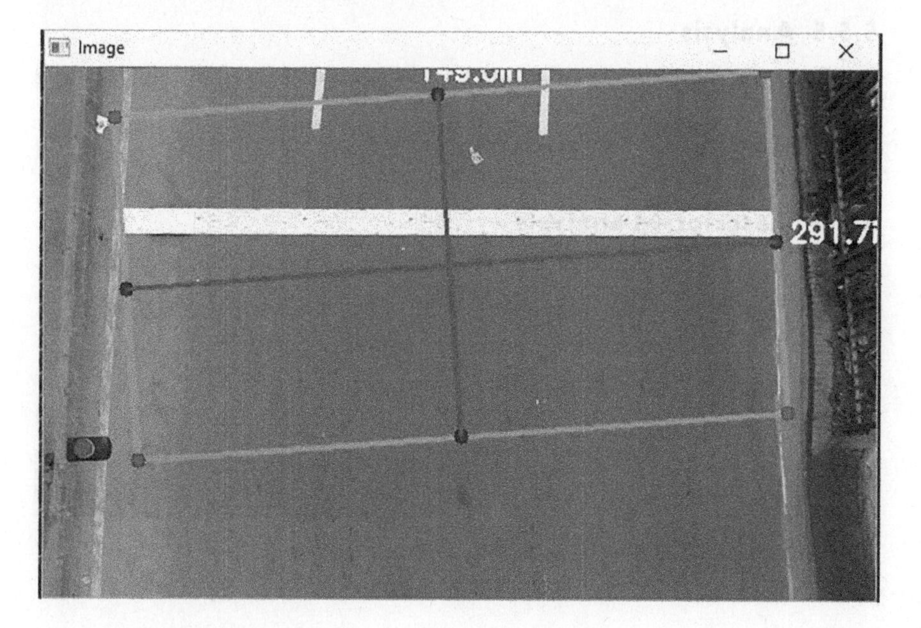

Figure 5.8 Road width detection.

Table 5.2 Actual object measurement versus output result

Actual measurement		*Generated result*	
Reference object	20.47*11.8 inch	Reference object	20.5*11.4 inch
Road width	154 inch	Road width	151.3 inch

acquiring the result we captured the same image from a different distance. The descriptive analysis of our experiment is analyzed and shown in the graphs and tables.

5.5.7 Object distance calculation from different distance

For more acceptance of this approach the Object Measurement has been experimented and displayed in the below table from different camera distance in this section.

See Figure 5.9 and Table 5.3.

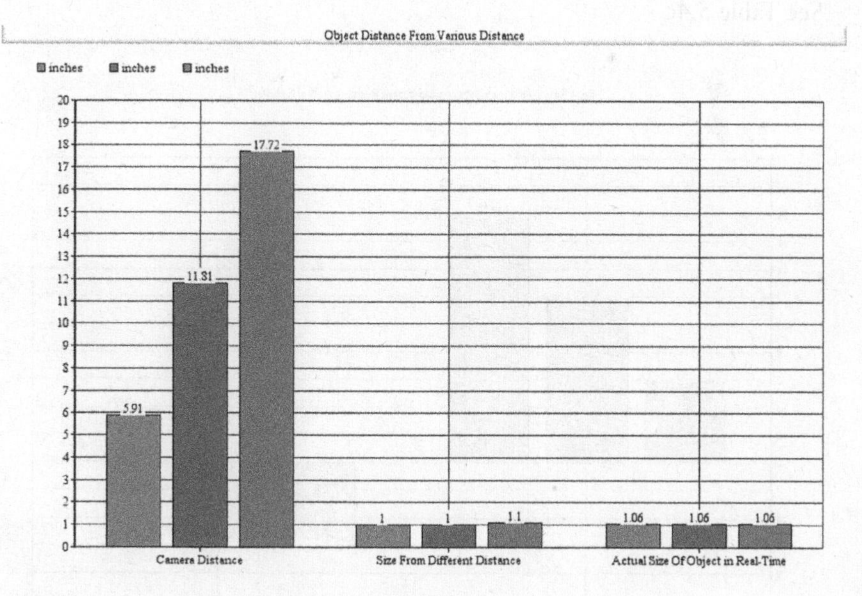

Figure 5.9 Object distance calculation graph from a different distance.

Table 5.3 Object distance calculation from different distance

Camera distance (inch)	*Size of different objects (inch)*	*Size of real objects (inch)*
5.910	1, 1, 0.9, 1.1, 1	1.060, 1.060, 0.87, 1.14, 1.02
11.810	1.1, 1, 0.9, 1.1, 1	1.060, 1.060, 0.87, 1.14, 1.02
17.720	1.1, 1.1, 0.9, 1.1, 1	1.060, 1.060, 0.87, 1.14, 1.02

5.5.8 Error rate analysis

The image below describes the error rate of this thesis work. The project is to be 96% accurate. With the increasing distance, the effect of the distance does not interrupt object distance very much (Figure 5.10).

5.5.9 Google map distance

This section shows the sample of distance measurement using Google Map built-in scale.

See Figure 5.11.

5.5.10 Error rate comparison

In this section we have differentiated the measurement error rate on the basis of Field Data, Google Map Data and data using our proposed method.

See Table 5.4.

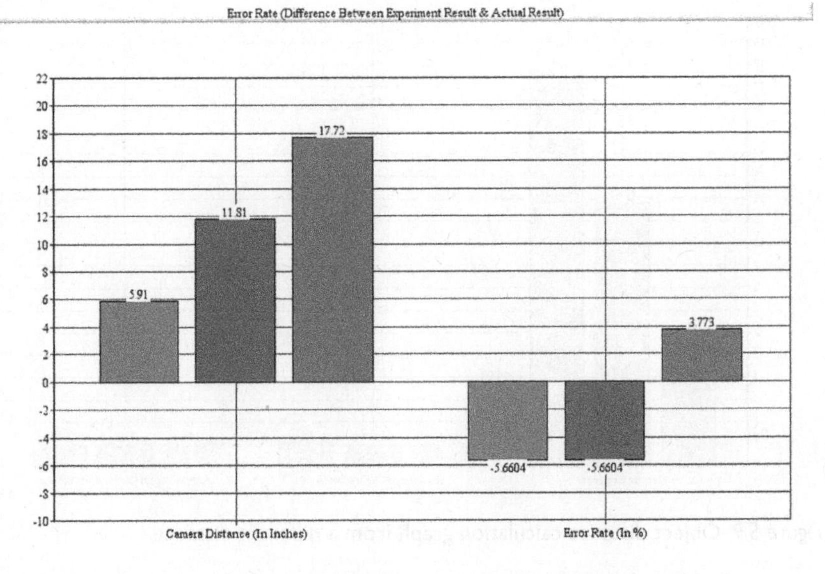

Figure 5.10 Error rate analysis between experiment result and actual result.

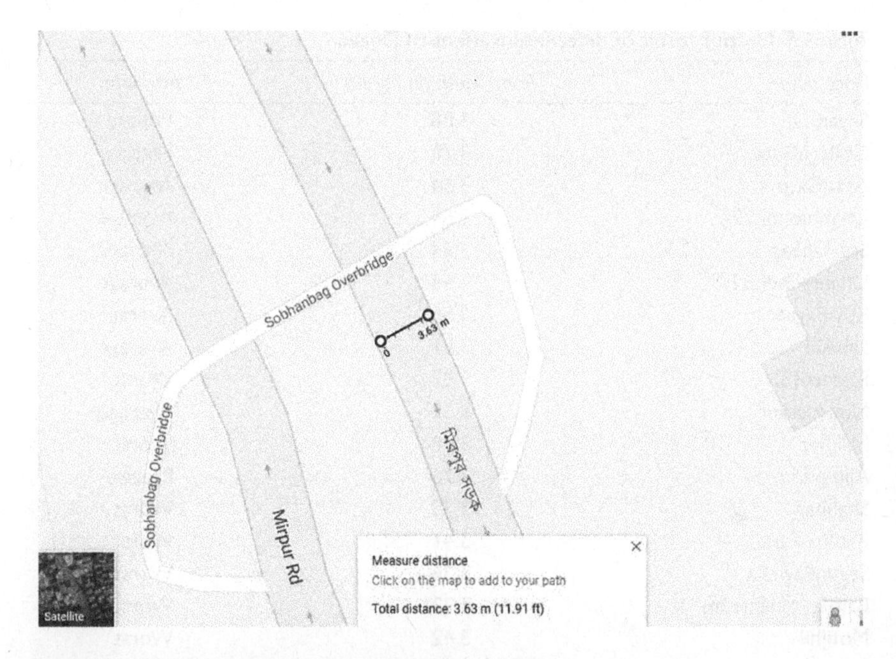

Figure 5.11 Road width detection using Google map.

5.5.11 Some data analyzed around Dhaka

Here we have included some road-width values of different places and applied different data mining algorithms to find out the aesthetics of Dhaka urban life.

The data are compared with the standard road length (3.7 m) [1] of the United States of America. We defined three attributes worst, beneath, precise to find out the condition of the road. We classified them as precise if the roads are above 3.7 m, beneath if the roads are between 3.5 and 3.69 m, and worst if the roads are less than 3.5 m in length. Here we have provided a table to represent all the data (Table 5.5).

All the data have been collected from real-life images and using the proposed methodology. Once we applied "Naïve Bayes" to all these data, we found out that if we consider Dhaka based on these main roads, the result comes worst.

Table 5.4 Error rate of the project

Real data	*Google*	*Our proposed method*
3.92 m	3.63 m	3.84 m
0%	–7.398%	–2.0408%

Table 5.5 Output result of different locations of Dhaka

Place name	*Road width (in meter)*	*Attribute*
Shyamoli	3.88	Precise
CollegeGate	3.76	Precise
AsadGate	3.54	Average
Dhanmondi 27	3.54	Average
SobhanBag	3.64	Average
Dhanmondi 32	3.54	Average
KolaBagan	3.56	Average
LabAid	3.51	Average
ScienceLab	3.42	Worst
NewMarket	3.52	Average
Nilkhet	3.48	Worst
Ajimpur	3.76	Precise
Shahbag	3.38	Worst
Ramna Park	3.41	Worst
SegunBagicha	3.36	Worst
Baitum Mukarram	3.22	Worst
Motijhil	3.42	Worst
Arambag	3.59	Average
Rajarbag	3.70	Precise
Kakrail	3.6	Average
Poribag	3.08	Worst

We also tried to match this data with authorized and verified road information in government records to find out the real condition and corruption. But that information was inaccessible due to proper authority issues.

5.6 PART 6: CONCLUSION

The summary of this thesis work ended up with an accuracy of 96%. We would like to add that we have tried to capture the images with a 90° view. But as it is captured by hand, it can't be said a perfect 90° angle. But the methodology we used is quite suitable for a cost-effective solution.

After the finishing of this project, we can come on some epilogues. First of all, we are now aware of the limitations. The calibration is an important issue in distance measurement from images. On the other hand, this finding also gives us the current condition of roads (a glimpse of urban life) of Dhaka. This data may be used in various fields.

Throughout the whole procedure in the research project, we have tried to focus on the main roads of Dhaka. The actual data fetched from the real-life image can provide lots of statistical analysis of urban planning, interior design, and event management. In this field, there are lots of future scopes as it is a very cost-effective procedure in the prospect of our country.

This thesis project is based on 2D images. Nowadays 3D images are being commonly available to all. The measurement concept is totally different in 3D from that in 2D. Our future target is to work with 3D images.

The reference object will always not be available in all the images. So, we tend to measure the object size without a reference object in the future. As Google street-view images do not provide updated images of locations, the issue of getting image data using API calling remains unresolved. But after all these issues, we will develop an application using this methodology to provide ease of use.

REFERENCES

1. Standard Road Length https://en.wikipedia.org/wiki/Lane, last accessed on 20-07-2018 at 4.53pm.
2. Learn about Population of Dhaka at http://www.worldpopulationreview.com/world-cities/dhaka-population/, last accessed on 30-07-2018 at 9:27am.
3. N. Jiang, Z. Jiang "Distance measurement from single image based on circles" *2007 IEEE International Conference on Acoustics, Speech and Signal Processing - ICASSP '07*, 15–20 April 2007.
4. K.A. Rahman, M.S. Hossain, M.A.-A. Bhuiyan, T. Zhang, M. Hasanuzzaman, H. Ueno "Person to camera distance measurement based on eye-distance" *2009 Third International Conference on Multimedia and Ubiquitous Engineering*, 4–6 June 2009.
5. L. Pu, R.T. Hsiao-Chun Wu, K. Yan "Novel object-size measurement using the digital camera" *2016 IEEE Advanced Information Management, Communicates, Electronic and Automation Control Conference (IMCEC)*, 3–5 October 2016.
6. MeasuringPlanarObjectswithaCalibratedCamerahttp://www.mathworks.com/help/vision/examples/measuring-planar-objects-with-a-calibrated-camera.html, last accessed on 27-07-2018 at 9:20am.
7. Know about JMicroVision http://www.linuxsoft.cz/en/sw_detail.php?id_item=11345#down last accessed on 27-07-2018 at 9:35am.
8. Know about Image Acquisition http://buzztech.in/image-acquisition-in-digital-image-processing/ last accessed on 27-07-2018 at 9:48pm.
9. M.S. Nixon and A.S. Aguado. *Feature Extraction and Image Processing.* Academic Press, 2008, p. 88.
10. S.E. Umbaugh (2010). *Digital Image Processing and Analysis: Human and Computer Vision Applications with CVIPtools* (2nd ed.). Boca Raton, FL: CRC Press. ISBN 978-1-4398-0205-2.
11. H.G. Barrow and J.M. Tenenbaum (1981) "Interpreting line drawings as three-dimensional surfaces", *Artificial Intelligence*, vol 17, no. 1–3, pp. 75–116.
12. T. Lindeberg (2001) [1994], "Edge detection", in *Hazewinkel, Michiel, Encyclopedia of Mathematics*, Springer Science+Business Media B.V./Kluwer Academic Publishers, ISBN 978-1-55608-010-4.

13. T. Lindeberg (1998) "Edge detection and ridge detection with automatic scale selection", *International Journal of Computer Vision*, vol 30, no. 2, pp. 117–154.
14. Know about Canny Edge Detection https://www.slideshare.net/louiseantonio58/image-processing-based-intelligent-traffic-control-systemmatlab-gui last accessed on 27-07-2018 at 9:35am.
15. Learn About Dilation & Erosion https://www.mathworks.com/help/images/morphological-dilation-and-erosion.html last accessed on 27-07-2018 at 10.01pm.
16. Learn About Dilation & Erosion https://www.cs.auckland.ac.nz/courses/compsci773s1c/lectures/ImageProcessing-html/topic4.htm#refs last accessed on 27-07-2018 at 10.01pm.
17. Learn about Contour Detection, available at http://www.imageprocessingplace.com/downloads_V3/root_downloads/tutorials/contour_tracing_Abeer_George_Ghuneim/intro.html/ last accessed on 20-07-2018 at 11.42am.
18. Learn about ImageJ at https://imagej.nih.gov/ij/docs/index.html, last accessed on 27-07-2018 at 9.42pm.

Chapter 6

Security and privacy issues in VANET

A comprehensive study

Aqeel Khalique, M. Afshar Alam, Imran Hussain, Safdar Tanweer, and Tabrej A. Khan

Jamia Hamdard

CONTENTS

6.1 INTRODUCTION

To understand the vehicular ad hoc network (VANET), we must look at the technologies that work behind the abstraction of "smart computing" and have molded the world into the one we recognize today. These technologies can be used separately for different fields or combined for usage in completely unrelated areas. At the heart, there is a network of interconnected devices that work together to share or compute data. The network is made up of different

DOI: 10.1201/9781003156789-6

nodes that are connected to the internet or the cloud. The cloud is just a metaphor for the internet. With the technological enhancements with remarkable research in the field, the quality of life and need for information and communication technologies (ICT) have impacted positively [1]. However, to increase the quality of life, technology must be sustainable as well [2]. Sustainability is a must as if we continue using the resources at the current life, technology any consideration or regard to the environment or the reserves for the future, we will quickly run out of the resources or deplete the supply drastically which will have a negative impact on future generations.

The connectivity provided by the World Wide Web showed the world how effective and resourceful the internet is. Although the connection in Web 1.0 was fairly limited, with the introduction of Web 2.0, the content creators and consumers are more connected than ever before. The notion of interconnectivity has been introduced mainly due to the interactions over online social media platforms, information systems, content-sharing platforms, streaming services. [3].The evolution from these technologies and the need for interconnectivity is a network of connected devices that share data and other information. Their network is known as the Internet of Things (IoT). These systems or devices may be anything that has a capability of connection with other devices such as mobile devices such and other sensors to even light bulbs and refrigerators. As mentioned, earlier, with the given technological enhancements, the enhancements, the increases. However, this should take sustainability as a major factor [4]. Without technologies like ICT and However, this would not be possible at the rate it is now.

From outcome a network that was able to communicate and automate vehicular experience depending upon the requirements of the user and environment. This network is known as VANET and is used for various purposes depending upon the requirement. Until recently, the area of vehicular activities was inside a realm of mechanical engineering. However, with the reducing rates of production cost and increasing computational capabilities, the vehicles have become more than just "wheels-on-road" intended for getting from point A to point B. The capabilities now include numerous processors interconnected with the system to provide better automation and task capabilities, the global positioning system, radars, cameras, sensors, and even the technologies used in other systems like event data recorders [5]. Most of the research in the emerging area of VANET is focused on collision avoidance applications to providing advertisements and infotainment services [6,7]. VANET will improve quality of life by decreasing congestion and hence reducing travel time and fuel/electricity usage. Figure 6.1 presents a detailed VANET layout.

In the coming years, development in autonomous vehicles is being anticipated as the vehicle will be able to interact with other road vehicles, alert other vehicles about the road conditions, and provide drivers and manufacturers with performance details. When a vehicle is stopped abruptly, vehicles behind the stopped one will know about it right away due to the

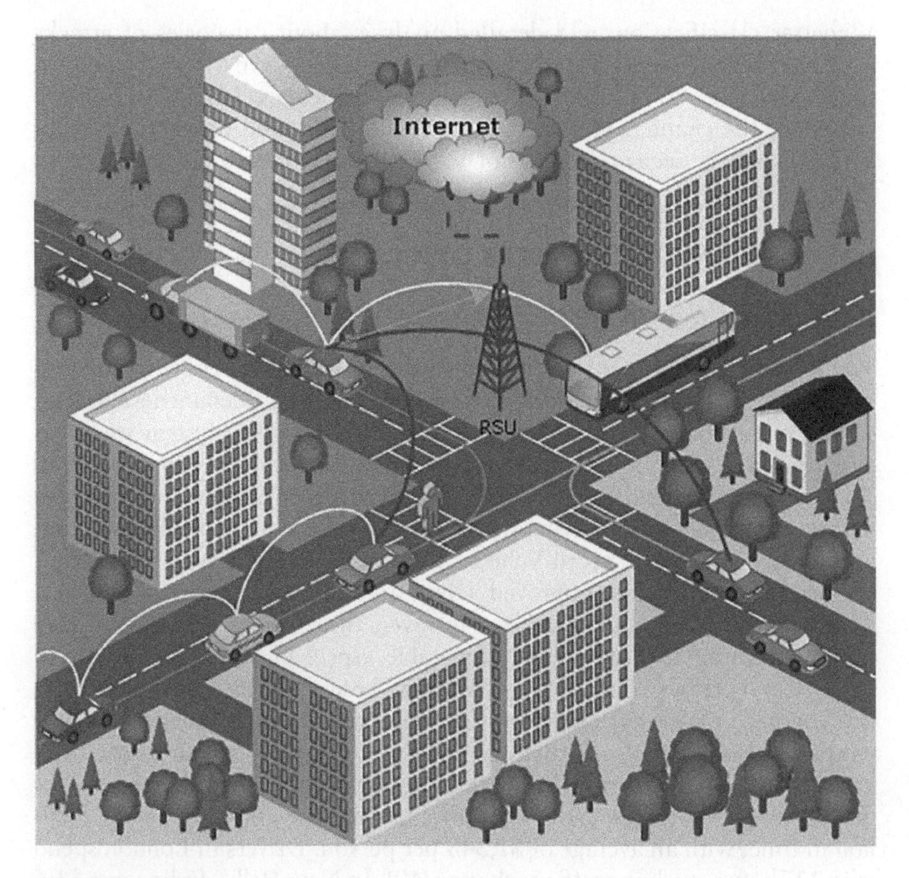

Figure 6.1 VANET.

low latency provided by the newer networking technologies and also avoid collisions ultimately saving human lives.

Years, development is a type of MANET, specifically applicable to vehicles and their connectivity. The connectivity of vehicles in VANET is primarily through the wireless network. Nowadays, VANET is specifically used to provide several additional features to the driver and passengers. These additional features include but are not limited to safety, comfort to drivers, communication among vehicles, etc. VANET is primarily concerned with sharing data generated from different vehicles.

In this chapter, we discuss security and privacy issues related to VANET in detail. Section 6.2 describes the evolution, need, growth, the effect of 5G, and standards related to VANET. Section 6.3 describes the network, architecture, communication models, components, characteristics, and applications of VANET. Section 6.4 describes security aspects viz. need, attributes, and security requirements of VANET. Furthermore, attacks in

Vignetter classification, and detailed analysis about categories of attacks are presented. Section 6.5 describes privacy challenges in VANET and possible approaches to overcome these privacy challenges. Lastly, Section 6.6 concludes the chapter. Hence, the chapter is a comprehensive study about VANET and the security aspects of VANET in particular.

6.2 VANET: EVOLUTIONARY DISCUSSION

6.2.1 Need for VANET

The World Health Organization published a global status report in 2009 on road safety covering the situation of road safety for 178 countries. Among several causes of accidents or road crashes, lack of timely system warnings and driver's exhaustion was the common causes for accidents and crashes [8]. Therefore, recommendations from the nearby vehicles' observations will play an important role in improving safety for the vehicle users [9]. Hence, connected cars or the Internet of Vehicles (IoV), as they are termed in VANET, are needed in providing improved safety and driving experience and also in reducing accidents and crashes. However, these connected vehicles may pose an additional threat to the vulnerable aspect of VANET resulting in security and privacy challenges in VANET. These challenges are discussed in this chapter in later sections. The population of vehicles on the road is also increasing exponentially leading to massive traffic problems in major cities around the world. Statistically, traffic congestion in the United States makes drivers lose 97 hours per year in traffic, costing the country over $87 billion in time, with an average of $1,348 per person. Drivers in London spend over 227 hours stuck in traffic each year [10]. In New Delhi, India, over $1.6 million worth of fuel is wasted every day by vehicles stuck in traffic [11]. Therefore, traffic congestion can lead to fuel wastage, excessive time delays, economic losses, and environmental pollution and hence affect travel, economy, business etc. These are several reasons which give birth to the need for VANET in the present and upcoming future requirement of transportation.

6.2.2 Growth of VANET

VANET mainly consists of connected vehicles having access to the internet for sharing data with smart devices inside the car and with smart objects outside the car. Smart objects outside the car include other cars, traffic lights, road sensors, roadside units (RSUs), etc. A connected car is an important entity of the VANET. Connected cars have huge computing power and data processing of 25 GB/h. Connected cars can optimize their operation and maintenance prioritizing the convenience and comfort of passengers. Connected cars use onboard sensors and internet connectivity for data processing and data transmission. In 2019, major automobile manufacturers

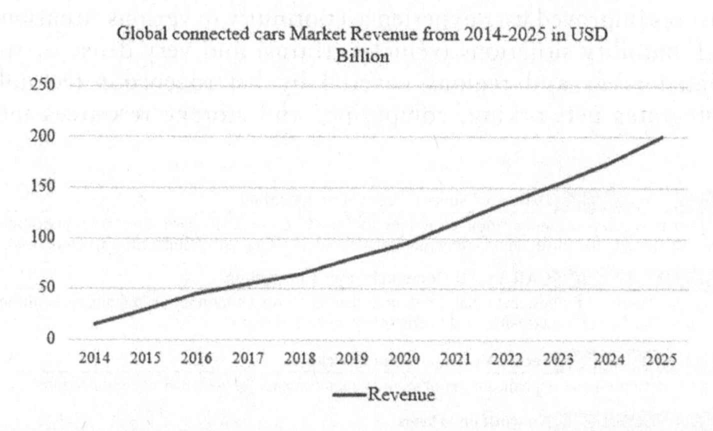

Figure 6.2 Global revenue of connected cars from 2014 to 2025.

are focusing on the growth of connected cars. These major automobile manufacturers are General Motors, BMW, Audi, Mercedes Benz, Tesla, Volkswagen, Jaguar, Porche and Nissan. By 2020, we may have 380 million connected cars (IoV) on the roads [12]. By 2021, close to 90 million connected cars will be shipped according to Business Insider [13]. The global growth of connected cars can be seen in Figure 6.2 [14].

6.2.3 Use of 5G in VANET

Presently the industry is exploring and researching the fifth-generation broadband networking (5G) [15] as well as its role in further enhancing applications of IoT, IoV, etc. Such industries include manufacturers, academics, pharmaceuticals, military, and vehicular sectors studying the future of autonomous vehicles. The industries are exploring the resiliency, affordability, performance, and protection of the 5G technology. The switch to 5G would certainly alter our everyday experiences with technology.

Network optimization is another significant aspect of the 5G infrastructure, in addition to increased efficiency and lower latency. This management of the network is enabled by network slicing and can be linked to many virtual networks depending on the form of the service requested. For example, notification messages and other associated security resources require a network connection with lower latency, fast transfer rate with a higher bandwidth [16].

The main features of 5G include the following:

- Ubiquitous connectivity for the huge number of users or devices uninterruptedly
- Very low latency up to a few milliseconds for critical systems, real-time applications, and services with zero delay tolerance
- High-speed gigabit connectivity

5G ensures improved user experience continuity in various situations such as high-mobility situations (vehicles, trains) and very dense or sparsely populated areas and regions covered by heterogeneous technologies. 5G integrates networking, computing, and storage resources into one

DSRC **Dedicated Short Range Communication**
- 802.11p-based wireless communication technology for highly secure, high-speed direct communication between vehicles and the surrounding infrastructure, without involving any cellular infrastructure band.

C2C-CC **CAR 2 CAR Communication Consortium**
- Develop specification of robust and reliable solutions that allow for a continuous and seamless evolution of required functionalities for increasing road traffic safety and efficiency

SEVECOM **Secure Vehicular Communication**
- Provide full definition and implementation of security requirements for vehicular communications

NoW **Network on Wheels**
- Solve technical key questions on the communication protocols & data security for C2C communications

DIRICOM **Intelligent Design of Wireless Communication Networks**
- Solve wireless networks design problems using intelligent techniques

SEISCIENTOS **Adaptive Ubiquitous Services in Vehicular Contexts**
- Framework for communication and infrastructure to provide dedicated services to end-user in ubiquitous vehicular environments.

WiSafeCar **Wireless traffic Safety network between Cars**
- Develop a reliable wireless traffic service platform to improve traffic safety, avoid traffic accidents, and provide variety of new type of services to vehicles.

MARTA **Mobility and Automotion through Advanced Transport Networks**
- ITS safety, efficiency, and sustainability

COOPERS **CO-OPerative SystEms for Intelligent Road Safety**
- Development of telematics applications between car industry and infrastructure operators

CVIS **Cooperative Vehicle-Infrastructure Systems**
- Design, develop and test new technologies needed to allow vehicles to communicate with each other and with the nearby roadside infrastructure

EVITA **E-Safety Vehicle Intrusion Protected Applications**
- Design, to verify, and to prototype building blocks for automotive on-board networks against tampering and protecting sensitive data

GST **Global System for Telematics**
- EU-funded Integrated Project is creating an open and standardised end-to-end architecture for automotive telematics services.

GeoNet **Geographic addressing and routing for vehicular communications**
- Implement and test a networking mechanism as a standalone software module

iTETRIS **Integrated Wireless and Traffic Platform for Real-Time Road Traffic Management Solutions**
- Develops an evaluation platform for large-scale, long-term simulations of cooperative traffic management applications.

PReVENT **PReVENTive and Active Safety Applications**
- Develop, test and evaluate safety related applications, using advanced sensor and communication devices integrated into on-board systems for driver assistance

SAFESPOT **Smart Vehicles on Smart Roads**
- Autonomous vehicle based safety systems are limited by the field of view of their sensors

SIMTD **Safe Intelligent Mobility**
- Test V2V and V2I communication innovations for traffic safety and mobility

Figure 6.3 Standards/projects related to VANET.

programmable and unified infrastructure. Therefore, scalable and sustainable network infrastructure that expands the coverage and decreases latency and can play a key role in improving future telecommunications networks.

VANET would benefit from 5G technology in terms of reduced latency for the device-to-device communication and transmission of alerts with improved data rates with a high-frequency spectrum. The 5G technology would also increase the number of concurrent connections by maximizing energy efficiency along with capacity and coverage. 5G technology enhances the quality of service with intelligent agents to manage routing and resource allocations with decreased complexity and provides better security control overall.

6.2.4 Standards/projects related to VANET

VANET standards are now being well-identified and accepted by different consortia. As described earlier, there are more standards/projects which are undergoing or have been widely adopted by different consortia globally. Figure 6.3 shows a list of a few standards/projects of VANET [17].

6.2.5 Existing literature related to VANET

There exist numerous works of literature related to VANET and their enabling technologies. The exhaustive literature survey for VANET is beyond the scope of the current topic. However, we list a few prominent research papers and summarize their work published in 2016 relevant to our scope of VANET, in particular, which discusses security attacks in VANET. Table 6.1 shows a summarized representation of a few research papers [18–23]:

Table 6.1 Summary of preexisting reviews on VANET

Authors	*Summary*
Jana et al. [18]	The paper analyses vulnerable security threats and suggested mechanisms to overcome these attacks.
Mishra et al. [19]	The paper discusses security issues and challenges.
Deepak et al. [20]	The paper discusses routing and security mechanisms.
Azees et al. [21]	The paper discusses characteristics and security challenges It also describes the taxonomy of security mechanisms.
Kaur et al. [22]	The paper discusses security and risk assessment.
Hasan et al. [23]	The paper discusses research issues and the vulnerability of security measures.

6.3 VANET: TECHNOLOGICAL DISCUSSION

6.3.1 Network

VANET is a network consisting of entities that can be divided into the following three domains:

- **Mobile domain**: consisting of two subparts namely vehicle domain (cars, buses, etc.) and mobile device domain (handheld devices, smartphones, navigation system, etc.).
- **Infrastructure domain**: consisting of two subparts namely roadside infrastructure domain (RSUs and traffic lights) and central infrastructure domain (traffic and vehicle management center).
- **Generic domain**: consisting of two subparts namely internet infrastructure domain and private infrastructure domain.

In VANET, routing protocols of MANET are not used. VANET architecture is generally categorized into the following three categories:

- Cellular and WLAN network consists of fixed gateways and WiMAX/WiFi which are used for internet connectivity, routing, and getting traffic information.
- Vehicles and fixed gateways consist of pure ad hoc connections between vehicles and gateways.
- Hybrid architecture is a combination of infrastructure and ad hoc networks.

6.3.2 Components

VANET has different components involved which are described below:

- **Trusted authority**: It is responsible for commissioning and maintaining the entire VANET and registering RSUs, onboard units (OBUs), and vehicles. It also ensures security management of VANET and verifies vehicle authentication ID and OBU ID. It consumes high power and a huge memory size. In case of any malicious activity, it can reveal OBU ID and other details. It also identifies potential attackers to expose them in the VANET.
- **Roadside unit (RSU)**: It connects to the cloud or internet which allows the application unit from several vehicles to connect to the internet. RSU consists of a radio-frequency antenna with a long range and high power, processor, memory, wireless and wired interface to connect with OBUs, RSUs, and the internet. It can increase the coverage area of OBUs by forwarding data. RSUs are installed along the roads near intersections, locations of high vehicle density, etc. Communication standards followed are IEEE 802.11, and all four IEEE 1609 protocols [24].

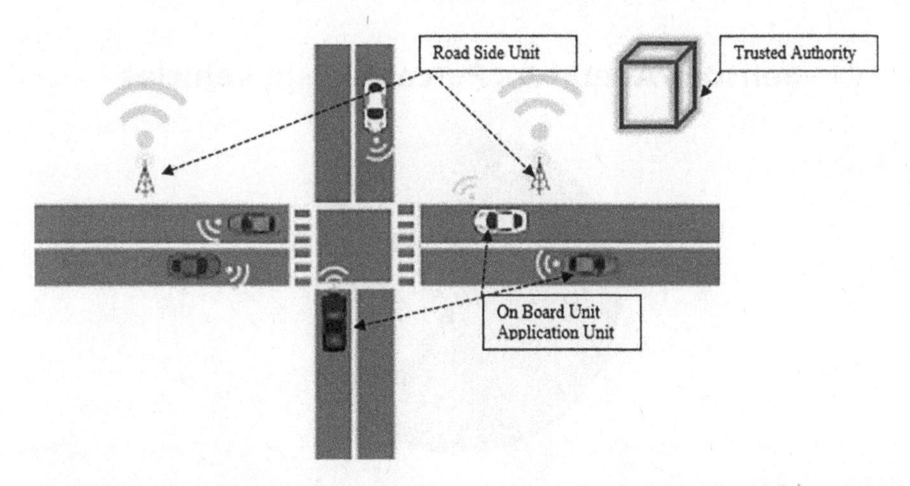

Figure 6.4 Typical VANET components.

- **Onboard unit (OBU)**: Applications may reside in the OBU. OBU is a peer device that uses the services. It has numerous sensors to collect data and process the required information. It is also responsible for sending information as messages to connected vehicles or RSUs. Typical structure for OBUs consists of a transceiver, radio-frequency antenna, wireless channel, processor, memory, user interface, connectivity (such as USB and Bluetooth), the Global Positioning System (GPS) sensor, etc. Communication standards followed are IEEE 802.11-2012, IEEE 1609.1, 1609.2, 1609.3 and1609.4 [25].
- **Application unit** (AU): Vehicles equipped with AU provided by the automobile manufacturer. The AU is responsible for providing an interface to the user of the vehicle. Typical applications such as navigation, traffic information, and infotainment, and messages from other vehicles/cloud can be displayed along with important information.

Figure 6.4 shows the VANET components in a typical VANET environment.

6.3.3 Sensors in connected vehicles

In VANET, vehicles and other roadside infrastructure are equipped with various sensing, computing or processing features, and wireless networking capabilities for connectivity. Ambient intelligence senses the immediate surroundings of vehicles and is also intelligent to take quick runtime critical decisions for safe driving. Today, modern connected vehicles are equipped with approximately 100 sensors onboard, and this number is expected to exceed over 200 by the end of the year 2022 [26]. These sensors are diverse in types and their nature and generally include other smart devices such as radar sensors, accelerometers, GPS, LIDARs, video

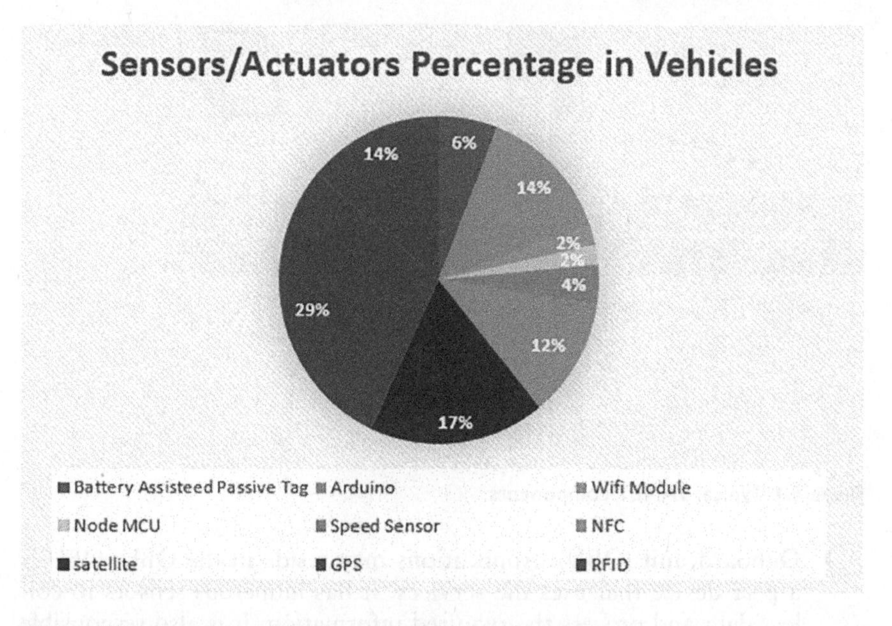

Figure 6.5 Percentage contribution of sensors in vehicles.

cameras, ultrasonic sensors, gyroscopes, infrared sensors, and odometry sensors. Therefore, a massive amount of data is generated by these sensors for analysis and processing for intelligent decision-making. Statistically, Intel estimated, connected vehicles generate 40 TB of data for every 8h of their driving [27]. This huge amount of data may flood the communication network and lead to considerable delay and disruption in the network services leading to the overall decline in quality of service and quality of experience of vehicle users. Therefore, to mitigate such a network management overhead, it is essential to deal with this massive volume of vehicular data observations or results at a smaller level. Thus, only the tasks which cannot be handled by the local compute and storage can be transferred to the backend servers. Figure 6.5 shows types of sensors and their percentage of presence generally present in connected vehicles used in VANET [28].

6.3.4 Communication

In VANET, different types of communication and connectivity are shown in Figures 6.6 and 6.7. Generally, there are the following four types of communication model (V2X) exist in VANET:

- V2V: Vehicle to vehicle.
- V2I: Vehicle to infrastructure (such as RSU).

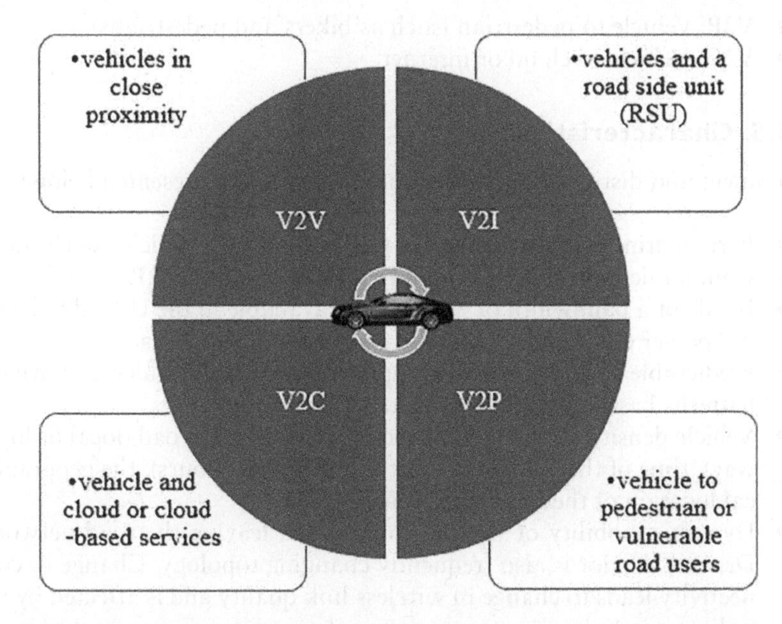

Figure 6.6 Type of communication in VANET.

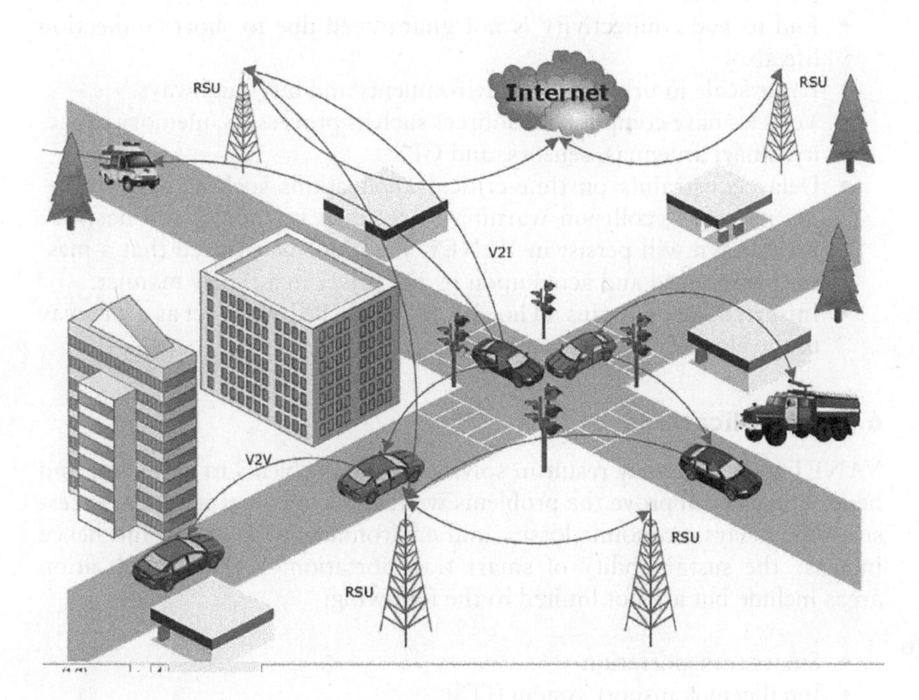

Figure 6.7 Communication in a typical VANET environment.

- V2P: Vehicle to pedestrian (such as bikers and pedestrians).
- V2C: Vehicle to cloud or internet.

6.3.5 Characteristics

Prominent and distinct characteristics of VANET are presented below:

- Participating nodes are vehicles, OBUs, and RSUs which can be static.
- Communication types are mainly V2V, V2R, V2I, V2P.
- Bands of a bandwidth of 75 MHz are available in the United States.
- No power constraint because of rechargeable batteries.
- Predictable mobility due to compliance with traffic rules and driving patterns based on road topology.
- Vehicle density on the road depends on the type of road (local or highway), time of the day (peak or non-peak traffic hours), the geographical location of the road (city or urban)
- Dynamic mobility of vehicles joining and leaving the road network. Driver behavior is also frequently changing topology. Change in connectivity leads to change in wireless link quality and is affected by the radio communication range.
- Connection life remains short due to road conditions, traffic conditions, traffic.
- End to end connectivity is not guaranteed due to short connection life also.
- Large scale in urban dense environments and busy highways.
- Vehicles have computing resources such as processors, memory capacity, smart antennas, sensors, and GPS.
- Delay constraints on time-critical applications such as early warning messages, collision warnings, pre-crash warnings, and network breakdown will persist in VANET. It cannot be ensured that a message is received and acted upon by the driver in a timely manner.
- Infrastructure remains ad hoc because only RSUs can act as a gateway to the cloud or the internet.

6.3.6 Applications

VANET will ultimately result in solving traffic problems to an extent and hence may also improve the problems we face such as fuel wastage, excessive time delays, economic losses, and environmental pollution and hence increase the sustainability of smart transportation. VANET application areas include but are not limited to the following:

- Smart transportation
- Intelligent transport system (ITS)

- Active road safety applications
- Smart traffic light system
- Parking guidance system
- Dynamic traffic management system
- Infotainment applications
- Vehicle monitoring and navigation

6.4 VANET: SECURITY ASPECTS

6.4.1 Need for Security

The primary concern of VANET includes safety, security, and privacy. VANET's mobility, dynamic topology, and frequent wireless links breakages make it difficult to provide adequate security and privacy mechanisms. The vulnerability of VANET is more due to the nature of a self-centralized network. Lack of security in VANET exposes them to different types of attacks that can have serious consequences for the safety of drivers on the road including loss of life. Another challenge is users' privacy as information exchanged between vehicles contains private information about the driver and the vehicle. Therefore, only legitimate and true information should be communicated over VANET. As vehicles become smarter, security becomes an ever more challenging issue for VANET. It is reported that 100% of 2016 vehicle models contain wireless technology and 60% of vehicles for the same year have internet access. Vulnerabilities in VANET increase as vehicles can join and leave the network without any security measures [29]. Therefore, a security attack on VANET can have a huge impact on human lives and assets.

6.4.2 Security requirements

Security requirements that describe more concretely what measures and controls must be prescribed to assure the security of each of the application domains. Generally, the primary goals of network security can be described in terms of three fundamental security properties:confidentiality, integrity, and availability. Though these three requirements form the basis, we list some other requirements including these three requirements. The list below describes the security requirement in detail and covers almost all aspects of security in VANET.

- **Availability:** Availability of network services in real-time and availability of resources
- **Authentication:** Verifying the authenticity of data transferred from V2V, V2I, and I2V along with verifying the identity of a user or device

- **Authorization:** Validating whether the user or device has rights or privileges to access a resource or service
- **Integrity:** Verifying that the transmitted messages are forwarded to the correct locations and without modification
- **Non-repudiation:** Non-denying any act of communication took place among different entities
- **Accountability:** Ensuring that every action can be traced back to a single user or device
- **Reliability:** Ensuring consistent intended behavior of the entire VANET
- **Privacy:** Ensuring that information should not be exposed to malicious or misbehaved vehicles
- **Physical security:** Denying any unauthorized physical access to devices, resources, or vehicles in VANET
- **Tracking of vehicle:** Ensuring that tracking the identity of the vehicle sending or receiving the messages is feasible in VANET
- **Scalability:** VANET should be scalable in accepting the number of additional vehicles without affecting the system performance
- **Access control:** Access policies must be designed and executed by legitimate users or vehicles in VANET
- **Encryption:** Enciphering transmitted messages so that only the intended recipient can decipher them
- **Secure booting:** Enabling device or vehicle to check every software installation or update using digital signature or checksums. Ensuring only authorized software must be running and it should be checked when the device is first powered on, or whenever it restarts
- **Secure updates:** Ensuring that security patches from service providers must be legitimate using digital signatures so that patches cannot be intercepted, extracted, and modified
- **Backup:** Ensuring archiving or backup mechanisms must be in place in the event of data loss
- **Efficiency:** Increasing the efficiency of VANET by minimizing the overhead and reducing the computation and delays

6.4.3 Security objectives

Based on the above security requirements, we can formulate rigid security objectives that must be adhered to or comply by all entities of VANET. These security objectives are listed below:

- Data in transit must be enciphered (confidentiality, privacy).
- Sensors must respond to queries from authorized entities only (authenticity, authorization).

- No modification to transmitted or received messages should be allowed (integrity).
- The origin of every message must be verified (message authenticity).
- Availability of services of VANET and its components must available to authorized entities (availability, reliability).
- No tampering of sensing devices and infrastructure should be allowed (physical security).
- Ensuring that no entity can deny participating in any communication (NR, accountability).

6.4.4 Attacks classification

Different types of attacks are possible in VANET and these attacks can be classified into different categories. We present a classification of attacks based on the following:

- Membership
- Activity
- Intentions
- Classes

This classification based on membership, activity and intention is presented in Figure 6.8 and classification based on attack classes is presented in Figure 6.9, respectively. Figure 6.10 (a–f) presents the security attributes

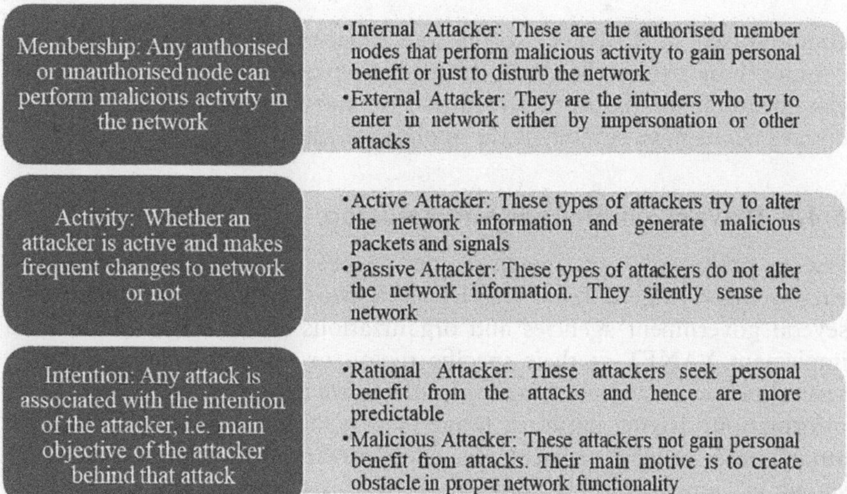

Membership: Any authorised or unauthorised node can perform malicious activity in the network

•Internal Attacker: These are the authorised member nodes that perform malicious activity to gain personal benefit or just to disturb the network

•External Attacker: They are the intruders who try to enter in network either by impersonation or other attacks

Activity: Whether an attacker is active and makes frequent changes to network or not

•Active Attacker: These types of attackers try to alter the network information and generate malicious packets and signals

•Passive Attacker: These types of attackers do not alter the network information. They silently sense the network

Intention: Any attack is associated with the intention of the attacker, i.e. main objective of the attacker behind that attack

•Rational Attacker: These attackers seek personal benefit from the attacks and hence are more predictable

•Malicious Attacker: These attackers not gain personal benefit from attacks. Their main motive is to create obstacle in proper network functionality

Figure 6.8 Classification of attacks in VANET based on membership, activity, and intention.

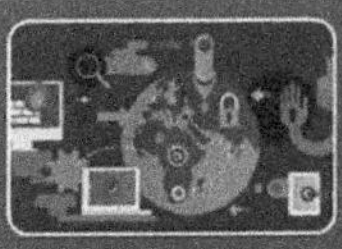

Network Attacks

• These are the most serious attacks. The whole network got affected from this. They are the direct attacks over functionality of network and node.

Application Attacks

• These types of attacks are primarily concerned with the information being shared and with the application being served.

Timing Attacks

• These attacks perform alteration in time slots of messages to add some delay.

Social Attacks

• All such messages or attacks that create emotional imbalance in other drivers come into this category. In this class of attacks unethical messages are sent to vehicles that disturb the driver and hence results into driving disruption, loss of other prerequisites of security system.

Monitoring Attacks

• In these attacks, attacker silently monitors and tracks the whole system and can perform malicious activities based on those observations. All passive attacks come into this category.

Figure 6.9 Classification of attacks in VANET based on classes.

compromised in different attacks and security mechanisms requires to counter these attacks. From the detailed comparison and analysis of attack classification, we further analyzed that the most vulnerable class in which attacks occur most severely is the application class.

6.4.5 Privacy challenges and solution

Privacy preservation is a much-required aspect in VANET; however, privacy is most vulnerable in VANET [30]. Due to this high vulnerability, several government agencies and organizations are finding it difficult to implement VANET in their specific region or country. In vulnerability, several privacy issues are ranging from vehicle details, passenger or driver information, driving patterns, usage of personal devices, communication among other vehicles or users, etc. These privacy challenges or attacks are shown in Figure 6.11. The solutions to privacy challenges are not concrete or one shoe fits all. The solution may vary due to devices, vehicle types, and several other factors such as a country's policies for security and privacy. In VANET, privacy may be classified into three categories: data privacy, identity privacy, and location privacy (see Figure 6.12).

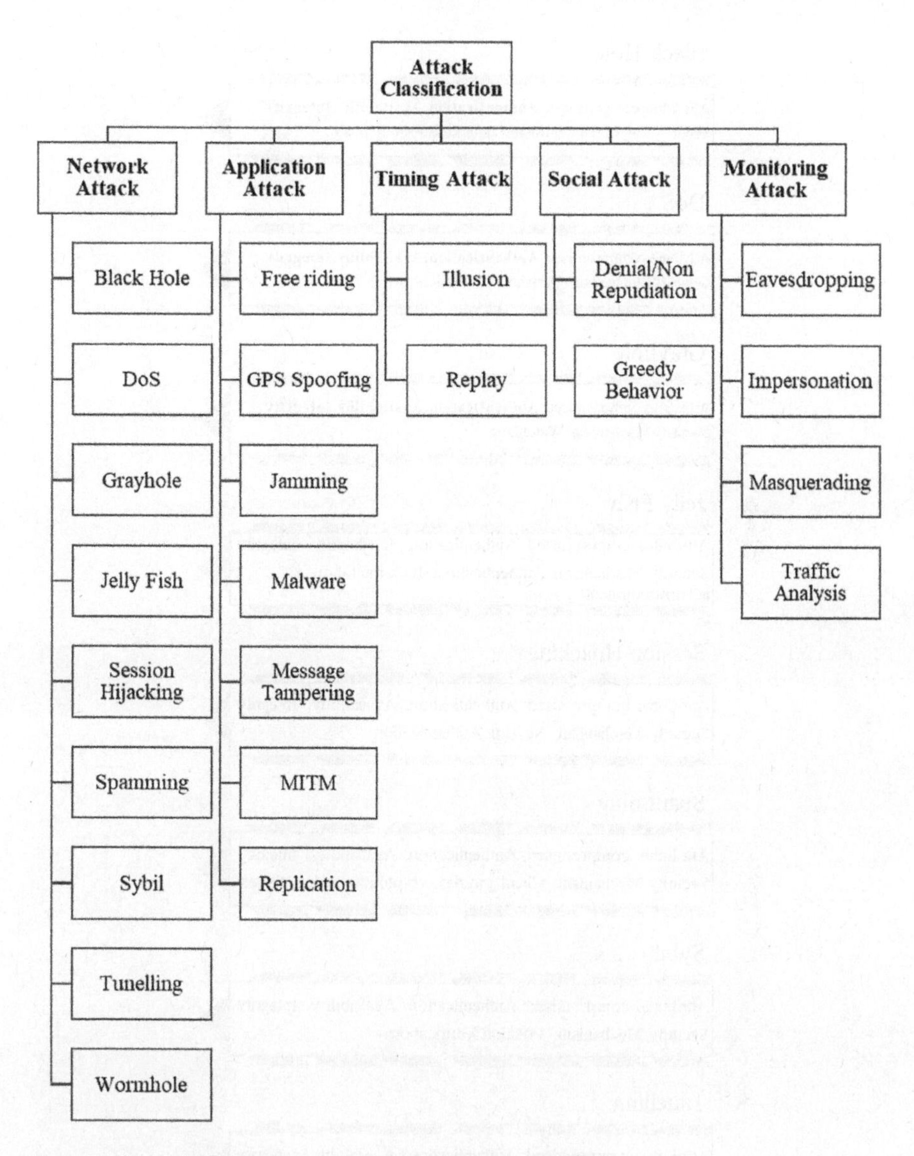

Figure 6.10 (a) Different attacks in VANET based on classes.

(Continued)

Black Hole
Attributes compromised: Authentication, Availability, Integrity
Security Mechanism: Intrusion Detection System (IDS)
DoS
Attributes compromised: Authentication, Availability, Integrity
Security Mechanism: Threshold Limitation
Grayhole
Attributes compromised: Authentication, Availability, Integrity
Security Mechanism: WatchDog
Jelly Fish
Attributes compromised: Authentication, Availability, Integrity
Security Mechanism: Authentication, IDS, multi-hop acknowledgments
Session Hijacking
Attributes compromised: Authentication, Availability, Integrity
Security Mechanism: Session Authentication
Spamming
Attributes compromised: Authentication, Availability, Integrity
Security Mechanism: Client puzzles, cryptographic techniques
Sybil
Attributes compromised: Authentication, Availability, Integrity
Security Mechanism: Position Verification
Tunelling
Attributes compromised: Authentication, Availability, Integrity
Security Mechanism: Signal strength distribution
Wormhole
Attributes compromised: Authentication, Availability, Integrity
Security Mechanism: Data authentication

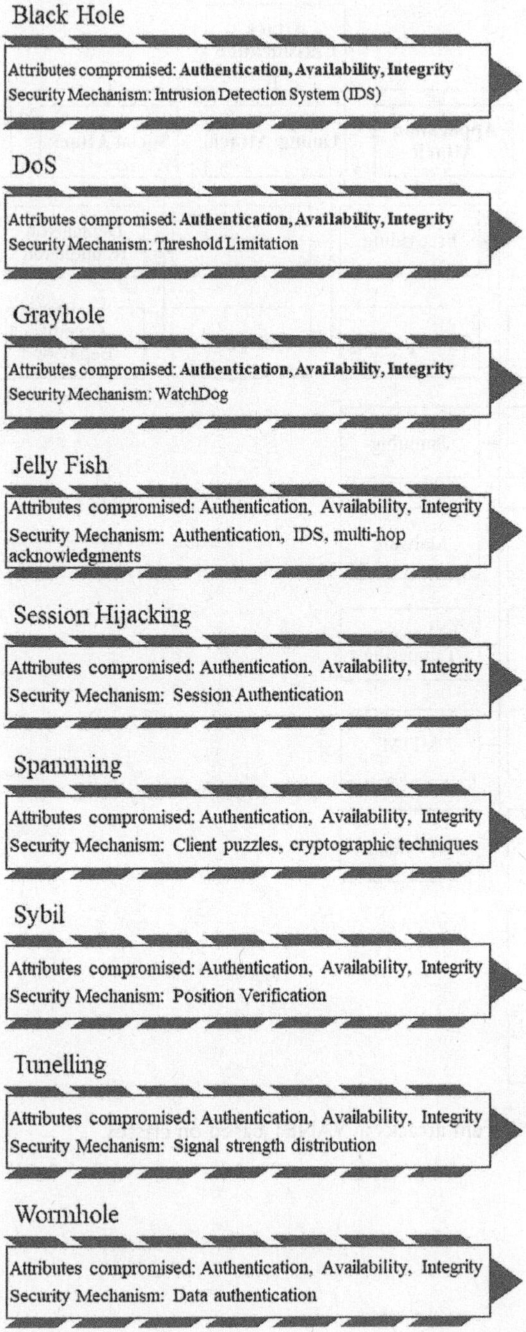

Figure 6.10 (b) Different attacks in network class, compromised attributes, and potential security mechanisms.

(*Continued*)

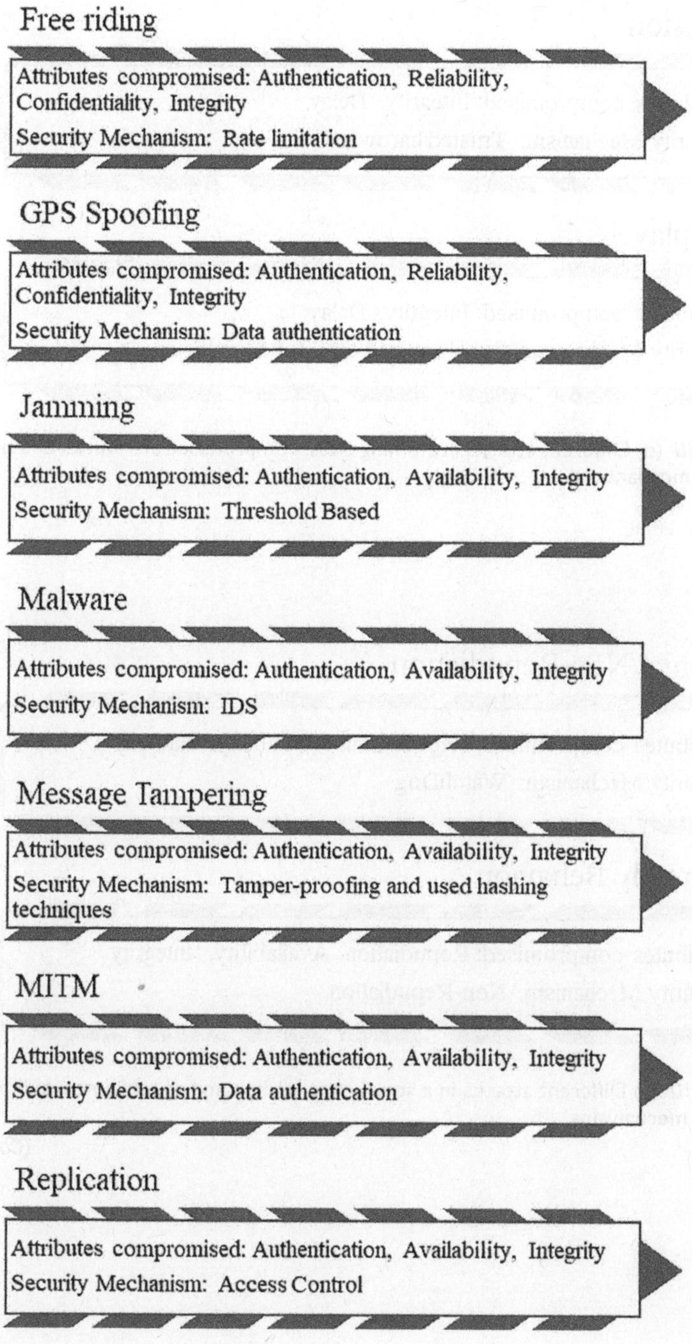

Free riding

Attributes compromised: Authentication, Reliability, Confidentiality, Integrity

Security Mechanism: Rate limitation

GPS Spoofing

Attributes compromised: Authentication, Reliability, Confidentiality, Integrity

Security Mechanism: Data authentication

Jamming

Attributes compromised: Authentication, Availability, Integrity

Security Mechanism: Threshold Based

Malware

Attributes compromised: Authentication, Availability, Integrity

Security Mechanism: IDS

Message Tampering

Attributes compromised: Authentication, Availability, Integrity

Security Mechanism: Tamper-proofing and used hashing techniques

MITM

Attributes compromised: Authentication, Availability, Integrity

Security Mechanism: Data authentication

Replication

Attributes compromised: Authentication, Availability, Integrity

Security Mechanism: Access Control

Figure 6.10 (c) Different attacks in an application class, compromised attributes, and potential security mechanisms.

(Continued)

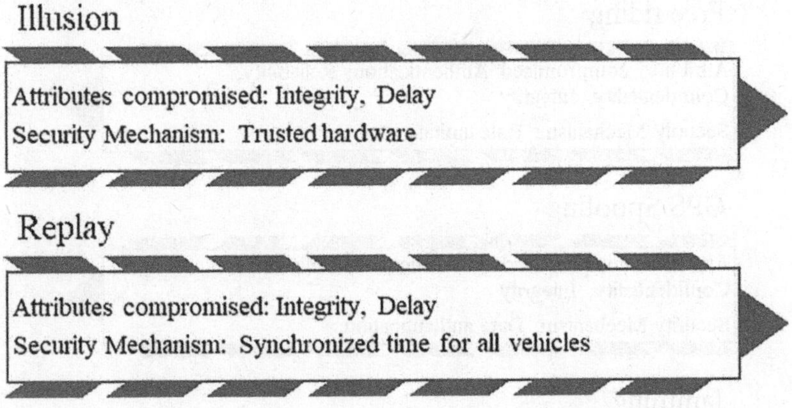

Illusion

Attributes compromised: Integrity, Delay
Security Mechanism: Trusted hardware

Replay

Attributes compromised: Integrity, Delay
Security Mechanism: Synchronized time for all vehicles

Figure 6.10 (d) Different attacks in a timing class, compromised attributes, and potential security mechanisms.

Denial/Non-Repudiation

Attributes compromised: Repudiation, Availability, Integrity
Security Mechanism: WatchDog

•Greedy Behavior

Attributes compromised: Repudiation, Availability, Integrity
Security Mechanism: Non-Repudiation

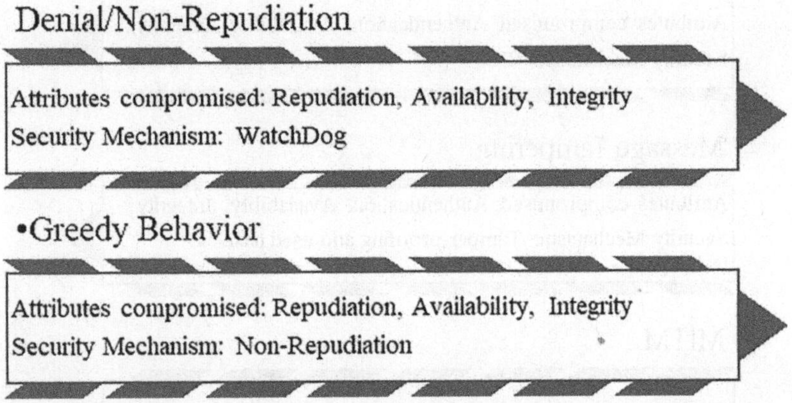

Figure 6.10 (e) Different attacks in a social class, compromised attributes, and potential security mechanisms.

(Continued)

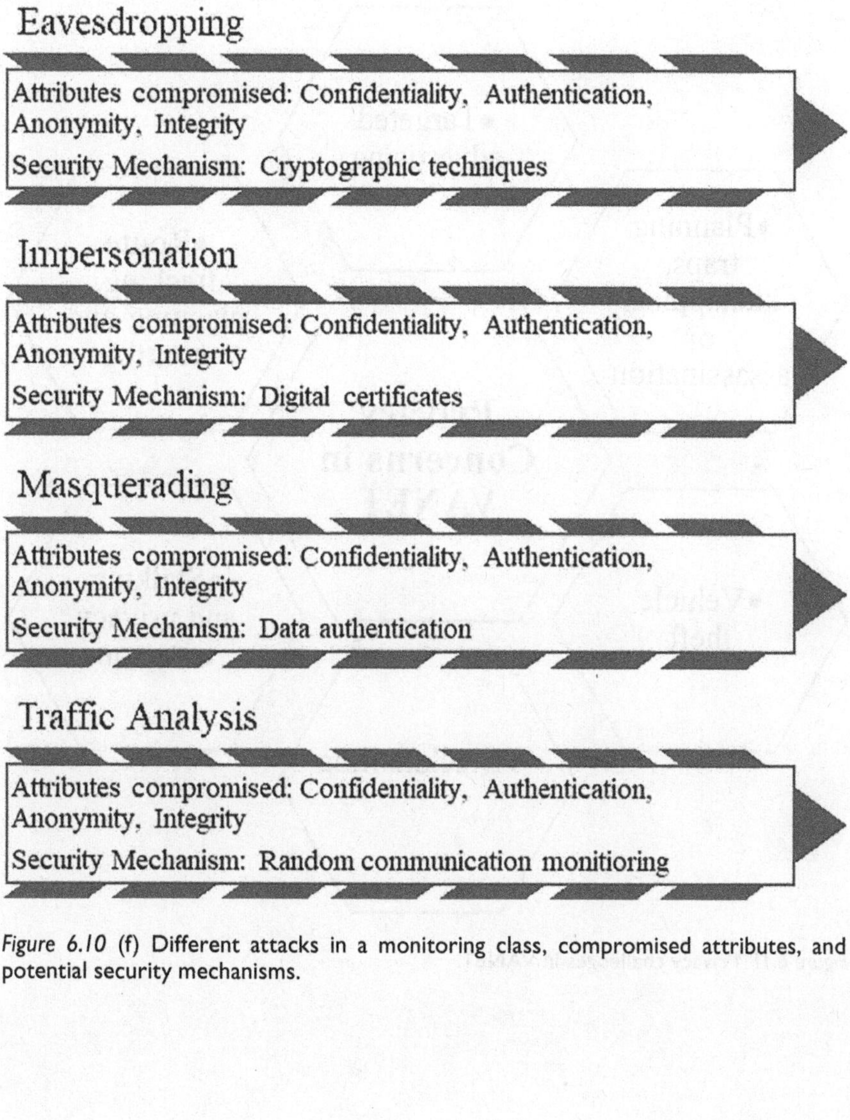

Eavesdropping

Attributes compromised: Confidentiality, Authentication, Anonymity, Integrity

Security Mechanism: Cryptographic techniques

Impersonation

Attributes compromised: Confidentiality, Authentication, Anonymity, Integrity

Security Mechanism: Digital certificates

Masquerading

Attributes compromised: Confidentiality, Authentication, Anonymity, Integrity

Security Mechanism: Data authentication

Traffic Analysis

Attributes compromised: Confidentiality, Authentication, Anonymity, Integrity

Security Mechanism: Random communication monitioring

Figure 6.10 (f) Different attacks in a monitoring class, compromised attributes, and potential security mechanisms.

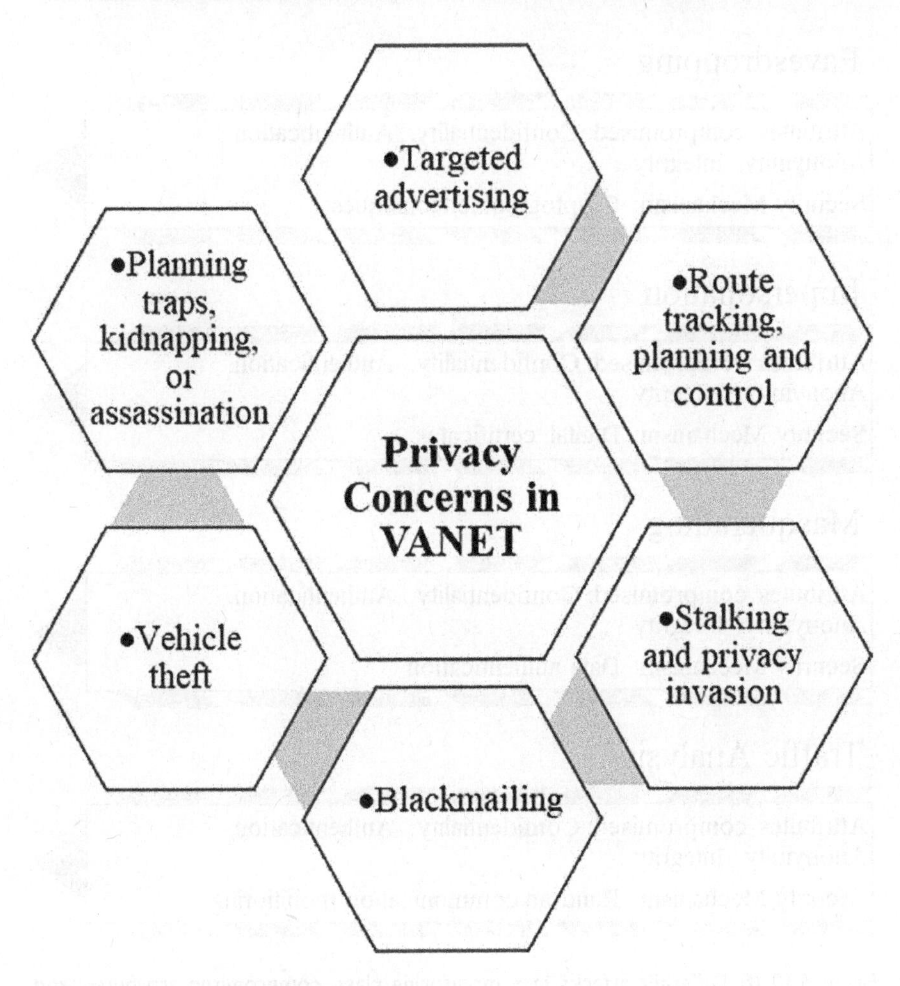

Figure 6.11 Privacy challenges in VANET.

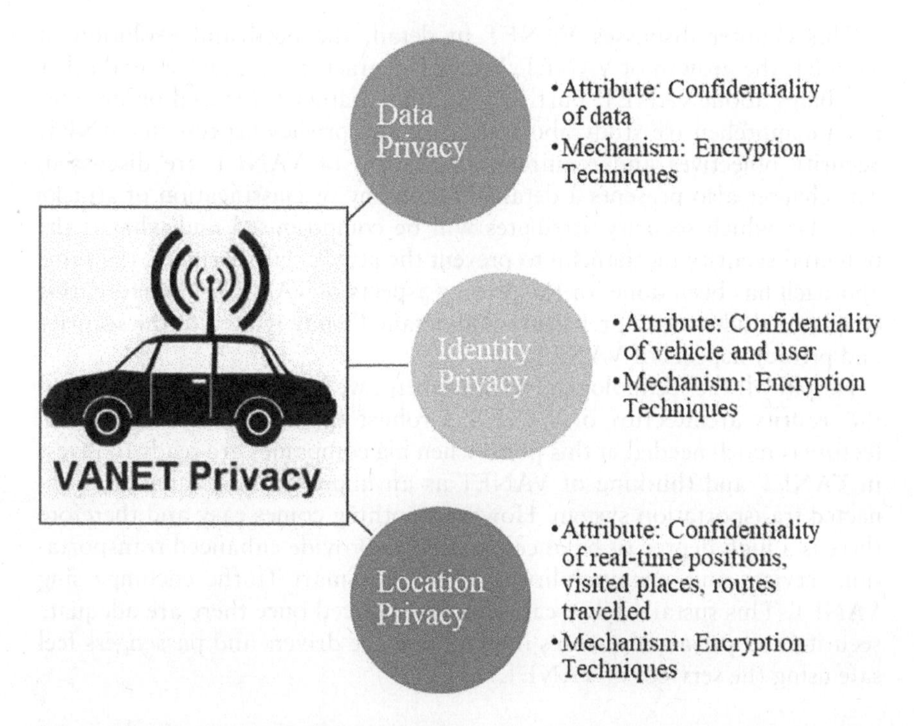

Figure 6.12 Privacy classification in VANET, compromised attributes, and a potential mechanism for privacy-preserving.

6.5 CONCLUSION

VANET is growing as it offers better services and safety. Several countries like Europe are implementing VANET and making regulations/standards for the deployment of VANET. VANET is not only limited to vehicles, but it opens the gateway for ITS, IoV, cloud-enabled connected cars, and even smart traffic. The most vulnerable aspect of VANET is the security and privacy of vehicles and passengers traveling. The most essential requirement of ITS, IoV, and VANET is safety on the roads. This vulnerability is the result of increased connectivity and unique services provided to drivers and passengers. The cost of obtaining enhanced user-based services and cloud connectivity poses an increased risk of numerous attacks by malicious vehicles or users. These attacks may lead to loss of life, property damages, vehicle loss, etc. These losses are a burden on several insurance providers. Hence, it is very important to provide and implement adequate security and privacy mechanisms.

This chapter discusses VANET in detail, the need and evolution of VANET, the growth of VANET, VANET characteristics, and several other attributes about VANET. Furthermore, this chapter is focused on presenting a comprehensive study about security and privacy aspects in VANET. Security objectives and requirements in light of VANET are discussed. The chapter also presents a detailed taxonomy or classification of attacks and also which security attributes will be compromised and what is the potential security mechanism to prevent the attack. Furthermore, the same approach has been done for the privacy aspects of VANET. Therefore, this chapter includes a comprehensive and detailed study related to the security and privacy aspects of VANET.

Despite all the technological advancement, we lack in standardization of the security architecture of VANET. A robust and resilient security architecture is much needed at this point when big companies are ready to invest in VANET and thinking of VANET as an improved and enhanced connected transportation system. However, nothing comes easy and therefore there is a high degree of balance required to provide enhanced transportation services and sustainability of ITS, IoV, Smart Traffic encompassing VANET. This sustainability can only be balanced once there are adequate security and privacy measures in place and the drivers and passengers feel safe using the services of VANET.

REFERENCES

1. Moghe, U., Lakkadwala, P., & Mishra, D. K. (2012). Cloud computing: Survey of different utilization techniques. *6th IEEE International Conference on Software Engineering*, CONSEG 12, 1–4.
2. Paiva, S., Ahad, M., Zafar, S., Tripathi, G., Khalique, A., & Hussain, I., (2021). Privacy and security challenges in smart and sustainable mobility. *SN Applied Sciences*, 2(7), 1–10.
3. Harris, A. & Rea, A. (2009). Web 2.0 and virtual world technologies: a growing impact on IS education. *Journal of Information Systems Education*, 2(2), 137–144.
4. Khalique, A. et. al. (2020). Wearable computing and its applications: an approach towards sustainable living. *Green Automation for Sustainable Environment*, *1*(1), 51–64, Taylor & Francis.
5. Raya, M., & Hubaux, J. (2007). Securing vehicular ad hoc networks. *Journal of Computer Security*, *15*(1), 39–68.
6. Baiocchi, A., & Cuomo, F. (2013). Infotainment services based on push-mode dissemination in an integrated VANET and 3G architecture. *Journal of Communications and Networks*, *15*(2), 179–190.
7. Nafis, M. T., & Khan, M. H. (2018). IoT enabled traffic control model using raspberry Pi. *International Journal of Advanced Research in Computer Science*, *9*(3), 157–160.

8. Global status report on road safety, World Health Organization, Geneva, 2009. [Online]. Available: https://www.who.int/violence_injury_prevention/road_safety_status/2009/web_version.pdf [Accessed: 03- Jan-2021].
9. Safe movement of vehicles at workplaces, [Online]. Available: https://www.commerce.wa.gov.au/sites/default/files/atoms/files/safe_movement_of_veh.pdf [Accessed: 03- Jan-2021].
10. Capital's drivers spent 227hours stuck in traffic in 2018 | London Road Safety Council,*Londonroadsafetycouncil.org.uk*, 2021. [Online]. Available: https://londonroadsafetycouncil.org.uk/capitals-drivers-spent-227-hours-stuck-in-traffic-in-2018/#:~:text=London%20has%20been%20named%20as, stuck%20in%20traffic%20in%202018.&text=Not%20only%20is%20London%20ranked, and%20sixth%20in%20the%20world. [Accessed: 03- Jan-2021].
11. Mahmood, Z. (2020). *Connected Vehicles in the Internet of Things*, Springer Nature Switzerland, 3–18, 223–251.
12. Connected Vehicles|FPT Software by IoV Platform Services, 2021. [Online]. Available: https://www.fpt-software.com/service/connected-vehicles/. [Accessed: 03- Feb-2021].
13. How 5G & IoT technologies are driving the connected smart vehicle industry, *Business Insider*, 2021. [Online]. Available: https://www.businessinsider.com/iot-connected-smart-cars. [Accessed: 03- Feb-2021].
14. Global connected cars market size, share, analysis & forecast 2025, *Adroitmarketresearch.com*, 2021. [Online]. Available: https://www.adroit-marketresearch.com/industry-reports/connected-cars-market. [Accessed: 03- Feb-2021].
15. Hayes, M., & Omar, T. (2019). End to End VANET/ IoT communications A 5G smart cities case study approach. *IEEE International Symposium on Technologies for Homeland Security (HST)*, 1–5.
16. A guide to 5G network security insight report, 2021. [Online]. Available: https://www.ericsson.com/en/security/a-guide-to-5g-network-security. [Accessed: 03- Feb-2021].
17. VANET/ITS initiatives/projects | VANET/ITS Website (NEO),Neo.lcc.uma.es, 2021. [Online]. Available: https://neo.lcc.uma.es/staff/jamal/vanet/index.html%3Fq=node%252F3.html. [Accessed: 03- Feb-2021].
18. Jana, B., Mitra, S., & Poray, J. (2016). An analysis of security threats and countermeasures in VANET. *International Conference on Computer, Electrical & Communication Engineering*, ICCECE, India, 1–6.
19. Mishra, R., Singh, A., & Kumar, R. (2016). Routing and security analysis in vehicular ad-hoc networks (VANETs). *International Conference on Electrical, Electronics, and Optimization Techniques*, 1050–1055.
20. Kumar, D., & Rishi, R. (2016). Routing and security analysis in vehicular ad-hoc networks (VANETs). *IEEE 1st International Conference on Power Electronics, Intelligent Control and Energy Systems*, ICPEICES, India 1–5.
21. Azees, M., Vijayakumar, P., & Jegatha, D. (2016). Comprehensive survey on security services in vehicular ad-hoc networks. *IET Intelligent Transport Systems*, *10*(6), 379–388.

22. Kaur, M., Martin, J. & Hu, H. (2016). Comprehensive view of security practices in vehicular networks. *International Conference on Connected Vehicles and Expo*, ICCVE, USA, 19–26.
23. Hasan, A., Hossain, M.S., & Atiquzzaman, M. (2016). Security threats in vehicular ad hoc networks. *International Conference on Advances in Computing, Communications and Informatics*, ICACCI, India, 404–411.
24. T. Program, ITS standards program | Fact Sheets | ITS standards fact sheets,*Standards.its.dot.gov*, 2021. [Online]. Available: http://www.standards.its.dot.gov/Factsheets/Factsheet/80. [Accessed: 03- Feb-2021].
25. Status of the dedicated short-range communications technology and applications,2015. [Online]. Available: https://rosap.ntl.bts.gov/view/dot/3575/dot_3575_DS1.pdf?
26. On the way to an intuitive connected vehicle: HMI technology trends,*Medium*, 2021. [Online]. Available: https://medium.com/@infopulseglobal_9037/on-the-way-to-an-intuitive-connected-vehicle-hmi-technology-trends-c71d7be8ef1f?source=---------36-------------------. [Accessed: 03- Feb-2021].
27. "For self-driving cars, there's big meaning behind one big number: 4 terabytes",*Intel Newsroom*, 2021. [Online]. Available: https://newsroom.intel.com/editorials/self-driving-cars-big-meaning-behind-one-number-4-terabytes/#gs.s5kbpa. [Accessed: 03- Feb-2021].
28. Guerrero-Ibáñez, J., Zeadally,S., & Contreras-Castillo, J., (2018). Sensor technologies for intelligent transportation systems, *Sensors*, *18*(4), 1212–2018.
29. Sheikh, M., Liang, J., & Wang, W. (2020). Security and privacy in vehicular ad hoc network and vehicle cloud computing: a survey. *Wireless Communications and Mobile Computing*, *2020*, 1–25.
30. Alfadhli, S., Lu, S., Fatani, A., Al-Fedhly, H., & Ince, M. (2020). SD2PA: a fully safe driving and privacy-preserving authentication scheme for VANETs. *Human-centric Computing and Information*, *10*(1) 1–25.

Chapter 7

Parametric optimization of liquid flow control process using evolutionary algorithms

Pijush Dutta and Madhurima Majumder
Global Institute of Management & Technology

Asok Kumar
Vidyasagar University

Korhan Cengiz
Trakya University

Rishabh Anand
HCL Technologies Limited

CONTENTS

7.1 INTRODUCTION

The liquid flow control process is a complex technique and has many influencing factors such as types of the flow sensor, pipe diameter, and characteristics of the liquid properties [1,2]. It is one of the common, real, nonlinear, and large time delay industrial processes. To obtain the optimal parameter by the conventional controller is a difficult task.

DOI: 10.1201/9781003156789-7

Many researchers have carried out many advanced control strategies for the liquid flow control process such as neural network control model [3,4], fuzzy logic controller [5,6], genetic algorithm [7], hybrid GA-ANN model [8], ANFIS model [9], hybrid metaheuristic optimization techniques like FPA-ANN [10] and improved versions of original elephant swarm water search algorithm [11] and also the liquid flow control process with non-linear, multivariable, boundary conditions and fluctuation of complex features. However, there is still scope for improving the results. Therefore, the estimation of a highly accurate model for describing the liquid flow control process is still an open problem to us.

Due to the restriction of process parameters and interaction between the input process variable and the response it is difficult to obtain the real-time quality response from the system. In such cases, metaheuristics or bioinspired optimization techniques can effectively solve the real-time predictive problem for online measurements. Due to global optimization methods and scale well to higher-dimensional problem-solving capability, we propose three different evolution optimization techniques. To represent the nonlinear model of a liquid flow process, we use analysis of variance (ANOVA).

From the simulation results, it is seen that proposed optimization techniques are effective and feasible to meet the real-time control requirements of the liquid flow control process. Modeling of the liquid flow control is described in Section 7.2. The mathematical description is briefly introduced in Section 7.3. The proposed methodology is described in Section 7.4. Results & discussion and conclusions are presented in Sections 7.5 and 7.6, respectively.

7.2 MODELING OF THE LIQUID FLOW CONTROL PROCESS

The experimental work is carried out with the flow & level measurement and control setup [10] (model no. WFT-20-I; see Figure 7.1).

For this work, a total of 134 sample data have been collected, which consist of four independent variables: sensor output voltage, pipe diameter, liquid (water) conductivity, and viscosity. Among these 17 datasets are utilized for the testing purpose (see Table 7.1).

7.3 MODELING OF THE LIQUID FLOW PROCESS

Due to nonlinear characteristics of liquid flow rate and liquid level process, it is very difficult to determine the process input variables such as pipe diameter, pipe diameter, and change in liquid properties to achieve the optimum liquid flow rate and liquid level using conventional controller techniques. Hence, we need computational intelligence tools. In this work, we have used ANOVA [11,12] to describe the mathematical relationship

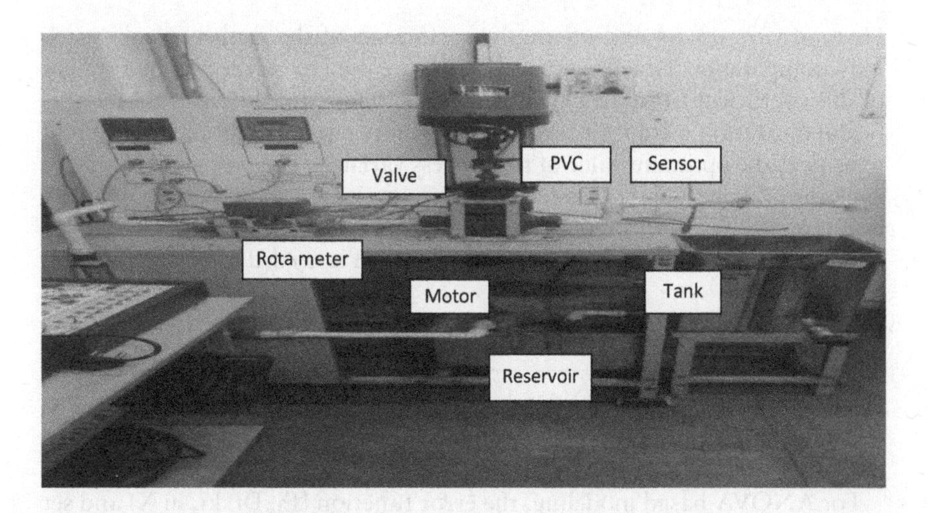

Figure 7.1 Experimental setup for liquid flow rate measurement [10].

Table 7.1 Experimental datasets for the liquid flow control process [10]

Sensor output	*Diameter*	*Conductivity*	*Viscosity*	*Flow rate*
0.218	0.024	0.606	0.8982	0.0008
0.221	0.025	0.616	0.7797	0.0008
0.225	0.025	0.616	0.8982	0.0016
0.232	0.025	0.597	0.7797	0.0016
0.234	0.02	0.615	0.8982	0.0024
0.237	0.027	0.622	0.7797	0.0024
0.238	0.03	0.6065	0.7254	0.0024
0.239	0.025	0.616	0.8982	0.0032
0.241	0.027	0.622	0.7797	0.0032
0.245	0.024	0.6065	0.7254	0.0032
0.247	0.024	0.616	0.8982	0.004
0.247	0.025	0.622	0.7797	0.004
0.25	0.025	0.6065	0.7254	0.0048
0.256	0.025	0.616	0.8982	0.0048
0.254	0.024	0.622	0.7797	0.0056
0.259	0.03	0.606	0.7254	0.0064
0.265	0.027	0.622	0.7797	0.0072

between input variables and the response of the liquid flow control process. In the mathematical model flow rate (F) can be expressed in terms of sensor output (E), pipe diameter (D) and liquid properties as follows:

$$F = \mu 1 \cdot E^{\mu 2} \cdot D^{\mu 3} \cdot k^{\mu 4} \ldots n^{\mu 5} \tag{7.1}$$

Here, $\mu 1$, $\mu 2$, $\mu 3$, $\mu 4$ and $\mu 5$ are the coefficients of the mathematical model. Now computational intelligence techniques are used to get the optimum values of this coefficient from the experimental dataset. Three different metaheuristics optimization techniques are applied as a computational intelligence tool to achieve the optimum value of the process variable. The process fits with the experimental one and the difference between measured and simulated flow rate is minimized. So, in this research, we used root mean square error (RMSE) as an objective function for the metaheuristic which is needed to be minimized.

$$\mathrm{RMSE}(X) = \frac{\sum_{i=1}^{N} \sqrt{f(E,D,K,X)^2}}{N} \tag{7.2}$$

where N is the number of the experimental data, X is the set of the estimated parameters.

For ANOVA-based modeling, the error function (Ei, Di, ki, ni X) and set of parameters X can be expressed as

$$(Ei, Di, X) = \mu 1 \cdot E^{\mu 2} \cdot D^{\mu 3} \cdot k^{\mu 4} \cdot n^{\mu 5} - F \tag{7.3}$$

$$X = \{\mu 1, \mu 2, \mu 3, \mu 4, \mu 5\} \tag{7.4}$$

where F is the experimental data.

7.4 PROPOSED METHODOLOGY

Before elaborating the proposed methodology [13] basic differential evolution (DE), particle swarm optimization (PSO), and genetic algorithm (GA) are elaborated initially.

7.4.1 Differential evolution (DE) calculation

It is a stochastic, populace-based improvement calculation for taking care of nonlinear advancement issues [14]. The calculation was presented by Storn and Price in 1996 [15]. A fundamental variation of the DE calculation works by having a populace of applicants' arrangements (called specialists). These specialists are moved around in the inquiry space by utilizing straightforward numerical formulae to join the places of existing operators from the populace. The new position of a specialist is an improvement, and it is acknowledged and frames some portion of the populace; generally, the new position is just disposed of. The procedure is rehashed and by doing so it is trusted, however not ensured, that a palatable arrangement will inevitably be found.

7.4.2 Particle swarm optimization (PSO)

PSO is a heuristic technique inspired by the interaction of a group of animals [16]. The process starts with random variables, each of which represents the possible solution to the optimization problem. Each of the variables represents two parameters: position and velocity. The position of the variable indicates the deviation from the ideal optimum solution, while the velocity indicates the motion of a variable around the search space. The social behavior of PSO is represented by other two parameters: Pbest and Gbest. Pbest identifies the individual best position of the particle, while Gbest identifies the global best position of the swarm. PSO has been effectively utilized in different research fields [17–22] such as renewable energy, nanoparticles, mathematical modeling, and nuclear energy field.

7.4.3 Genetic algorithm (GA)

GA has a place with a group of computational models propelled by Darwin's generation and natural selection hypothesis [23–25]. GA utilizes the essential propagation administrators, for example, hybrid and transformation to deliver the hereditary organization of the populace. Some trade and reordering of chromosomes, creating posterity that contains a mix of data from each parent, is frequently alluded to as hybrid due to the route strands of chromosomes traverse during the trade. Decent variety in the populace is accomplished by change. A regular hereditary calculation methodology makes the accompanying strides. The populace of applicants' answers (for the enhancement errand to be tackled) is introduced. New arrangements are made by applying hereditary administrators (change as well as hybrid). The wellness (how great the arrangements are) of the subsequent arrangements is assessed, and a reasonable choice procedure is then applied to figure out which arrangements will be kept up for the people to come. The methodology is then iterated. After a few cycles or ages, the calculation unites to the best person that speaks to the ideal answer for the current issue. Hereditary algorithms have been effectively utilized as an improvement device in operational research [15], multidisciplinary streamlining approach [14,16], the board [23], coordination [24], and aviation applications [25].

7.5 RESULT ANALYSIS

The parameters setting for every calculation in the examination is portrayed as pursues:

1. For GA, crossover percentage (pc) is 0.7, extra range factor for crossover (γ) is 0.4, mutation percentage (pm) is 0.3, and mutation rate (mu) is 0.1.

Table 7.2 A comparative study based on computational time

Method	*Average computational time (seconds)*
DE	146.1244
PSO	165.7174
GA	119.0112

2. For DE, the mutation factor (F) is 0.5, the crossover rate (C) is 0.9, and the maximum iteration number is 200.
3. For PSO, inertia weight (w) is 1, inertia weight damping factor (wdamp) is 0.99, personal learning coefficient ($c1$) is 1.5, and global learning coefficient ($c2$) is 2 according to the earlier work.

For all the algorithms, we choose the maximum iteration number as 5,000 and population as 100. For a liquid flow model, search space is confined to five-dimensional ($\mu1$, $\mu2$, $\mu3$, $\mu4$, $\mu5$) function optimization problems to look through ideal estimations as shown in equation 7.4. The search range [7] for the optimization of the liquid flow-based model is (–15, 15).

7.5.1 Computational efficiency test

Computational time is one of the major criteria for evaluating the effectiveness of the bioinspired optimization technique applied in a particular process of parametric optimization. In this subsection, we have taken the average execution time taken by each algorithm for each of the problems for a fixed number of iteration 5,000, population 100 and 10 times run the program. Table 7.2 shows a comparative study based on average execution time. It has been observed that the GA-based ANOVA model performed best by means of average computational time (Table 7.2).

7.5.2 Accuracy test

The accuracy test is another statistical criteria that indicate that the calculated value is how close to the measured value under the different experimental conditions. In accuracy test we have used two error indicator indexes: mean absolute error (MAE) and mean absolute percentage error (MAPE) to measure the error between experimental and the simulated current data, as defined in equations 7.5 and 7.6:

$$\text{IAE} = |\,F\text{measured} - F\text{calculated}\,| \tag{7.5}$$

$$\text{MAPE} = \frac{1}{n}\sum_{i=1}^{n} \frac{\text{IAE}}{F\text{measured}} \tag{7.6}$$

Moreover, MAE can be defined as:

$$\text{MAE} = \frac{\sum_{i=1}^{n} \text{IAE}_i}{n} \tag{7.7}$$

where, n is the number of the experimental dataset, and F_{measured} and $F_{\text{calculated}}$ are the experimental and estimated values of liquid flow rate. The best optimization algorithm has always produced the least RMSE for all different runs. The coefficient of the nonlinear models is obtained from DE, PSO, and GA MATLAB® codes as shown in Table 7.3.

The prediction error can be calculated using RMSE which can be defined as follows:

$$\text{RMSE} = \sqrt{\sum_{m\,i=1}^{m} \frac{X_{\text{exp}} \times X_{\text{cal}}}{X_{\text{exp}}} \times 100\%}$$

$$\text{Accuracy} = (100 - \text{RMSE})\%$$

Here, X_{exp} is the experimental value X_{cal} is the calculated value and m is the number of training data. DE offers the least MAPE, while PSO gives the least MAE (see Table 7.4). PSO optimization has the least RMSE error and maximum accuracy (Table 7.5). Figure 7.2 shows the relative errors vs. different liquid flow rate measurements for DE, PSO, and GA-based modeling, respectively. It can be seen that the proposed DE optimization has the least relative error than other optimization techniques.

Comparative study between calculated values is obtained from GA, DE, and PSO with experimental values of the outputs as shown in Figure 7.3. GA optimization provides a better calculated flow rate with respect to experimental flow rate. Figure 7.4 represents the graph between deviation ($=X_{\text{exp}}-$ Xcal) and experimental flowrate where the deviation is minimum between the flowrate of Xexp 250–500 Lpm DE have the least deviation with flowrate.

Table 7.3 **Estimated optimal parameters by using DE, PSO, and GA-based modeling of the liquid flow control process**

Method	*μ1*	*μ2*	*μ3*	*μ4*	*μ5*
DE	13.9440	11.3317	−1.3593	−5.0136	−0.7676
PSO	15.00	10.0466	−1.0550	−3.6296	−0.6247
GA	15.00	8.4074	−0.7271	−1.3846	−0.7300

Table 7.4 **A comparative study based on mean absolute error (MAE) in DE, PSO, and GA**

Method	*Mean absolute percentage error (MAPE)*	*Mean absolute error (MAE)*
DE	15.79	0.040
PSO	16.60	0.036
GA	20.41	0.039

Table 7.5 A comparative study based on root mean square error (RMSE) and accuracy

Method	*RMSE*	*Accuracy*
DE	0.0481	99.9519
PSO	0.0442	99.9558
GA	0.0492	99.9508

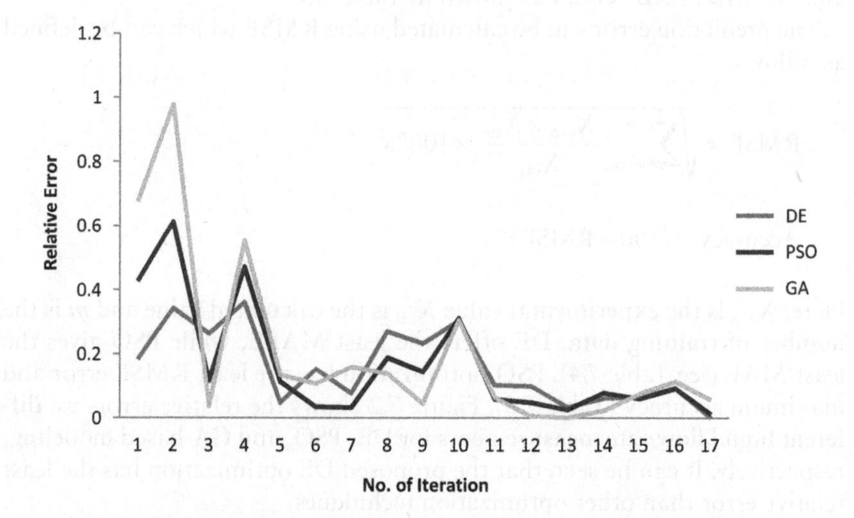

Figure 7.2 Relative errors for GA, PSO, and DE-based modeling of the liquid flow control process.

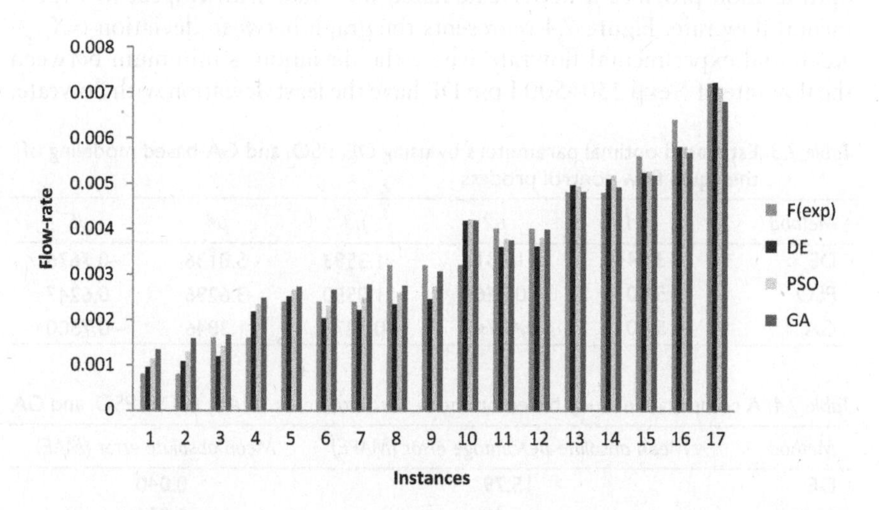

Figure 7.3 Comparisons of the characteristics of the experimental data and estimated liquid flow rate using DE, PSO, and GA-based model.

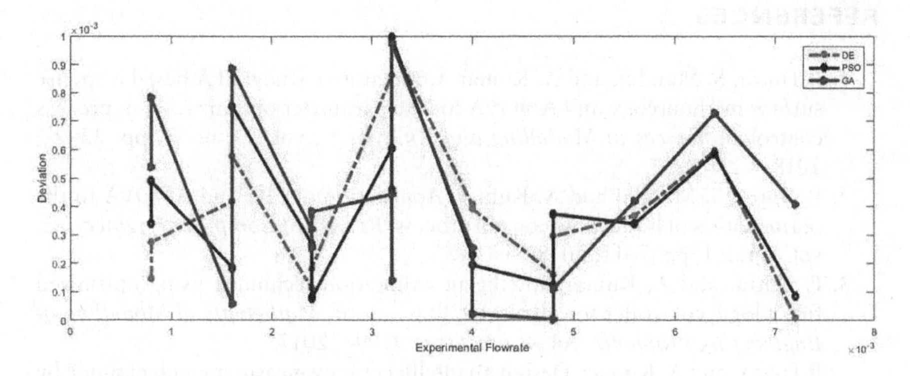

Figure 7.4 Deviation vs. experimental flow rate.

7.6 CONCLUSIONS

Modeling and optimization of liquid level and flow control in a process industry is an interesting task for researchers. In this study, our aim is to optimize the power equation model of ANOVA which is used to make a nonlinear relation between the input and output of the process variables.

In the next step, we need to find out the optimal values of the coefficient of the NN models using some suitable optimization techniques so that the estimated liquid flow rate fits best with the experimental results. In the final step, we perform different statistical analysis which is shown in Section 7.5. From the result analysis we get the following conclusion: (a) in respect to computation time GA is more effective; (b) in respect to MAE, RMSE, and accuracy PSO show the more effective output and (c) DE shows the better result in respect to MAPE and relative error. So, all the evolution algorithms do not provide accurate results in all statistical result analysis aspects. However, both algorithms can predict the liquid flow rate with satisfactory accuracy of more than 99.95%.

More accurate modeling can be designed by considering the other type of contact-type liquid flow sensor and further tuning can be done by metaheuristic optimization techniques to achieve better stability and accuracy are the future aspect.

COMPLIANCE WITH ETHICAL STANDARDS

Conflict of interest: The authors declare that they have no conflict of interest.

Human and animal rights: This article does not contain any studies with human participants or animals performed by any of the authors.

REFERENCES

1. P. Dutta, S. Mandal, and A. Kumar, Comparative study: FPA based response surface methodology and ANOVA for the parameter optimization in process control, *Advances in Modelling and Analysis C*, vol. 73, no. 1, pp. 23–27, 2018.
2. P. Dutta, S. Mandal and A. Kumar, Application of FPA and ANOVA in the optimization of liquid flow control process, *Review of Computer Engineering*, vol. 5, no. 1, pp. 7–11, 2018.
3. P. Dutta, and A. Kumar, Intelligent calibration technique using optimized fuzzy logic controller for ultrasonic flow sensor, *Mathematical Modelling of Engineering Problems*, vol. 4, no. 2, pp. 91–94, 2017.
4. P. Dutta, and A. Kumar, Design an intelligent flow measurement technique by optimized fuzzy logic controller, *Journal Europen des Systmes Automatiss*, vol. 51, pp. 89–107, 2018.
5. P. Dutta, and A. Kumar, Study of optimizated NN model for liquid flow sensor based on different parameters. *International Conference on Materials, Applied Physics & Engineering*, 2018.
6. K.V. Santosh, and K.V. Roy, An intelligent flow measurement technique using ultrasonic flow meter with optimized neural network, *International Journal of Control and Automation*, vol. 5, no. 4, pp. 185–196, 2012.
7. P Dutta, A Kumar, Modelling of Liquid flow control system using optimized genetic algorithm. Communicated.
8. P. Dutta, and A. Kumar, Design an intelligent calibration technique using optimized GA- ANN for liquid flow control system, *Journal Europen des Systmes Automatiss*, vol. 50, no. 4–6, pp. 449–470, 2017.
9. P. Dutta, and A. Kumar, Application of an ANFIS model to optimize the liquid flow rate of a process control system, *Chemical Engineering Transactions*, vol. 71, pp. 991–996, 2018.
10. P. Dutta, and A. Kumar, Modeling and optimization of a liquid flow process using an artificial neural network-based flower pollination algorithm, *Journal of Intelligence System*, 2018, Doi: 10.1515/jisys-2018-0206.
11. P. Dutta, S. Mandal, and A. Kumar, Modeling of liquid flow control process using improved versions of elephant swarm water search algorithm, *SN Applied Sciences*, vol. 1, p. 886, 2019, Doi: 10.1007/s42452-019-0914-5.
12. S.A. Glantz, B.K. Slinker, and T.B. Neilands, *Primer of Applied Regression & Analysis of Variance*, McGraw-Hill, New York, 2016.
13. L.F. Gonzalez, E.J. Whitney, K. Srinivas, and J. Periaux, Optimum multidisciplinary and multi-objective wing design in CFD using evolutionary techniques. *International Conference on Computational Fluid Dynamics 3*, Toronto, Canada, July 2004.
14. V. Arunachalam, *Optimization using Differential Evolution, Water Resources Research Report no.60*. Facility for Intelligent Decision support Department of Civil and environmental Engineering, Ontario, July 2008.
15. R. Stron and K. Price, Differential Evolution –a Simple and efficient heuristic for global optimization over continuous spaces, *Journal of Global Optimization*, vol. 11, no. 4, pp. 341–359, 1997.

16. J. Kennedy and R.C Eberhart, Particle swarm optimization. *Proceedings of IEEE International Conference on Neural Network*, Piscataway, NJ, pp. 1942–1948, 1995.
17. I.M. de Mendonça, I.C.S. Junior, and A.L.M. Marcato, Static planning of the expansion of electrical energy transmission systems using particle swarm optimization, *International Journal of Electrical Power and Energy Systems*, vol. 60, pp. 234–244, 2014.
18. T.-D. Liu, T.-E. Fan, G.-F. Shao, J.-W. Zheng, and Y.-H. Wen, Particle swarm optimization of the stable structure of tetrahexahedral Pt-based bimetallic nanoparticles, *Physics Letters A*, vol. 378, no. 40, pp. 2965–2972, 2014.
19. U. Aich and S. Banerjee, Modeling of EDM responses by support vector machine regression with parameters selected by particle swarm optimization, *Applied Mathematical Modelling*, vol. 38, no. 11–12, pp. 2800–2818, 2014.
20. C.-J. Chou, C.-Y. Lee, and C.-C. Chen, Survey of reservoir grounding system defects considering the performance of lightning protection and improved design based on soil drilling data and the particle swarm optimization technique, *IEEJ Transactions on Electrical and Electronic Engineering*, vol. 9, no. 6, pp. 605–613, 2014.
21. B.A. Lee, B.S. Kim, M.S. Ko, K.Y. Kim, and S. Kim, Electrical resistance imaging of two-phase flow with a mesh grouping technique based on particle swarm optimization, *Nuclear Engineering and Technology*, vol. 46, no. 1, pp. 109–116, 2014.
22. P. Thakral and A.K. Bakhshi, Computational atomistic blueprinting of novel conducting copolymers using particle swarm optimization, *Journal of Computer-Aided Molecular Design*, vol. 28, no. 2, pp. 111–122, 2014.
23. S.S. Chaudhury, Application of genetic algorithm in production and operations management: a review, *International Journal of Production Research*, vol. 43, No. 19, pp. 4083–4101, October 2005.
24. D. Kalyanmoy, and S. Tiwari, Multi-objective optimization of a leg mechanism using genetic algorithm, *Engineering Optimization*, vol. 37, No. 4, pp. 325–350, June 2005.
25. C.K. Chang, Genetic algorithms for project management, *Annals of Software Engineering*, vol. 11, pp. 107–39, 2001.

Chapter 8

Internet of Things (IoT)-based industrial monitoring system

Syeda Florence Madina, Md. Shahinur Islam, and Fakir Mashque Alamgir
East West University

Mohammad Farhan Ferdous
Japan–Bangladesh Robotics and Advance Technology Research Center (JBRATRC)

Muhammad Arif
University of Lahore

CONTENTS

DOI: 10.1201/9781003156789-8

8.1 PART 1: INTRODUCTION

The importance of environment monitoring exists in many aspects. The environment is required to be monitored to maintain the healthy growth in crops and to ensure a safe working environment in industries, and so on. Due to technological growth, the process of reading the environmental parameters becomes easier compared to the past days.

The Internet of Things (IoT) is rapidly increasing technology. IoT is the network of physical objects or things embedded with electronics, software, sensors, and network connectivity, which enables these objects to collect and exchange data. In this project, we are developing a system which will automatically monitor the industrial atmosphere, evaluate stored data, and make decisions using the concept of IoT. IoT has given us a promising manner to construct effective industrial system and applications with the aid of the use of wireless gadgets, androids, and sensors. The main contribution of this project is that it provides IoT in industries with various sensors and control units [1].

In recent years, a wide range of industrial IoT applications have been developed and deployed. In our project, we use different types of sensors which will help us to gather data and inform about the current atmosphere where this project will be implemented. DHT11 features a temperature and humidity sensor complex with a calibrated digital signal output. By using the exclusive digital-signal-acquisition technique and temperature and humidity sensing technology, it ensures high reliability and excellent long-term stability.

The sensitive material of the MQ2 gas sensor is SnO_2 with lower conductivity in clean air. When the target combustible gas exists, the sensor's conductivity is higher along with the gas concentration rising. Gas sensor (MQ2) has high sanctity to liquified petroleum gas (LPG), propane, and hydrogen and can also be used for methane and other combustible steam; it is of low cost and suitable for different applications [2].

Passive infrared (PIR) sensors use heat emitted to detect the place of an object or a living creature. Humans and warm-blooded animals emit heat and also infrared rays. This ray cannot be seen with the naked eye because of its low frequency. Infrared rays' frequency is between 3T and 430T, while visible light frequency is between 430T and 750T [3]. IoT refers to the rapidly growing network of connected objects that can collect and exchange data using embedded sensors. It is nowadays finding profound use in each sector and also plays a key role in the proposed environmental monitoring system. IoT converging with cloud computing offers a novel technique for better management of data coming from different sensors, collected and transmitted by low power, NodeMCU and ESP8266 [4].

8.1.1 Background

The proposed project is implemented for monitoring the gas, temperature, motion, etc., and informs the connected people about the environmental condition through smartphone or portable device and give an alert about the harmful conditions.

Bangladesh has taken an initiative to adopt the latest technological innovations, and some companies have already set up IoT laboratories in the country. Different countries have already adopted this technological innovation and 30 billion IoT devices are expected to be in the world by 2020, helping to improve performance [4].

IoT is a technology that connects machines to machines, giving city-dwellers better performance and improving technological efficiency. This technological advancement will help find easy solutions for various problems, uploaded official website.

At present, Bangladesh has many sectors where IoT technology is used such as smart buildings, industrial automation, smart grids, water management, waste management, smart agriculture, telecare, intelligent transport systems, environment management, smart urban lighting, and smart parking.

The IoT industry is the next wave of internet technologies. IoT will fundamentally change how business and manufacturing are done worldwide. It continues to spread across the home and the enterprise and affects how we live and work every day. IoT adoption reached some 43% of enterprises worldwide by the end of 2016. Total investment between 2015 and 2020 will be $6 trillion among both consumer and industrial IoT markets, with industrial IoT leading the growth [4].

8.1.2 Project organization

In Section 8.2, we have discussed literature reviews of project relevance. In Section 8.3, we have discussed the hardware and software we used in this project. In Section 8.4, we have discussed the experimental studies and results that we have conducted for our project purpose. In Section 8.5, we have discussed the limitations and conclusion.

8.2 PART 2: LITERATURE REVIEW

8.2.1 "IoT" and its smart applications

This section discusses the concept of the project we are trying to implement, what kind of work has already been done before, and how we are different at. This chapter discusses the concept of the project we have implemented, what kind of work has already been done before, and how we are different and more improved than those projects. In every organization, there is always an information desk that provides information, advertisement messages, and many notifications to their customers and staff. The problem is that it requires some staff who is dedicated to that purpose and who must have up-to-date information about the offer's advertisement and the organization. Due to IoT, we can see many smart devices around us. Many people hold the view that cities and the world itself will be overlaid with sensing and actuation, many embedded in "things" creating what is referred to as a smart world. Similar work has been already done by many people around the world. In the literature review, IoT refers to intelligently connected devices and systems to gather data from embedded sensors, actuators, and other physical objects. IoT is expected to spread rapidly in the coming years adding a new dimension of services that improve the quality of life of consumers and productivity of enterprises, unlocking an opportunity. Now, mobile networks already deliver connectivity to a broad range of devices, which can enable the development of new services and applications. This new wave of connectivity is going beyond tablets and laptops; to connected cars and buildings; smart meters and traffic control; with the prospect of intelligently connecting almost anything and anyone. This is what the GSMA refers to as the "connected life" [5].

8.2.2 Health monitoring and management using IoT sensing with cloud-based processing: opportunities and challenges

In this paper, the authors proposed the health monitoring system and management using IoT sensing with cloud-based processing. They use a networking sensor to make possible the gathering of rich information indicative of our physical and mental condition. In this paper, they highlight the opportunities and challenges for IoT realizing this vision of future health care. Day-to-day technology will be improved to our benefit. But here, the main context is the

IoT system proposed. This is the most important thing. Recent years have seen arising interest in wearable sensors and today several devices are commercially available for personal health care, fitness, and activity awareness. Technologically, the vision presented in the preceding paragraph has been feasible for a few years now. Yet, wearable sensors have, thus far, had little influence on the current clinical practice of medicine. In this paper, we focus particularly on the clinical arena and examine the opportunities afforded by available and upcoming technologies and the challenges that must be addressed to allow integration of these into the practice of medicine. Here, we created the cloud and stored the data. We created the data accusation sensing transmission, data concentration cloud processing, and finally cloud processing visualized. But our cloud let's processing and visualization are connected between the internet. This model is the system architecture. It is impractical to ask physicians to pore over the voluminous data or analyses from IoT-based sensors. To be useful in clinical practice, the results from the Analytics Engine need to be presented to physicians in an intuitive format where they can readily comprehend the inter-relations between quantities and eventually start using the sensory data in their clinical practice. Finally, data gathered or inferred from IoT sensors span the complete spectrum of categories outlined in the previous paragraph and therefore an array of different visualization methodologies are required for effective use of the data. Eventually, this proposed system is effective for future technology effects [6].

8.2.3 IoT-based monitoring and control system for home automation

In this paper, the authors explained IoT-based monitoring and control systems for home automation. A home automation system will control lighting, climate, entertainment systems, and appliances. It may also include home security such as access control and alarm systems. The home automation system uses the portable device as a user interface. This project was aimed at controlling home appliances via smartphones using Wi-Fi as a communication protocol and raspberry pi as a server system. The user here will move directly with the system through a web-based interface over the web, whereas home appliances like lights, fans, and door locks are remotely controlled through the easy website [7].

8.2.4 Industrial automation using IoT

Automation is one of the increasing needs within industries as well as for domestic applications. Automation reduces human efforts by replacing human efforts with a self-operated system. The internet is one way of the growing platform for automation, through which new advancement is made through which on easily monitor as well control the system using the internet. As we are making use of the internet the system becomes secured and live data monitoring is also possible using the IoT system.

Using IoT in this paper, they are developing a system that will automatically monitor the industrial applications and generate alerts/alarms or make intelligent decisions using the concept of IoT. IoT is achieved by using local networking standards and remotely controlling and monitoring industrial device parameters by using Raspberry Pi and embedded web server technology. Raspberry Pi module consists of the ARM11 processor and real-time operating system whereas embedded web server technology is the combination of embedded device and internet technology. Using an embedded web server along with raspberry pi it is possible to monitor and control industrial devices remotely by using a local internet browser. They have developed new technologies that have allowed us to move from the first generation of the internet into the current transition into the fourth generation. This generation has been propelled by the concept of IoT [8].

8.3 PART 3: HARDWARE AND SOFTWARE

For this project, we used different types of hardware and software. NodeMCU, ESP82663, gas sensor, PIR sensor, DHT11 were the main hardware. Arduino IDE and Thinger.io studio were mainly used in this project.

8.3.1 Hardware

We used NodeMCU as the microprocessor for the system. To sense the LPG gas and to detect the motion uses, PIR sensor and MQ2 sensor are used. All the hardware components set up the generic breadboard.

8.3.1.1 NodeMCU

NodeMCU is an open-source IoT platform. It has firmware that runs espresso system ESP8266 Wi-Fi SOC and ESP-12 module-based hardware. By default, the word "NodeMCU" refers to firmware instead of development toys. The firmware uses the Lua scripting language (Figure 8.1).

NodeMCU is an open-source software and hardware development environment that is built on a very inexpensive system on a chip called ESP8266 [9].

8.3.1.2 ESP8266

The ESP8266 is a low-cost Wi-Fi microchip with a full TCP/IP stack and microcontroller capability produced by the manufacturer Espresso Systems in Shanghai, China. It is a highly integrated chip designed to provide full internet connectivity in a small package. ESP8266 is capable of functioning consistently in industrial environments, due to its wide operating temperature range. With highly integrated on-chip features and minimal external discrete component count, the chip offers reliability, compactness, and robustness (Figure 8.2).

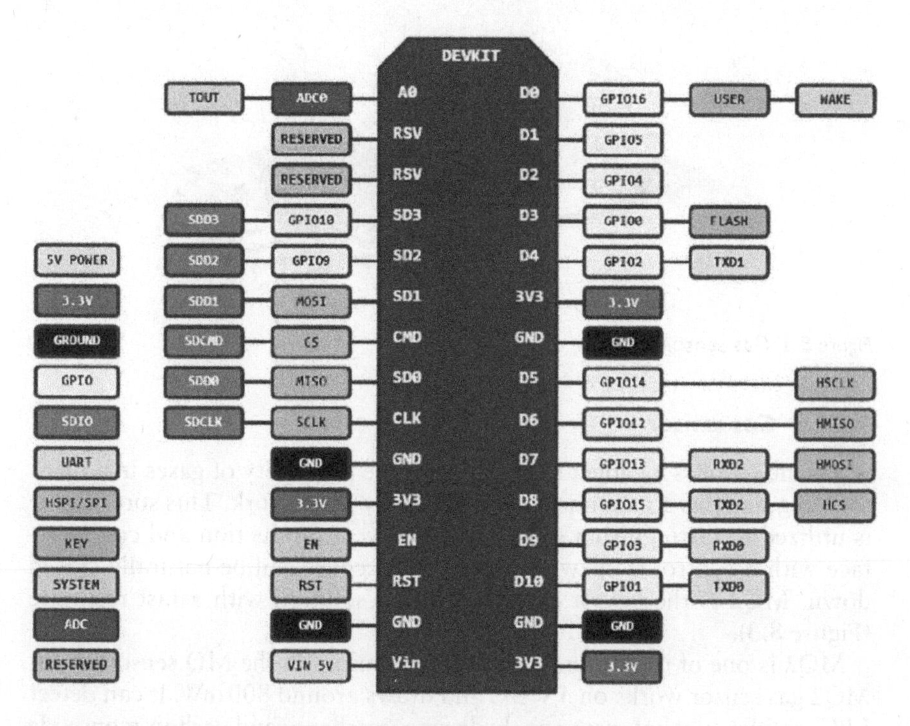

Figure 8.1 NodeMCU pin point diagram [9].

Figure 8.2 ESP 8266 module figure [10].

ESP8266 contains a built-in 32-bit low-power CPU, ROM, and RAM. It is a complete and self-contained Wi-Fi network solution that can carry software applications as a stand-alone device or connected with a microcontroller unit (MCU). The module has built-in AT Command firmware to be used with any MCU via COM port [10].8.

Figure 8.3 Gas sensor [11].

8.3.1.3 Gas sensor

A gas indicator is a gadget that identifies the proximity of gases in a territory, regularly as a major aspect of a security framework. This sort of gear is utilized to distinguish a gas spill or different production and can interface with a control framework; so, the procedure can be naturally closed down. MQ2 is the lowest cost monitoring solution with a fast response (Figure 8.3).

MQ2 is one of the commonly used gas sensors in the MQ sensor series. MQ2 gas sensor works on 5 V DC and draws around 800 mW. It can detect LPG, smoke, alcohol, propane, hydrogen, methane, and carbon monoxide concentrations anywhere from 200 to 10,000 ppm.

8.3.1.4 DHT11

The DHT11 is a basic, ultralow-cost digital temperature and humidity sensor. It uses a capacitive humidity sensor and a thermistor to measure the surrounding air and spits out a digital signal on the data pin (no analog input pins needed). It's fairly simple to use but requires careful timing to grab data (Figure 8.4).

They consist of a humidity sensing component, an NTC temperature sensor (or thermistor), and an IC on the backside of the sensor. For measuring humidity, they use the humidity sensing component which has two electrodes with moisture holding substrate between them.

8.3.1.5 PIR sensor

PIR sensors allow us to sense motion, almost always used to detect whether a human has moved in or out of the sensors range. PIR is made of a pyroelectric sensor which can be seen in Figure 8.5 as the round metal can with a rectangular crystal in the center can detect levels of infrared radiation.

The output can be used to control the motion of the door. Motion detection uses light sensors to detect either the presence of infrared light emitted

Figure 8.4 DHT11 sensor.

Figure 8.5 PIR sensor [12].

from a warm object or the absence of infrared light when an object interrupts a beam emitted by another part of the device [12].

8.3.1.6 Breadboard

A breadboard is a solder-less device for a temporary prototype with electronics and test circuit designs. Most electronic components in electronic

Figure 8.6 Power board.

circuits can be interconnected by inserting their leads or terminals into the holes and then making connections through wires where appropriate.

8.3.1.7 Power board

We design and build a board to control the power of the total device. This is the 10 V power supply. When a device is connected to the sensor for detecting data that time power consumption is needed (Figure 8.6).

8.3.1.8 MCP3008

Microchip Technology Inc.'s MCP3004/3008 analog-to-digital converter (ADC) devices are successive approximation 10-bit ADC with onboard sample and hold circuitry. The MCP3004 is programmable to provide two pseudo-differential input pairs or four single-ended inputs. The MCP3008 10-bit simple to advanced converter consolidates superior and low power utilization in a little bundle, making it perfect for inserted control applications. The MCP3008 highlights a progressive guess register design and an industry-standard SPI sequential interface, permitting 10-bit ADC ability to be added to any PIC microcontroller. The MCP3008 highlights 200k examples/second, eight information channels, low power utilization (5 nA run of the mill backup, 425 μA normal dynamic), and is accessible in 16-stick PDIP and SOIC bundles (Figure 8.7).

MCP 3008 ADC device is used for data acquisition where multiple analog sensors are present and multiple sensors are interfaces with this type of work.

8.3.1.9 Buzzer

A buzzer or beeper is an audio signaling device. Typical uses of buzzers and beepers include alarm devices, timers, and confirmation of user input such as a mouse click or keystroke (Figure 8.8).

Figure 8.7 MCP3008 [13].

Figure 8.8 Buzzer.

The vibrating circle in an attractive signal is pulled into the shaft by the attractive field. At the point when a wavering sign is traveled through the curl, it creates a fluctuating attractive field that vibrates the plate at a recurrence equivalent to that of the drive signal.

8.3.2 Software

In this project, we have used Arduino IDE to program NodeMCU. Arduino studio is used to connect to the web server. Here, a web server Thinger.io (https://thinger.io/) is used to it connect ESP8266.

8.3.2.1 Web server

A web server (thinger.io) has been used for this project. The sensor consumes the real-time data that transmits the data through the web server. Thinger.io platform is an open-source platform for IoT, it provides a ready-to-use scalable cloud infrastructure for connecting things. Some other important criteria differentiate IoT platforms between each other, such as scalability, customizability, ease of use, code control, and integration with 3rd party software, deployment options, and the data security level [14] (Figure 8.9).

The Cloud Console is related to the management front-end designed to easily manage devices and visualize their information in the cloud.

8.3.2.2 Arduino studio

The open-source Arduino Software (IDE) makes it easy to write code and upload it to the board. It runs on Windows, Mac OS X, and Linux (Figure 8.10).

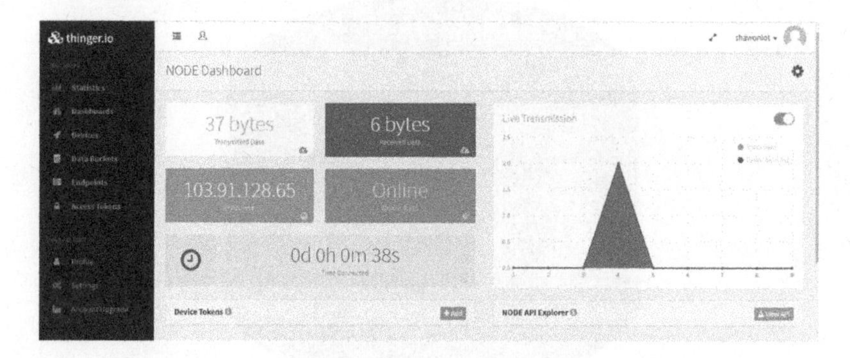

Figure 8.9 Web server.

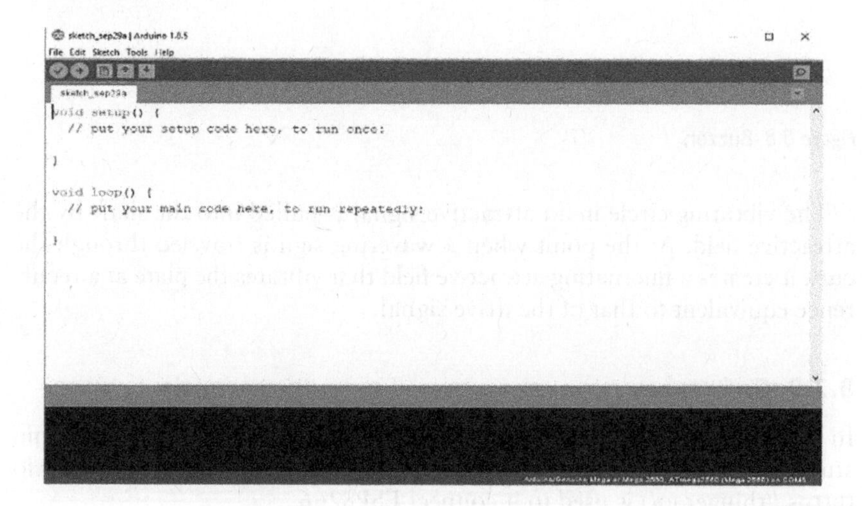

Figure 8.10 Arduino IDE.

8.4 PART 4: EXPERIMENTAL STUDIES AND RESULTS

We did some sensor analysis before implementing this project. We also review some other projects which were related to this project and taking some ideas and try to implement a new logic also cost-effective. We use different types of sensors for computing the environmental atmosphere and storing data. The required data will be stored on a website using IoT and the website is the Thinger.io server. (https://thinger.io/). By using those data from this website, users can make a decision. In our project, we used three sensors which we discussed in the previous section. Here, we represent and interpret those sensor data.

8.4.1 Sensor analysis

Sensors play a key role in modern industrial plant operations. Nevertheless, the information they provide is still underused. Extracting information from the raw data generated by the sensors is a complicated task, and it is usually used to help the operator react to undesired events, rather than preventing them. The complexity of modern industrial plants has called for equipment control automation that includes sensors for monitoring equipment behavior and remote-controlled valves to act upon undesired events. Plant automation physically protects plant integrity. However, it acts reacting to anomalous conditions. Equipment behavior set points within a working range are established and whenever the equipment behavior, such as equipment temperature, goes outside the designed range, an alarm is activated and a control equipment, such as a valve, is triggered to reset the equipment to the predefined working condition. The complexity of modern industrial plants has called for equipment control automation that includes sensors for monitoring equipment behavior and remote-controlled valves to act upon undesired events. Plant automation physically protects plant integrity. However, it acts by responding to abnormal conditions. Hardware conduct set focuses inside a working extent are built up and at whatever point the gear conduct for example hardware temperature goes outside the designed range an alert is initiated and a valve is activated to reset the hardware. The predefined working condition and the unpredictability of present-day mechanical plants have called for hardware control computerization that incorporates sensors for checking hardware conduct and remote-controlled valves to follow up on undesired occasions. Plant robotization physically ensures plant uprightness.

Sensor analytics is the statistical analysis of data that is created by wired or wireless sensors. A primary goal of sensor analytics is to detect anomalies. This is an insight that was gained by examining deviations from an established point of reference can have many uses, including predicting and proactively preventing equipment failure in a manufacturing plant, alerting a controller in an electronic intensive care unit when a patient's blood pressure drops, or allowing a data center administrator to make data-driven decisions about heating, ventilating and air conditioning.

Because sensors are often always on, it can be challenging to collect, store and interpret the tremendous amount of data they create. A sensor analytics system can help by integrating event monitoring, storage, and analytics software in a cohesive package that will provide a holistic view of sensor data. Such a system has three parts: the sensors that monitor events in real time, a scalable data store, and an analytics engine. Instead of analyzing all data as it is being created, many engines perform time series or event-driven analytics, using algorithms to sample data and sophisticated data modeling techniques to predict outcomes. These approaches may change. However, as advancements in big data analysis, object storage and processing technologies will make real-time analysis easier and less expensive to carry out.

8.4.1.1 Thinger.io overview

In this section, we are going to depict the general architecture of the new open-source platform for deploying data fusion applications by integrating Big Data, Cloud, and IoT technologies. Thinger.io is a new platform that is receiving interest from the scientific/technological community, finding projects in which this platform has been used successfully, including in the field of education. Thinger.io provides a ready-to-use cloud service for connecting devices to the internet to perform any remote sensing or actuation over the internet. It offers a free tier for connecting a limited number of devices, but it is also possible to install the software outside the cloud for private management of the data and devices connected to the platform, without any limitation.

This platform is hardware agnostic. So, it is possible to connect any device with the internet such as Arduino, Raspberry Pi, Sigfox, Lora solutions over gateways, or ARM. This platform has provided some out-of-the-box features such as device registry; bidirectional communication in real time, both for sensing or actuation; data and configuration storage, so it is possible to store time series data; identity and access management, to allow third-party entities to access the platform and device resources over REST/web socket APIs; third-party webhooks, so the devices can easily call other web services, send emails, SMS, push data to other clouds, etc. [12]. It also provides a web interface to manage all the resources and generate dashboards for remote monitoring. The general overview of this platform is available in Figure 8.11.

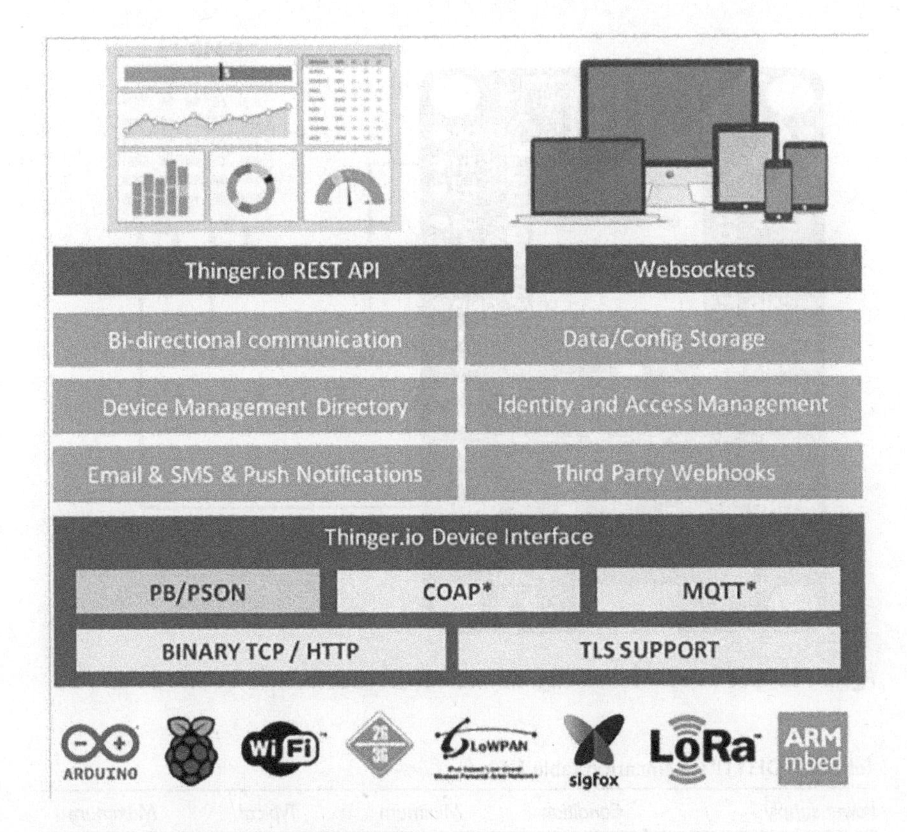

Figure 8.11 General overview of Thinger.io.

8.4.2 DHT11 sensor analysis

The DHT11 is a basic, ultralow-cost digital temperature and humidity sensor. It uses a capacitive humidity sensor and a thermostat to measure the surrounding air and spits out a digital signal on the data pin (no analog input pins needed). It's fairly simple to use but requires careful timing to grab data. Following pin, the interface is given below and two libraries will be required to run this code. Download the zip file, extract the same, and copy this in the Arduino library folder. The pin diagram of DHT11 with NodeMCU is given below. It connects the 3.3 V with NodeMCU. Some electrical characteristics of DHT11 are provided.

DHT11 Specification

- **Supply voltage:** +5 V
- **Temperature range:** 0°C–50°C error of±2°C
- **Humidity:** 20%–90% RH±5% RH error
- **Interface:** Digital

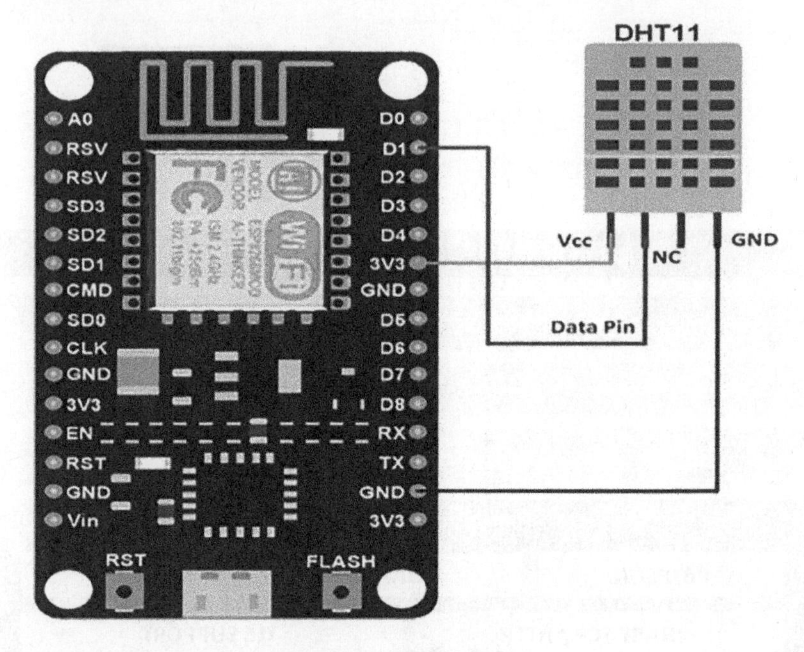

Figure 8.12 DHT11 connection diagram with NodeMCU.

Table 8.1 DHT11 specification table [15]

Power supply	*Condition*	*Minimum*	*Typical*	*Maximum*
Current supply	DC	3.3V	5V	5.5V
	Measuring	0.5 mA		2.5 mA
	Average	0.2 mA		1 mA
	Standby	100 μA		150 μA
Sampling period	Second	1		2

This graph shows the value of temperature and humidity (Figure 8.12 and Table 8.1). The temperature and humidity change in real time. DHT11 overall communication process along with real time is shown in Figure 8.13.

Communication process: Serial-interface (single-wire, two-way), the single-bus data format is used for communication and synchronization between NODEMCUand DHT11 sensor. One communication process is about 4 ms. Once the start signal is completed, DHT11 sends a response signal of 40-bit data and triggers a signal acquisition.

Data format: 8bit integral T data+8bit decimal T data+8bit checksum. If the data transmission is right, the checksum should be the last 8bit of "8bit integral data+8bit decimal data+8bit integral T data+8bit decimal T data".

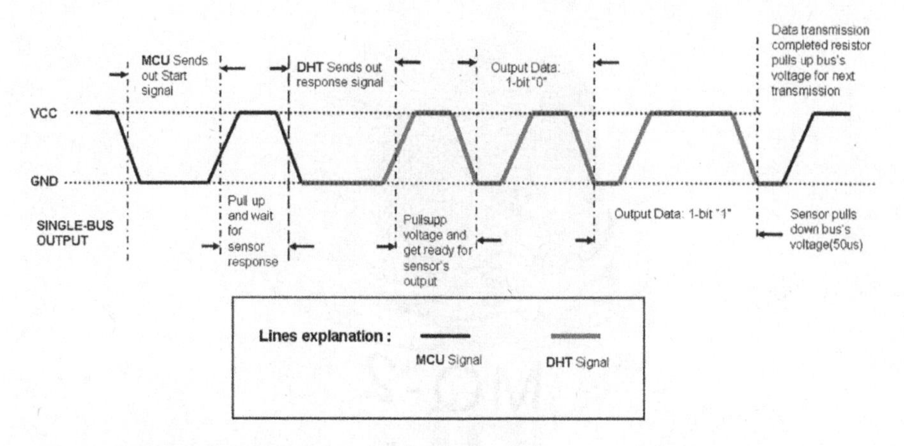

Figure 8.13 Overall communication process DHT11.

When MCU sends a start signal, DHT11 changes from the low-power-consumption mode to the running mode. Once the start signal is completed, DHT11 sends a response signal of 40-bit data and triggers a signal acquisition. Users can choose to collect (read) some data. Without the start signal from MCU, DHT11 will not collect temperature information spontaneously. Once data is collected, DHT11 will change to the low-power-consumption mode until it receives a start signal from MCU again.

8.4.3 MQ2 gas sensor analysis

MQ2 is one of the commonly used gas sensors in the MQ sensor series. It is a metal oxide semiconductor -type gas sensor also known as Chemi resistors as the detection is based upon a change of resistance of the sensing material when the gas comes in contact with the material. Using a simple voltage divider network, concentrations of the gas can be detected. Ratio and concentration are nonlinear (Figure 8.14).

The MQ2 gas sensor module detects gas leakage in homes and industries. They are sensitive to a range of gases and are used indoors at room temperature. The output of this sensor generates an analog signal and can read through an analog pin of the Arduino. But, NodeMCU has only one analog pin. So, we use the MCP3008 chip for observing the digital output. This sensor can be used in homes or factories for sensing gas leaks, suitable for gas, butane, propane, methane, alcohol, hydrogen, smoke, and other monitoring devices (Tables 8.2 and 8.3).

This sensor has specific criteria for detecting the gas leakage criteria. An output criterion is a high and low description. So some electrical specifications are given below.

Figure 8.14 MQ2.

Table 8.2 Complete specifications of the gas sensor (Model MQ2)

Operating voltage	+5V
Load resistance	20 kΩ
Heat resistance	33 Ω±5%
Heating consumption	<800 mW
Sensing resistance	10–60 kΩ
Concentration scope	200–10,000 ppm
Preheat time	Over 24 h

Table 8.3 Functional description of MQ2

Pin	*Description*	*Function*
GND	Ground	Connect to ground
DOUT	Digital output	1. Output signal: high • o No gas present • o LED's status: OFF 2. Output signal: low • o Gas present • o LED's status: ON
AOUT	Analog pin	This output voltage changes with the concentration of surrounding gas present. • Output voltage increase:increasing in concentration of surrounding gas. • Output voltage decrease:increasing in concentration of surrounding gas.
VCC	+5V	Connect to +5V

Figure 8.15 Pin connection with NODMCU.

Specification of MQ2

- Size: 35 mm × 22 mm × 23 mm (length × width × height)
- Main chip: LM393, ZYMQ2 gas sensors
- Working voltage: DC 5 V

Basic Characteristics of MQ2 (Gas/Smoke Sensor)

1. MQ2 gives a signal for output instruction.
2. It has dual signal output (analog output, and high/low digital output).
3. 0–4.2 V analog output voltage, the higher the concentration the higher the voltage.
4. MQ2 specially better sensitivity for gas, natural gas.
5. It has long service life and reliable stability.

NodeMCU connects to the MQ2 gas sensor. Gas sensor output shows the ADC converter (Figure 8.15).

This sensor gave real-time transmitted data through the web page which is connected to the thinger.io web page. When it detects the gas leakage, the output of this sensor's pick level becomes high; otherwise, pick level is low or remains in the steady state.

8.4.4 PIR sensor analysis

The construction and principle of operation of PIR detectors of a large detection range. An important virtue of these detectors is the highly efficient detection of slowly moving or crawling people. The PIR detector described here detects crawling people at a distance of 140 m. A high

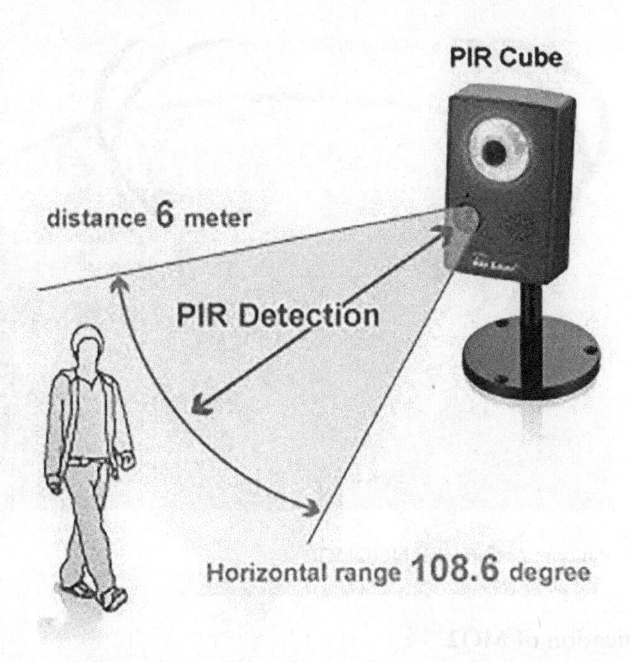

Figure 8.16 Object detection follows [15].

signal-to-noise ratio was obtained by using a large number of pyroelectric sensors, i.e., by using a large number of detection zones (channels). The original electronic system for the PIR detector is presented in which DC signal amplifiers from pyroelectric signals are used. In order to ensure large detection ranges, a new method of signal analysis was used. The main elements of security systems are PIR detectors. In general, detectors operating inside buildings have a small detection range, small ranges of working temperature, and relatively simple algorithms of intruder detection. A PIR sensor is an electronic sensor that measures infrared light radiating from objects in its field of view. They are most often used in PIR-based motion detectors. PIR sensors are commonly used in security alarms and automatic lighting applications. Moreover, to detect slowly moving or crawling people, the lower limit frequency of a transfer band of the PIR detector should be near zero. By fulfilling this condition, an increase in low-frequency noise occurs causing a decrease in the next detector's sensitivity. A PIR sensor is an electronic sensor that measures infrared light radiating from objects in its field of view. They are most often used in PIR-based motion detectors (Figures 8.16 and 8.17).

8.4.4.1 PIR working principle

The PIR sensor itself has two slots in it, each slot is made of a special material that is sensitive to infrared. When a warm body like a human or animal

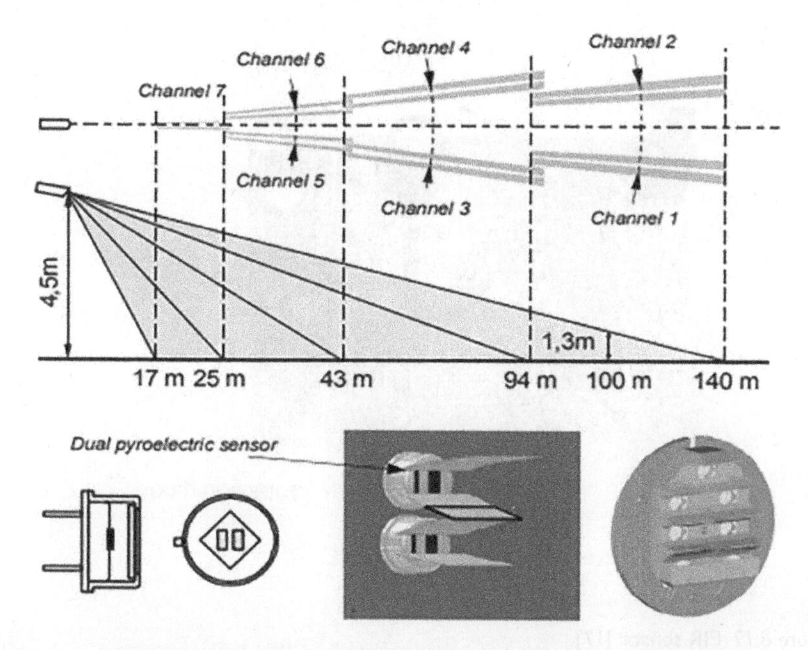

Figure 8.17 Detection zone of PIR detector in horizontal and vertical planes [16].

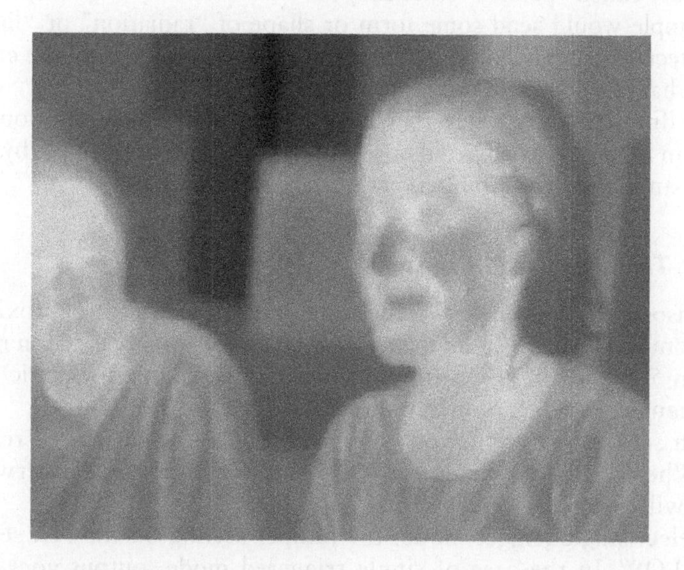

Figure 8.18 Example of infrared radiation [16].

passes through, it first intercepts one-half of the PIR sensor, which causes a positive differential change between the two parts. Note that the PIR just uses a relatively simple sensor and definitely not a camera (Figure 8.18).

Figure 8.19 PIR sensor [17].

PIRs are called "passive" since they are not assisted by any "helpers" that for example would send some form or shape of "radiation" or "light" to help detect. It's purely based on what the sensor can pick from the environment, what is being emitted by objects.

If the difference is too high then it will trigger – it detects "motion". This is done in a smart way to avoid false positives caused for example by a brief flash or an increase in room temperature (Figure 8.19).

8.4.4.2 The range of PIR sensor

PIR sensor detects a human being moving around within approximately 10 m from the sensor. It is an average value, as the actual detection range is between 5 and 12 m. PIR is fundamentally made of a pyroelectric sensor, which can detect levels of infrared radiation (Figure 8.20).

When sensors are connected with NodeMCU it observes the real-time data. When it detects any object, PIR output will be high; otherwise, its output will be low or steady state (Figure 8.21).

To select single trigger mode, the jumper setting on the PIR sensor is set on LOW. In the case of single triggered mode, output goes HIGH when motion is detected. After a specific delay, the output goes to LOW even if the object is in motion. The output is LOW for some time and again goes HIGH if the object remains in motion. This delay is provided by a user using the potentiometer. This potentiometer is on board of PIR

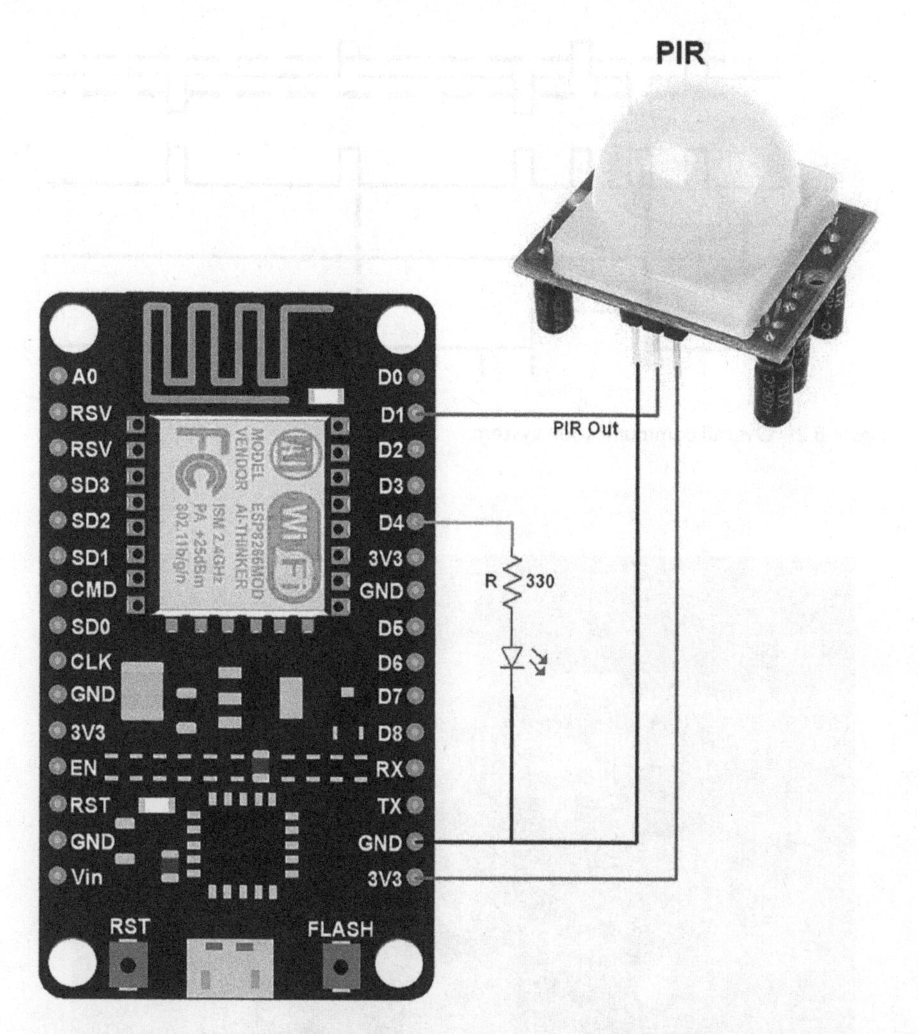

Figure 8.20 Pin connection with NodeMCU (PIR).

sensor module. In this way, the PIR sensor gives HIGH/LOW pulses if an object is in continuous motion. To select repeat trigger mode, the jumper setting on the PIR sensor is set on HIGH. In the case of Repeat Triggered Mode, Output goes HIGH when motion is detected. The output of the PIR sensor is HIGH until the object is in motion. When an object stops motion or disappears from the sensor area, the PIR continues its HIGH state up to some specified delay. We can provide this delay by adjusting the potentiometer. This potentiometer is on board of PIR sensor module. In this way, the PIR sensor gives HIGH pulse if an object is in continuous motion.

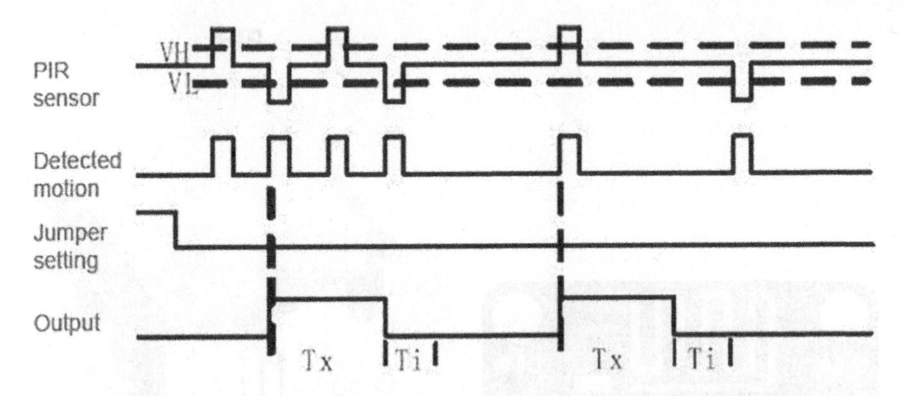

Figure 8.21 Overall communication system.

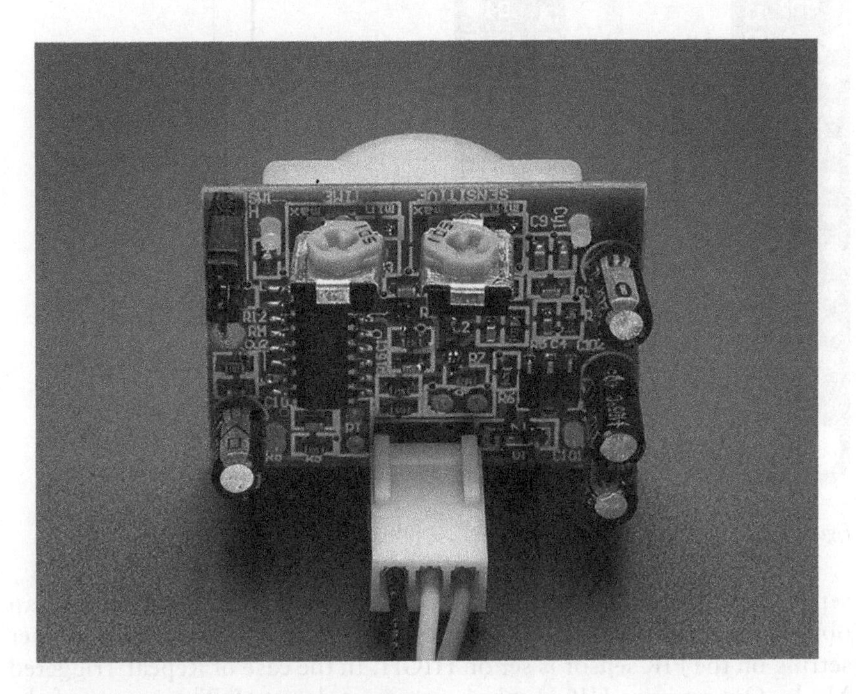

Figure 8.22 Sensitivity changing switch [17].

8.4.4.3 Changing sensitivity and delay time

There are two potentiometers on the PIR motion sensors board: sensitivity and time delay. It is possible to make PIR more sensitive or nonsensitive. The maximum sensitivity can be achieved up to 6 m. Time delay adjust potentiometer is used to adjust the time shown in the above timing diagrams. Clockwise movement makes PIR more sensitive (Figure 8.22).

8.4.5 Block diagram of this system

We assemble the hardware layout by using a sensor, Wi-Fi module, and others. The energy supply is mandatory for sensor reading. The whole power intake of the gadget is approximately 13–15 V. Now, connected the sensor and look at the information the Arduino studio serial displays. After observing the serial display facts, all the sensors are linked using an internet platform. When the internet console signal is inexperienced which means the hardware is connecting the web platform. Finally, all sensors and other modules are linked and we observed the transmitted sensor statistics online.

Now, the block diagram of our project is given in Figure 8.23.

Thinger.io can take real-time data from all three sensors simultaneously and update it on the web page. The data is entered into the storage channel of the website and displayed graphically to the user. Here the following representation of a one-time instance as the data is displayed to the user. The flow chart of this project is given in Figure 8.24.

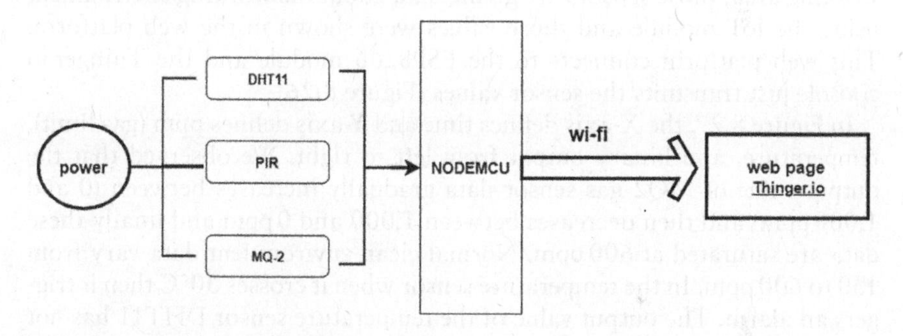

Figure 8.23 Project block diagram.

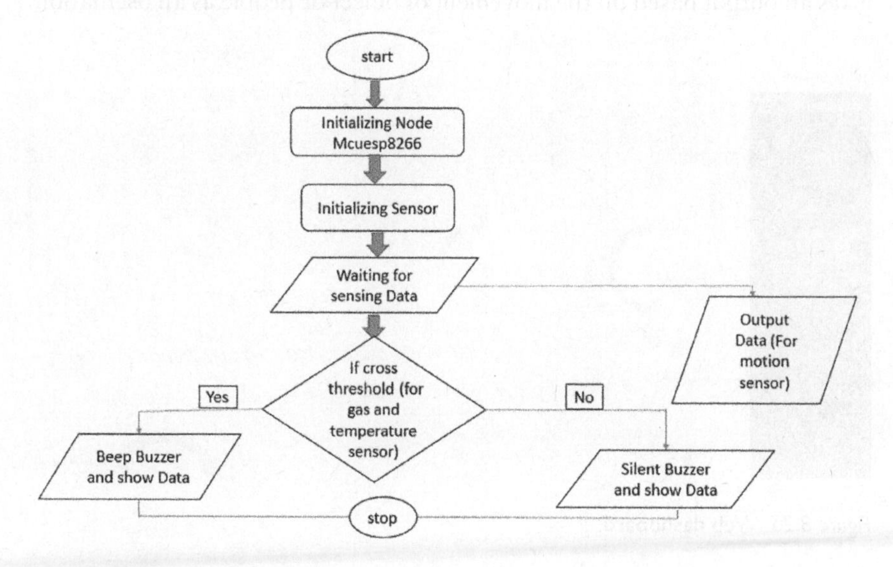

Figure 8.24 Flow chart.

8.4.6 Sensor recognition process

Our project consists of two parts. One is sensor reading and another is a web platform. All transmitted data is shown on the web platform. This is the most important thing in our project. Industry environment protection is an important sector for the security platform. We cannot monitor the whole working environment all the time manually. Using this platform, we can monitor the working environment and safety issues and can take proper steps regarding the worker's safety. This platform gives real-time data using an IoT-based module and shows data through the web console or platform. So the user can observe data from anywhere where the internet is available. Our project is mainly based on IoT. So, to monitor those types of workplaces we require an internet connection to observe data. Here showing the web platform visualization (Figure 8.25).

We use the Thinger.io web platform. This is the free web console. Data we stored from the different sensors which were implemented in the required working area, those sensors are giving data about the working environment using the IoT module and those values were shown in the web platform. This web platform connects to the ESP8266 module and the Thinger.io console just transmits the sensor values (Figure 8.26).

In Figure 8.27, the *X*-axis defines time and *Y*-axis defines ppm (gas limit), temperature, and binary output from left to right. We observed that the output value of MQ2 gas sensor data gradually increases between (0 and 1,000 ppm) and then decreases between 1,000 and 0 ppm and finally these data are saturated at 600 ppm. Normal clean environment data vary from 150 to 600 ppm. In the temperature sensor when it crosses 30°C then it triggers an alarm. The output value of the temperature sensor DHT11 has not changed much and PRI sensor monitors the whole work environment and gives an output based on the movement of object or people as an oscillation.

Figure 8.25 Web dashboard.

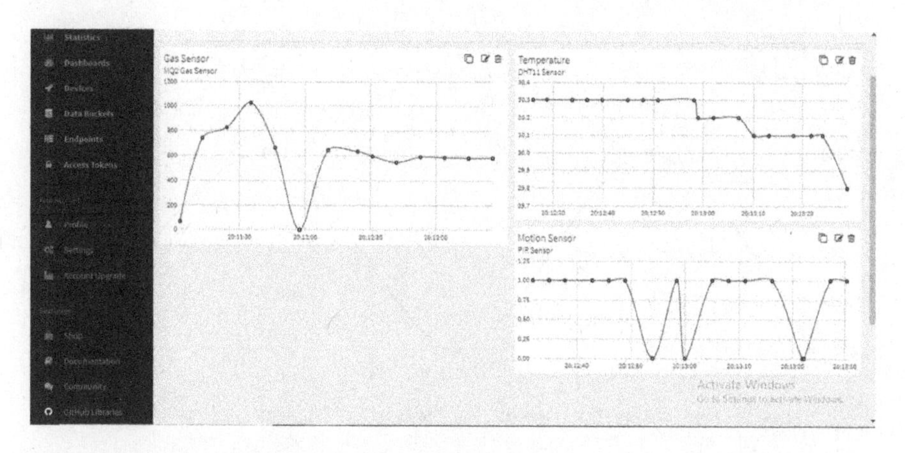

Figure 8.26 **Sensor transmitting values in a simple room.**

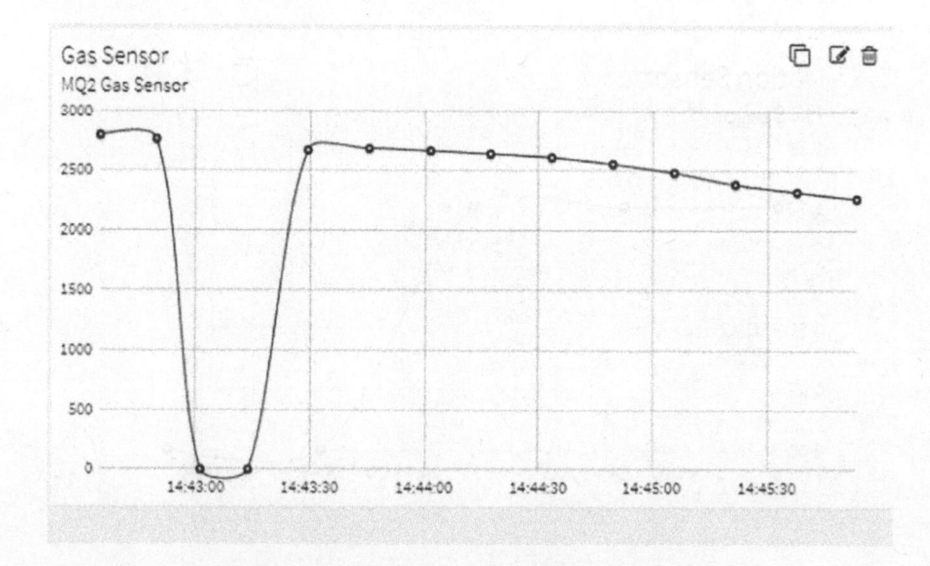

Figure 8.27 **Gas sensor transmitting values in a populated room.**

In Figure 8.27, the *X*-axis defines time and *Y*-axis defines ppm (gas limit). For a populated room, we observed that when the output value of MQ2 gas sensors stays above 1,000 ppm then it can be considered a normal clean environment. Dangerous environment sensed as soon as a drop below 1,000 ppm is observed. It can also be detected when a sudden drop in the gas limit is recorded. Such an instance triggers the alarm.

In Figure 8.28, the *X*-axis defines time and *Y*-axis defines temperature. Along with time, considering a fixed number of people in a room, the

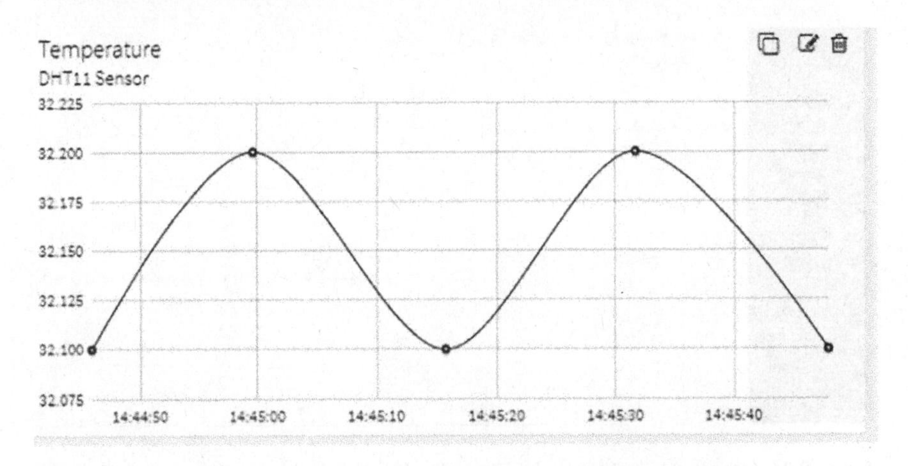

Figure 8.28 **Temperature sensor transmitting values in a populated room.**

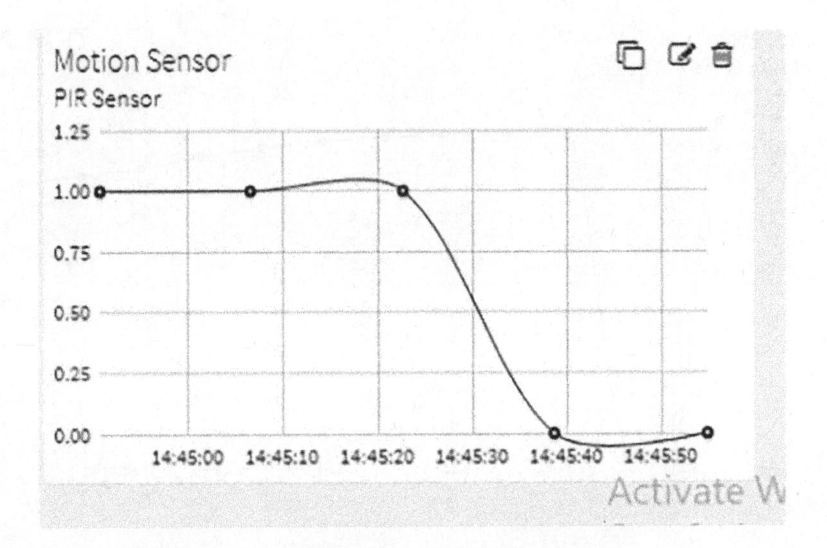

Figure 8.29 **Motion sensor transmitting values in a populated room.**

temperature remains fixed. As soon as the population increases with the entrance of new people in the room, the temperature increases, as seen at 14:45:00. When the population decreases again, the temperature goes back to normal value again as seen at 14:45:15.

In Figure 8.29, the *X*-axis defines time and *Y*-axis defines binary output. The PIR sensor records data with time, and when there is no motion in the room, the sensor reads a value of zero. When any motion is detected due to any object in the room, the binary value rises to 1. PRI sensors observe the whole working environment and give an output based on the movement of the object or human as an oscillation.

According to the atmosphere, we decided on three major levels for those sensors. MQ2 gas sensor is a gas sensor for this sensor below 100 ppm is lower level and below 1,500 ppm is the middle level and above 1,500 ppm is considered higher level and when working atmosphere gas level cross the middle level and entered into the higher level the buzzer will get on and give alert who monitor the working atmosphere [18]. Various companies have a different working atmosphere, chemical industry or pharmaceutical company mainly work in a clean environment, those type of environment are very gas sensitive.

Figure 8.30 represents the sensitivity characteristics of MQ2 for different gases in a different environment at 20°C temperature where humidity is 65% and O_2 concentration is 21%. And normal resistance RL= 5 kΩ, R0=sensor resistance at 1,000 ppm, H2 in the clean air. RS=sensor resistance at various concentration of gases [18].

The second sensor is DHT11 which is a temperature sensor and it also has three levels according to our working atmosphere. Some of the factories work on low temperature and some of the factories work on as usual

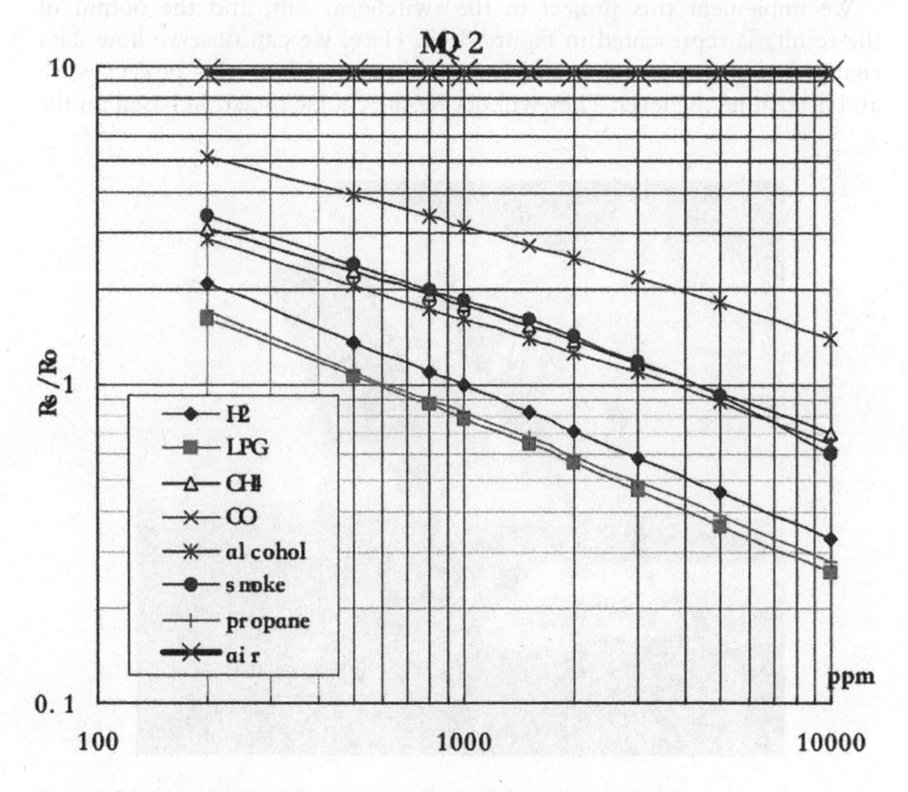

Figure 8.30 Sensitivity characteristics MQ2 for different gases [18].

normal temperature and most of the factories work on high temperatures like chemical company and welding company or any factory where different types of product are making used by a high-temperature machine that time we can set our three levels according to the working atmosphere. When the middle level is crossed and entered into the higher level of the system alert system, a buzzer will get active and give a signal to take proper action. Finally, the PRI motion sensor will not trigger any alarm but this sensor regularly monitors the working atmosphere movement and shows an oscillation in the output of the sensor.

Based on the interpretation of the diagrams, the operator decides his course of action. Signal lights present in the room will light up if a harmful environment is detected. The workers and operator would then determine the conditions based on the color of the lights.

8.4.7 Real-life implementation

We implemented this project in our department because we did not get permission to implement this project in a large factory. So the result analysis is based on data which we got from different implemented areas (Figure 8.31).

We implement this project in the switchgear lab, and the output of the results is represented in Figure 8.32. Here, we can observe how data changed along with the working atmosphere and how this project work and determine the level. They will decide the course of action based on the

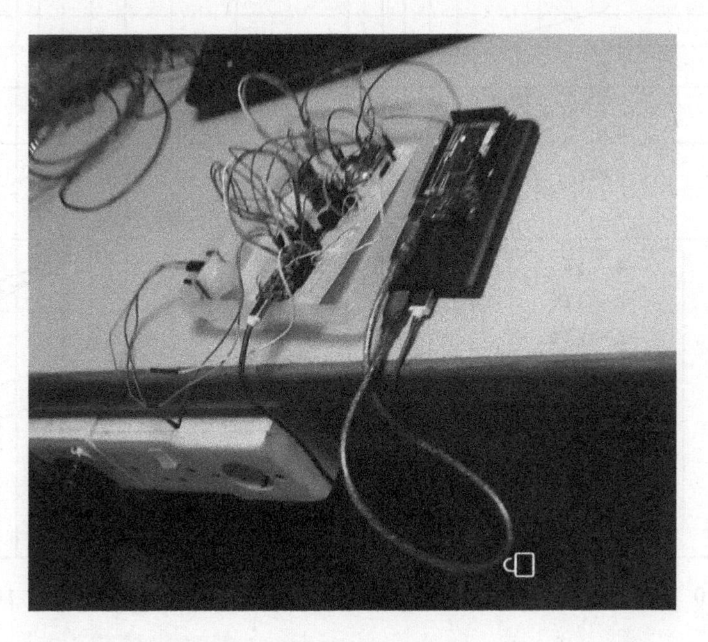

Figure 8.31 Real-life implementation in switchgear lab.

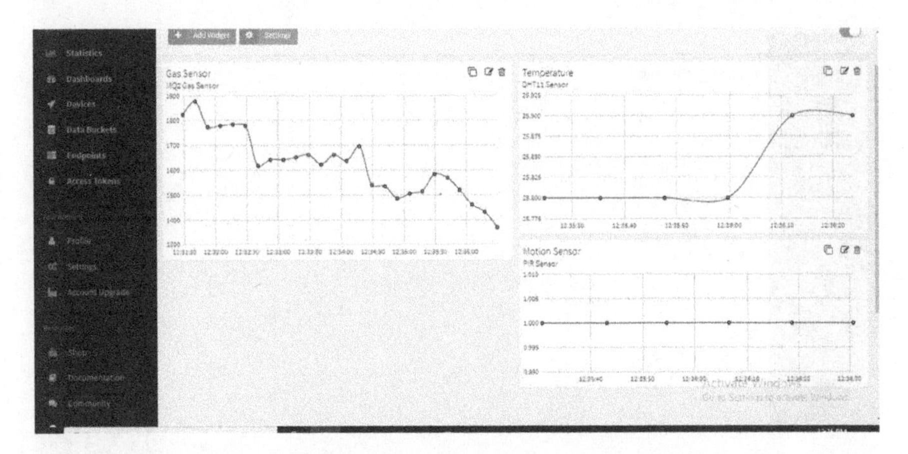

Figure 8.32 **Real-life implemented data in switchgear lab.**

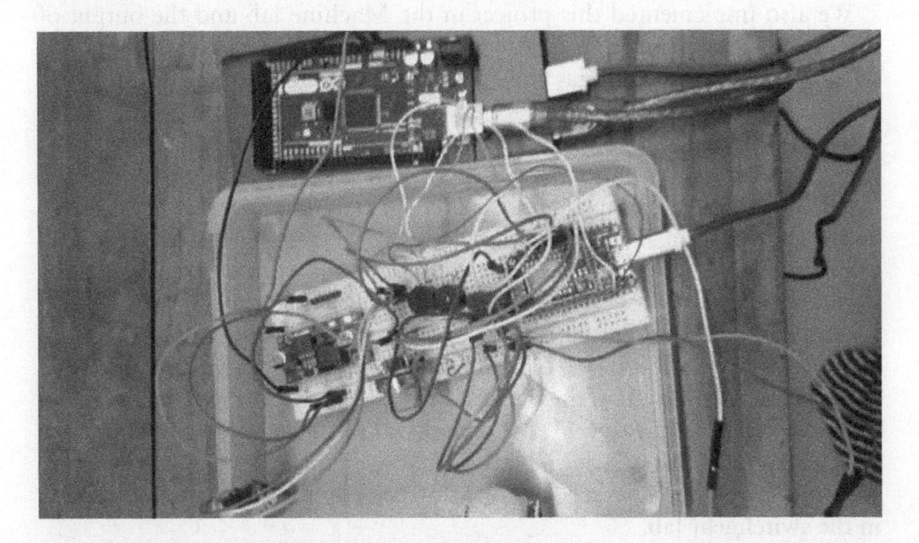

Figure 8.33 **Real-life implementation in machine lab.**

data and alarm. In these places, it represents the time in this plot X-axis and Y-axis for the first graph showing the gas sensor values has PPM count on its *Y*-axis (Figure 8.32). The second graph shows the temperature in Celsius on *Y*-axis, and the third graph shows motion detection by PIR sensors on *Y*-axis. The PIR sensor gives a binary output. It outputs 1 if some motion is detected and 0 if otherwise. The above figure shows that the MQ2 gas sensor gave the lowest value 1,075 ppm at a temperature of 25.4°C in the switchgear lab. The motion sensor observes the whole working environment and when a human being or any object is moving then the output of the PRI sensors shows an oscillation based on the movement (Figure 8.33).

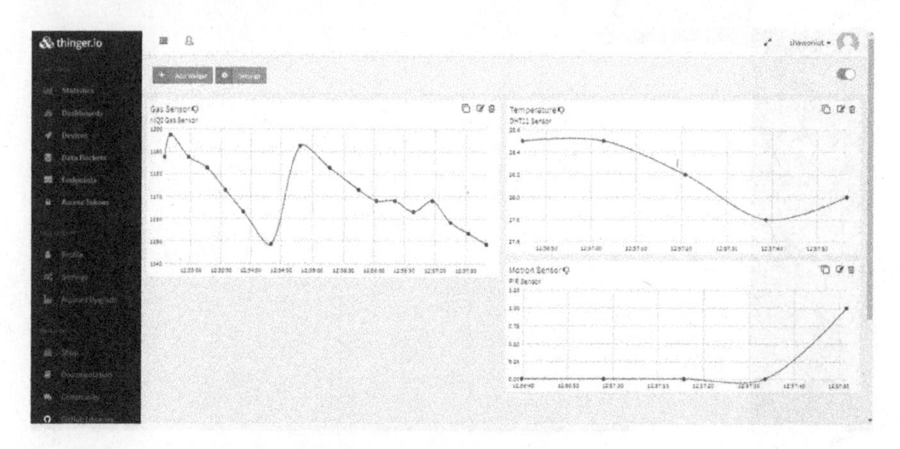

Figure 8.34 Real-life implemented data in machine lab.

We also implemented this project in the Machine lab and the output of this result is represented in Figure 8.34. In this figure, *X*-axis defines time and *Y*-axis defines ppm (gas limit), temperature, and binary output from left to right. We observe how data changed along with the working atmosphere and how this project work and determine the level. Along with the time (*X*-axis), the output of the MQ2 (*Y*-axis) is varying in ppm unit. Same as we saw a variation in the output of the temperature sensor in *Y*-axis along with time in *X*-axis. The temperature went down along with time after some time output of the temperature sensor became stable. At last, the motion sensor observed the whole working environment, and when a human being or any object was moving then the output of the PRI sensors showed an oscillation based on the movement. Then, they will decide the course of action based on the data and alarm. In this lab we did not get so much variation in the output, MQ2 and DHT11 gave quite similar values which we get in the switchgear lab. The above figure shows that the MQ2 gas sensor gave the lowest value of 1,150 ppm at a temperature of 28.2°C in the switchgear lab.

We have used various gases to test out the gas sensor by some experiments. The first gas used was carbon dioxide (CO_2). The normal ppm of CO_2 is 300–750 ppm. In the presence of this gas, if the ppm range is exceeded, the buzzer sounds an alarm.

In Figure 8.35, it is observed that the sensor crosses the specified range at a time around 17:52:00. The abnormality remains for a certain duration, and throughout this duration, the buzzer will sound. Throughout the rest of the time, normal condition prevails and the ppm remains within the specified range.

The next gas used was LPG. The normal ppm of LPG is 500–900 ppm. In the presence of this gas, if the ppm range is exceeded, the buzzer sounds an alarm.

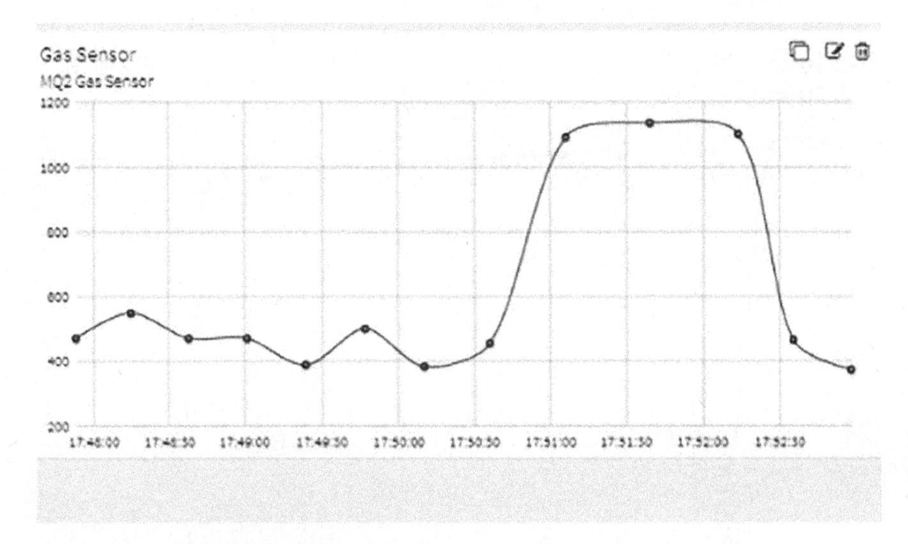

Figure 8.35 Implemented data for CO_2 gas.

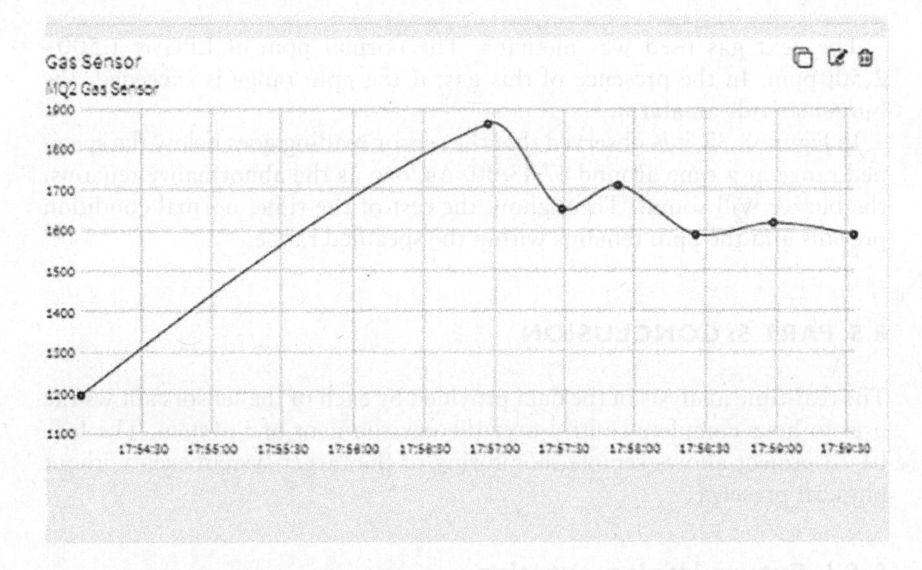

Figure 8.36 Implemented data for LPG gas.

In Figure 8.36, it is observed that the sensor crosses the specified range at a time around 17:57:00. As long as the abnormality remains, the buzzer will sound. Throughout the rest of the time, normal condition prevails and the ppm remains within the specified range.

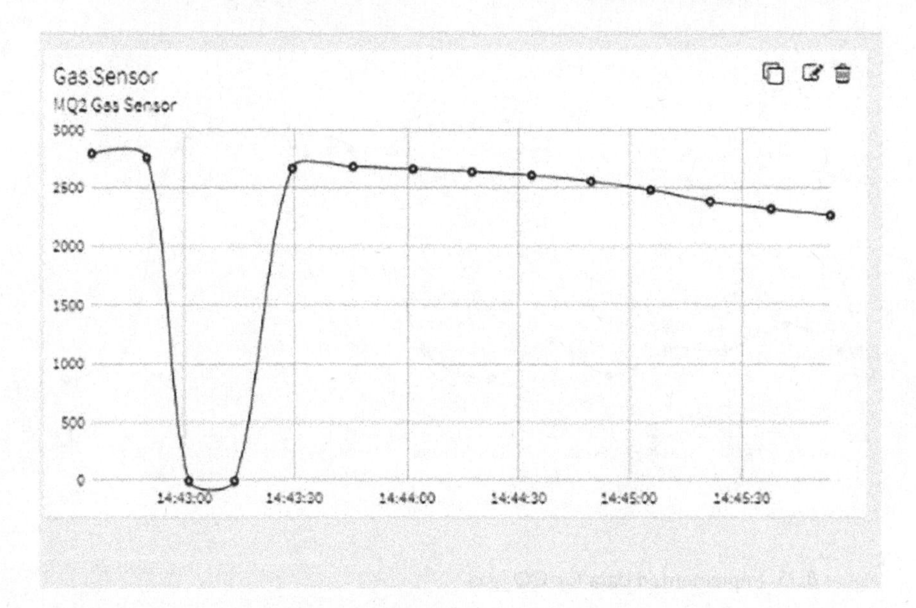

Figure 8.37 Implemented data for CH_4 gas.

The next gas used was methane. The normal ppm of LPG is 1,500–2,500 ppm. In the presence of this gas, if the ppm range is exceeded, the buzzer sounds an alarm.

In Figure 8.37, it is observed that the sensor reading goes below the specified range at a time around 17:43:00. As long as the abnormality remains, the buzzer will sound. Throughout the rest of the time, normal condition prevails and the ppm remains within the specified range.

8.5 PART 5: CONCLUSION

The real-time analysis of the data provided by each of the sensors allows the user to have complete control over the environment of a system. The IoT-based project allows remote monitoring of the target system even without physical presence.

8.5.1 Future implementation

- Integrating more sensors for more specific data acquisition and analysis
- Will be applicable in LPG gas services stations in an active situation
- This system will be used to provide monitoring services to rural areas at an affordable price

Our project can be considered as a platform to develop in the field of IoT on the monitoring system. In developing countries like ours, this kind of innovative and cost-effective project can improve the future of technology. So, we are looking forward to implementing the project in order to make an impact in the new era of technology.

8.5.2 Discussion

The industrial monitoring system is designed specially to address the cost, accuracy, and transparency problems in a highly secured approach. This system is more effective than the existing system since it uses an advanced controller for monitoring the environmental conditions and the controller collects the data from the sensor and those updated sensor values are written by the Arduino coding in a particular text file. Using Arduino studio, the value is read and updated on the webpage. Based on the collected value, the respective action will be carried out. Our web page is created by Thinger.io. The whole of this system is user-friendly because a user can see the environment information through the web page. Since this is a prototype design, we would like to build a professional one in near future based on our project concept.

REFERENCES

1. Shrouf, F., J. Ordieres, and G. Miragliotta. "Smart factories in Industry 4.0: A review of the concept and of energy management approached in production based on the Internet of Things paradigm." In *2014 IEEE International Conference on Industrial Engineering and Engineering Management*, pp. 697–701. IEEE, 2014.
2. Nayyar, A., V. Puri, and D.-N. Le. "A comprehensive review of semiconductor-type gas sensors for environmental monitoring." *Review of Computer Engineering Research* 3, no. 3 (2016): 55–64.
3. Yavuz, S.O., A. Taşbaşi, A. Evirgen, and K.A.R.A. Akay. "Motion detector with PIR sensor usage areas and advantages." *İstanbul Aydın Üniversitesi Dergisi* 4, no. 14 (1996): 7–16.
4. Deekshath, R., P. Dharanya, K.R. Dimpil Kabadia, G. Deepak Dinakaran, and S. Shanthini. "IoT based environmental monitoring system using Arduino UNO and Thingspeak." *IJSTE-International Journal of Science Technology & Engineering| ISSN (online): 2349–784X* 4, no. 9 (2018).
5. Sharma, V., and R. Tiwari. "A review paper on "IOT" & It's smart applications." *International Journal of Science, Engineering and Technology Research (IJSETR)* 5, no. 2 (2016): 472–476.
6. Díaz, M., C. Martín, and B. Rubio. "State-of-the-art, challenges, and open issues in the integration of Internet of things and cloud computing." *Journal of Network and Computer Applications* 67 (2016): 99–117.

7. Pavithra, D., and R. Balakrishnan. "IoT based monitoring and control system for home automation." In 2015 *Global Conference on Communication Technologies (GCCT)*, pp. 169–173. IEEE, (2015).
8. Deshpande, A., P. Pitale, and S. Sanap. "Industrial automation using Internet of Things (IOT)." *International Journal of Advanced Research in Computer Engineering & Technology (IJARCET)* 5, no. 2 (2016): 266–269.
9. Chandramohan, J., R. Nagarajan, K. Satheeshkumar, N. Ajithkumar, P. A. Gopinath, and S. Ranjithkumar. "Intelligent smart home automation and security system using Arduino and Wi-fi." *International Journal of Engineering and Computer Science (IJECS)* 6, no. 3 (2017): 20694–20698.
10. Cruz, T., J. Barrigas, J. Proença, A. Graziano, S. Panzieri, L. Lev, and P. Simões. "Improving network security monitoring for industrial control systems." In *2015 IFIP/IEEE International Symposium on Integrated Network Management (IM)*, pp. 878–881. IEEE, (2015).
11. Bustamante, L., M.A. Patricio, and J.M. Molina. "Thinger.io: An open source platform for deploying data fusion applications in IoT environments." *Sensors* 19, no. 5 (2019): 1044.
12. Micko, E.S. *PIR Motion Sensor Utilizing Sum and Difference Sensor Signals.* U.S. Patent 7,183,912, issued (February 27, 2007).
13. Li, Z., and G. Gong. "DHT-based detection of node clone in wireless sensor networks." In *International Conference on Ad Hoc Networks*, pp. 240–255. Springer, Berlin, Heidelberg, (2009).
14. Al-Haija, Q.A., H. Al-Qadeeb, and A. Al-Lwaimi. "Case study: Monitoring of AIR quality in King Faisal University using a microcontroller and WSN." *Procedia Computer Science* 21 (2013): 517–521.
15. Zhang, Z., X. Gao, J. Biswas, and J.K. Wu. "Moving targets detection and localization in passive infrared sensor networks." In *2007 10th International Conference on Information Fusion*, pp. 1–6. IEEE, 2007.
16. Muller, K.A., and H. Mahler. *Range Insensitive Infrared Intrusion Detector.* U.S. Patent 4,990,783, issued (February 5, 1991).
17. Prasad, S., P. Mahalakshmi, A.J. Clement, and R. Swathi. "Smart surveillance monitoring system using raspberry pi and pir sensor." *International Journal of Computer and Information Technologi* 5, no. 6 (2014): 7107–7109.
18. Ramya, V., and B. Palaniappan. "Embedded system for hazardous gas detection and alerting." *International Journal of Distributed and Parallel Systems (IJDPS)* 3, no. 3 (2012): 287–300.

Chapter 9

CFC-free dual-functional gadgets with solar power

Kaniz Amena
United International University

Mohammad Farhan Ferdous
Japan–Bangladesh Robotics and Advanced Technology Research Center (JBRATRC)

CONTENTS

DOI: 10.1201/9781003156789-9

9.1 PART 1: INTRODUCTION

Our solar-powered refrigerator with geyser can be used in any shop like tea stalls or the family work in provincial territories. This may decrease the fuel cost by taking out or lessening their requirement for wood, gas, and power to warm water. The objective of our venture is to plan a CFC-free solar-powered refrigerator and use the warmth that is synchronized by the cooler for use, which will course through the aluminum water blocks connected with the warmth sink.

It maintains a specific temperature to keep the freshness of the beverage. Simultaneously, it warms the water, which can be utilized for making tea, coffee, or soup. The entire gadget can be controlled by solar power or battery. Importantly, the gadget is affordable.

9.1.1 Background and literature review

Solar power is utilized in many nonindustrial nations. Solar-based boiling water, solar-powered cooler, solar-based cooking, and high-temperature handled warmth are utilized in numerous nations. Solar-based energy applies energy from the Sun as solar-based radiation for heat or to create power. Solar-based power age utilizes either photograph voltaic or warmth motors (concentrated solar-based force).

In Cherney's study, a basic solid sorption cycle with the working pair zeolite water was upgraded in the 1970s [1]. Since then, many types of research have been proved, theoretically and practically as well, but the cost of these gadgets is still making them noncompatible for markets. Consequently, the focal point of some examination is set on cost decrease and the expansion of the effectiveness of the machines, and promising outcomes have effectively been acquired [2].

In 1861, Mouchout developed the steam engine fully powered by solar energy. After that in 1883, an American inventor, named Charles Fritts introduced the very first solar cells using Selenium Wafers. After a century, by 1983, around 60% of the population started to heat the water using solar energy. But in that century, some of the people were using the stove to heated water by burning of wood or coal or gas was rarely available and pricy as well. To overcome these situations, a cost-efficient and safer way to heat water was established. The process of the solar water heaters was accepted by the huge market and started to fulfill the demand of everyday household needs (SHC annual report, 2001) [3].

9.2 PART 2: METHODOLOGY

Figure 9.1 shows that the point when we turned on the force of the framework, within the refrigerator begins cooling, the fans race to spread the

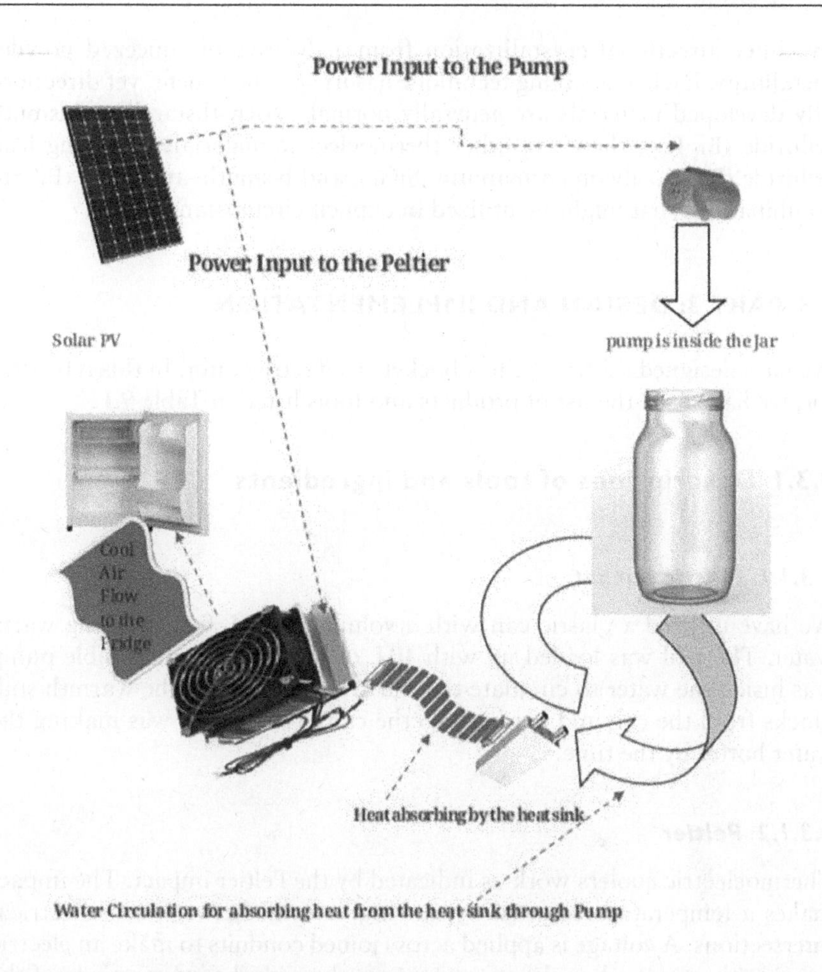

Figure 9.1 Visual representation of the system.

cooling inside the cooler. At outside, the pump works for the water moving through the aluminum blocks and saved into the source back. The water dissemination keeps on expanding the warmth of the water.

The cool side of Peltier will be cooling the inside of the refrigerator and the hot side of the Peltier will increase the heat of the water. These processes are caused for the properties of thermoelectric material. The heat difference of the two sides of Peltier causes an increase in power consumption.

9.2.1 Principle of thermoelectric material

The thermoelectric semiconductor material regularly utilized in the present TE coolers is an amalgam of bismuth telluride that has been reasonably doped to give singular squares or components having particular "N" and "P" qualities. Thermoelectric materials frequently are manufactured by either directional crystallization from a dissolve or squeezed powder metallurgy. Each assembling technique has its specific benefit, yet directionally developed materials are generally normal. Notwithstanding, bismuth telluride (Bi_2Te_3), there are other thermoelectric materials including lead telluride (PbTe), silicon germanium (SiGe), and bismuth–antimony (Bi–Sb) combinations that might be utilized in explicit circumstances.

9.3 PART 3: DESIGN AND IMPLEMENTATION

We have designed a 3 feet * 2 feet bucket-sized refrigerator. In this refrigerator, we have used the list of products and tools listed in Table 9.1.

9.3.1 Descriptions of tools and ingredients

9.3.1.1 Plastic bucket

We have utilized a plastic can with a volume of 12 L for gathering warm water. The pail was loaded up with 10 L of water. The submersible pump was inside the water to circulate the warm water through the warmth sink blocks from the can and got back to the can. This course was making the water hotter by the time.

9.3.1.2 Peltier

Thermoelectric coolers work as indicated by the Peltier impact. The impact makes a temperature contrast by moving warmth between two electrical intersections. A voltage is applied across joined conduits to make an electric flow. At the point when the current moves through the intersections of the two conductors, heat is eliminated at one intersection and cooling happens. Warmth is saved at the other intersection [4].

Table 9.1 List of products and tools

SL	*Product*	*Quantity*
1	Plastic bucket	1 pc
2	Peltier	4 pc
3	Heat sink fan	Small 2 pc, large 1 pc
4	Aluminum water block	Large 2 pc
5	Submersible pump	1 pc
6	Digital thermometer	2 pc
7	Cork sheets	Large 1 pc
8	Water pipes	3 feet
9	Thermal adhesive glue	1 pc
10	Plastic container	1 Pc, Big

The fundamental use of the Peltier impact is cooling. The Peltier impact can likewise be utilized for warming or control of temperature. For each situation, a DC voltage is required [4].

9.3.1.3 Heat sink fan

A heat sink fan is a functioning chilling arrangement used to chill coordinated circuits in PC frameworks, usually a central processing unit. As the name proposes, it is made out of a detached cooling unit (the heat sink fan). The heat sink is typically produced using a high-temperature conductive material like aluminum and copper, and the fan is a DC brushless fan, which is the standard utilized for PC frameworks [5].

9.3.1.4 Aluminum water block

These blocks braze parts into a whole. The internal flow channel is extrusion forming. It can be connected with an internal diameter 7–8 mm/0.27–0.31″ water pipe. The solid construction, aluminum material is a good choice for heat dissipation [6].

9.3.1.5 Submersible pump

Submersible pumps push the liquid to the surface rather than stream siphons which make a vacuum and depend upon air pressure. It utilizes compressed liquid from the surface to drive a water-powered engine downhole, as opposed to an electric engine, and is utilized in hefty oil applications with warmed water as the thought process liquid [7].

9.3.1.6 Digital thermometer

A digital thermometer works by sensing the heat by a heat sensor connected to the experimental object. It is very useful to know the accurate temperature of the object in form of decimals. So, the user can assume the small to smallest change in temperature for the experiment. In this way, the experiment will be more precise rather than analog ones.

9.3.1.7 Cork sheets

Cork sheets are easy to use for their low density and lightness. It is waterproofed, on the other hand, it is a good thermal insulator. Again, it is a high coefficient of friction.

9.3.1.8 Water pipe

We used 3 feet of water pipe with the same diameter as the pipes of heat sink block for water circulation from plastic buckets to aluminum water blocks.

9.3.1.9 Thermal adhesive glue

Thermal adhesive glue is a material that has low-temperature serviceability. It also consists of thermal cycling resistance which prevents heat transfer by thermal insulation.

9.3.1.10 Plastic container

For the cool chamber of the gadget, we made a closed compartment in a plastic container. We covered the walls of this container with cork sheets for thermal resistance.

9.4 PART 4: CONSTRUCTION OF THE STRUCTURE

Initially, we have cut the cork sheet in the size of the walls of the plastic container. Figure 9.2 shows that we attach the pieces of cork sheets with the thermal adhesive glue to the interior walls of the plastic container to make the closed compartment heat resistant. Then we attach the heat sink water blocks to the hot surface of the palters with thermal glue. Figure 9.3 shows that we also attach the heat sink fans to the cold surface of the Peltier with thermal glue. Now, for the attached materials we cut one wall including its cork sheet measuring the size of the Peltier. After that, we attach the partier-fan-aluminum block sets with that wall maintaining the precaution so that the outer heat cannot be entered into the closed chamber. Figure 9.4 shows that the aluminum water block was fitted on the outer side and the heat sink fan's side was fitted to the inner cool chamber.

Figure 9.2 Structural design of the gadget.

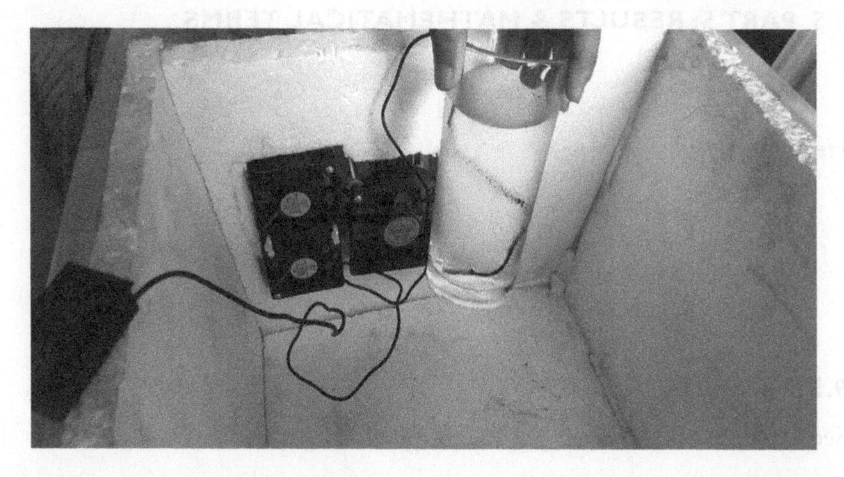

Figure 9.3 Cold chamber of the gadget.

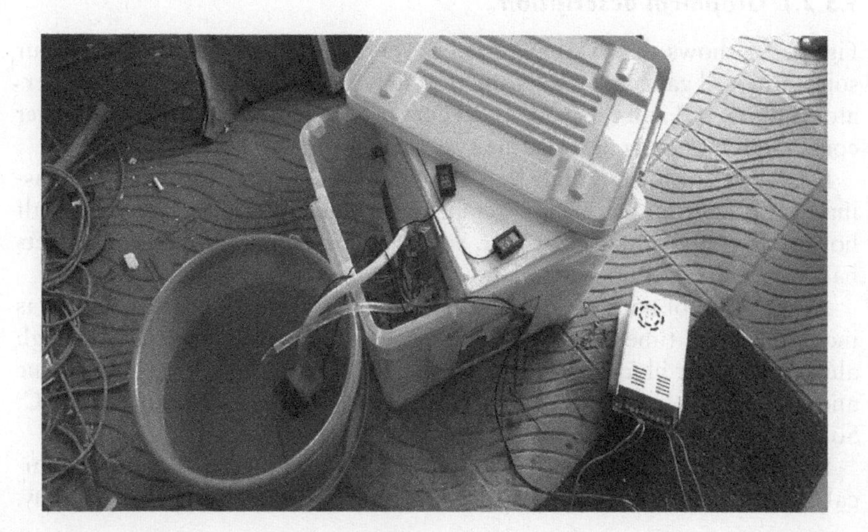

Figure 9.4 Experimental setup of the gadget.

9.4.1 Wiring

We simply interface the positive wires of all the Peltier together to the positive wire of the submersible pump and associate with the positive of the supply of the solar PV. At that point, we interface the negative wires of all the Peltier together to the negative wire of the submersible pump and associate with the negative of the supply of the solar PV.

9.5 PART 5: RESULTS & MATHEMATICAL TERMS

$$Q = mS\Delta\Theta$$

Here,

- Q=amount of heat
- m=mass of the heated object
- S=relative heat of the object
- $\Delta\Theta$=temperature difference

9.5.1 Results

See Table 9.2.

9.5.2 Graphical representation of the experimental data

9.5.2.1 Graphical description

Figure 9.5 shows that we have tried to represent the characteristics of our solar-powered gadgets by showing the curve of both data from the refrigerator (blue curve) and the geyser (red curve). We have also plotted the power consumption (green curve) of the system over time.

It is clear from the graph that the refrigerator temperature is decreasing over time from the normal temperature (26.3°C). Within one and half hours it dropped down to 10.9°C. So, the inner temperature of our gadgets has been decreased by 15.4°C.

At the point of normal temperature (26.3°C), the data of the geyser was increasing over time. The simultaneous circulation of the water through aluminum water blocks increases the temperature of the geyser. Within one and half hours, the temperature of the geyser has been increased to 50.5°C. So, the temperature of the geyser has increased by 24.2°C.

We would like to focus your attention on a key point of our graphical representation, which is very impressive at the view of the economy.

Table 9.2 Experimental data

Time (hour)	*Refrigerator temperature (°C)*	*Geyser temperature (°C)*	*Power consumption (W)*
0.25	26.3	26.3	91.69
0.5	17.1	33.7	85.62
0.75	13.2	38.8	83.58
1	11.5	43.3	81.78
1.25	10.9	47.6	80.26
1.5	10.9	50.5	79

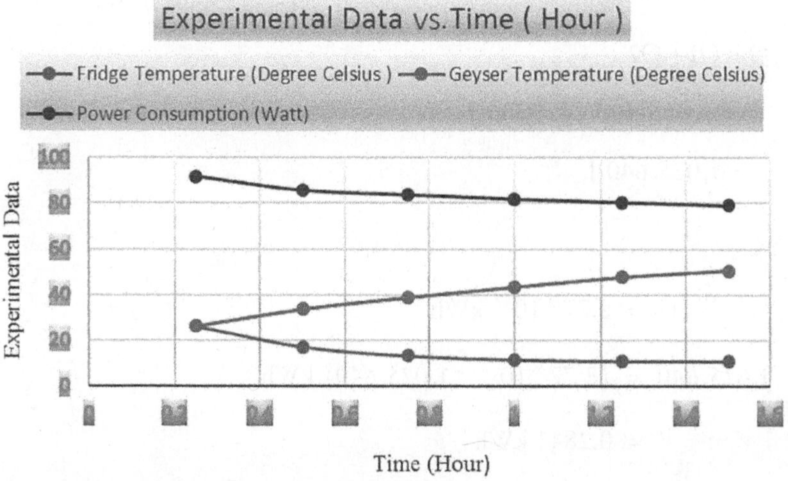

Figure 9.5 **Graphical representation of experimental data.**

The power consumption curve is decreasing over time from its starting point (91.69 W). It has decreased to 79 W, within just one and a half hours of the experiment. So, it has been decreased by 12.69 W over time. The more the system will run, the more the power consumption will be decreased.

9.5.3 Calculation

In a 1.5 h experiment, we find that our device increases 24.2°C (26.3°C–50.5°C) of 10 L water, whose specific heat is 4,200 J $kg^{-1}k^{-1}$.

So, the heat produced is

$$Q1 = mS\Delta\Theta$$

$$= 10 * 4,200 * (50.5 - 26.3)$$

$$= 1,016,400 \text{ J}$$

In a 1.5 h experiment, we find that our device decreases 20°C (26.3°C–10.9°C) of 250 mL water, whose specific heat is 2400 J $kg^{-1}k^{-1}$.

So, the heat produced is

$$Q2 = mS\Delta\Theta$$

$$= 0.25 * 2,400 * (26.3 - 10.9)$$

$$= 9,240 \text{ J}$$

So, the total energy is

$$Q = Q1 + Q2$$

$$= 1{,}016{,}400\ \text{J} + 9240\ \text{J}$$

$$= 1{,}025{,}640\ \text{J}$$

Now,

$$1\text{J} = 2.77 * 10^{-7}\ \text{kWh}$$

$$1{,}025{,}640 = \left(2.77 * 10^{-7} * 1{,}025{,}640\right)\ \text{kWh}$$

$$= 0.2841\ \text{kWh}$$

So, the solar power required is

$$\text{Solar Power} = 0.2841\,\text{kWh}/1.5\,\text{h}$$

$$= 0.1894\ \text{kW}$$

$$= 189.4\ \text{W}$$

So, the efficiency with respect to 250 W/h solar PV is

$$\eta = 189.4 / 250$$

$$= 0.7576$$

$$= 75.76\%$$

9.6 PART 6: COST AND COMPARISON

We are comparing our solar-powered gadgets with any other gadgets that work by electricity or gasoline. The proof that our gadgets are more cost-efficient than electricity and gasoline is given below:

- **Solar power over gasoline:**
  According to this paper [8], the average cost of gasoline per liter = \$1.4
  Now, produced energy of 1 L gasoline = 8.9 kWh
  So, price of 8.9 kW energy with respect to gasoline = \$1.4

That means, 1 kW costs = 1.4/8.9
= $0.15
According to Google [9], 1 unit (kWh) solar power costs = $0.08
Our, device consumes 0.1894 kW/h.
So, our device costs per hour = 0.1894*0.08
= $0.015
So, the cost ratio of gasoline to solar power =
Cost of gasoline per hour/cost of solar power per hour = $0.15/$0.015
= 10
From the above analysis, it is clear that our device costs ten times cheaper than burning gasoline.

- **Solar power over electricity:**
According to this paper [10], price of 1 unit (kWh) electricity = 13.31 cents
= $ 0.1331
So, the cost ratio of electricity to solar power =
Cost of electricity per hour/cost of solar power per hour = $0.1331/$0.015
= 8.87
From the above analysis, it is clear that our device costs 8.87 times cheaper than any other gadget that runs with electricity.

9.6.1 Graphical representation of the cost comparison

Figure 9.6 shows that we have plotted the above cost analysis into a column chart. In this chart, we can see that the use of solar power is way more cost-efficient than that of gasoline and electricity. Our appliance will be a revolutionary invention to the rural locality as well as urban areas.

9.7 PART 7: DISCUSSION

9.7.1 Future of the gadget

We are planning to implement IoT into our gadgets to make them programable and controllable. After reaching a certain temperature the gadget will be turned off automatically to prevent unnecessary power consumption. Then after a certain increase/decrease of temperature, the gadget will be turned on automatically.

9.7.2 Conclusion

We have tried to make a 2 in 1 home appliance using solar energy to prevent the cost of electricity or fuel. It is designed for the shops and hospitals of

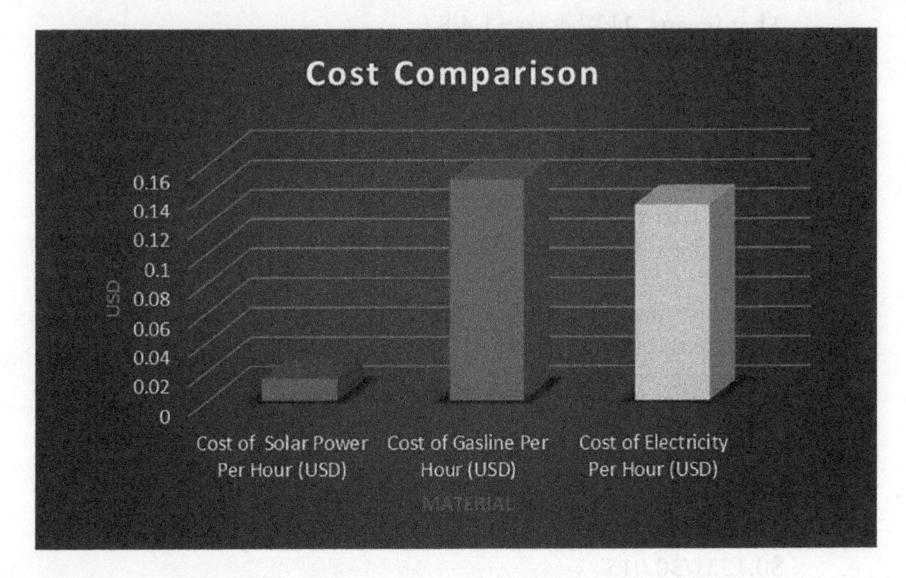

Figure 9.6 Chart of cost comparison

remote areas where the supply of electricity is not sufficient enough to reach the demand of the locality. It can be used in tea stalls of those areas for the preservation of ice cream and cold drinks, on the other hand, the warm water can be used for making tea, coffee, soup, etc. In the future, we will add more features to the gadget.

REFERENCES

1. D. I. T Cherney (1978) *Solar Energy Application of Natural Zeolites, in Natural Zeolite: Occurrence, Properties and Use.* Oxford: Pergamon Press, 479 p.
2. R. Z. Wang and R. G Oliveira *Adsorption Refrigeration - An Efficient Way to Make Good Use of Waste Heat and Solar Energy.* Shanghai: Oliveira Institute of Refrigeration and Cryogenics, Shanghai Jiao Tong University, 23 p.
3. T. S. Vinubhai, R. J. Vishal and K. Thakkar, Solar water heating systems, *Conference: National Conference on Emerging Vista of Technology in 21st Century,* At: Parul Institute of Technology, Limda, Vadodara, 8 p.
4. T. P. Sonwani, A. Kumar, G. C. Sahu, K. L. Sahu, Thermoelectric refrigerator using Peltier module, *International Research Journal of Engineering and Technology (IRJET),* Shri Rawatpura Sarkar University, Raipur, Chhattisgarh, India.
5. https://www.techopedia.com/definition/5291/heat-sink-and-fan-hsf.
6. https://www.electronics.com.bd/Aluminum-Water-Cooler-Block-80x40x12mm.
7. https://en.wikipedia.org/wiki/Submersible_pump.

8. https://www.globalpetrolprices.com/USA/gasoline_prices/.
9. https://www.solar.com/learn/solar-panel-cost/.
10. https://www.electricchoice.com/electricity-prices-by-state/#:~:text=The%20average%20electricity%20rate%20is,is%2013.31%20cents%20per%20kWh.

Chapter 10

Blockchain technology in intelligent transportation systems

Arshil Noor, Md Tabrez Nafis, and Samar Wazir
Jamia Hamdard

CONTENTS

10.1 INTRODUCTION

Recently, an intelligent transportation system (ITS) has arisen as an approach to improve transportation frameworks' effectiveness, travel well-being, vehicle security, and the information given to drivers and travelers. These targets can be cultivated using methods, frameworks, and gadgets that permit information gathering, correspondence, investigation, and dispersion among moving people and vehicles, foundations, and administrations.

DOI: 10.1201/9781003156789-10

Rewording this definition, a progression of key ideas clarifies circumstances and results that lead to the acknowledgment of this proposition work. "Progressed applications" catches the broadest meaning of the framework planned to get this work. An assortment of utilizations and conventions are utilized related to acquiring a high level of use regard to what we are utilized to for the traditional transportation organization. The understanding of "without encapsulating intelligence" draws out the genuine quintessence of a foundation of such a kind. The association cycle between two clients or a client and a vehicle or foundation should incorporate the most unconceivable presence of a human middle person. Human knowledge should be changed in artificial intelligence (AI) when this one can be completely trusted by anybody. This artificial intromission prompts the making of "creative administrations" that exploit quicker handling and better exhibitions. When there are no delegates, customary cycles become lighter. Henceforth, by the blend of these lighter cycles, new developments may happen in what is viewed as a norm. At last, the idea of "more brilliant use" gives help to introduce the other significant subject of this proposition, because the descriptor "smart and advance" accepts explicit importance in this unique situation.

Blockchain technology is developed by some engineers and researchers, while they were working on the digital currency later became famous as bitcoins. As blockchain technology works in the distributed networks environment, it was utilized in the bitcoins, that as digital currency as cryptographic numerical apparatuses. This technology of blockchain was first introduced in a paper that was published very secretly which was based on cryptography and its mailing and ordering in 2008 [1]. Since then, an incredible turn of events, which has been completed on introductory ideas, has prompted the formation of many dispersed and dynamic blockchains. The blockchain idea includes various types of information and is convoluted to the point that many great personalities have expressed in public that this particular technology is unbelievably amazing and will change the world. Since then, organizations and associations have been utilizing blockchain technology-based frameworks, and numerous researchers and monetary specialists anticipate that great innovations should be fixated on this creative idea, particularly in the logistic area [2–5]. A few creators have even characterized this innovation as troublesome regarding the transportation field [6].

10.2 BLOCKCHAIN

Blockchain is a very special type of technology, as it is a distributed ledger technology that is an extraordinary kind of distributed data set over the network (Figure 10.1). In this technology, every PC 'hub' or part of the organization stores an indistinguishable 'record' or data set. This data

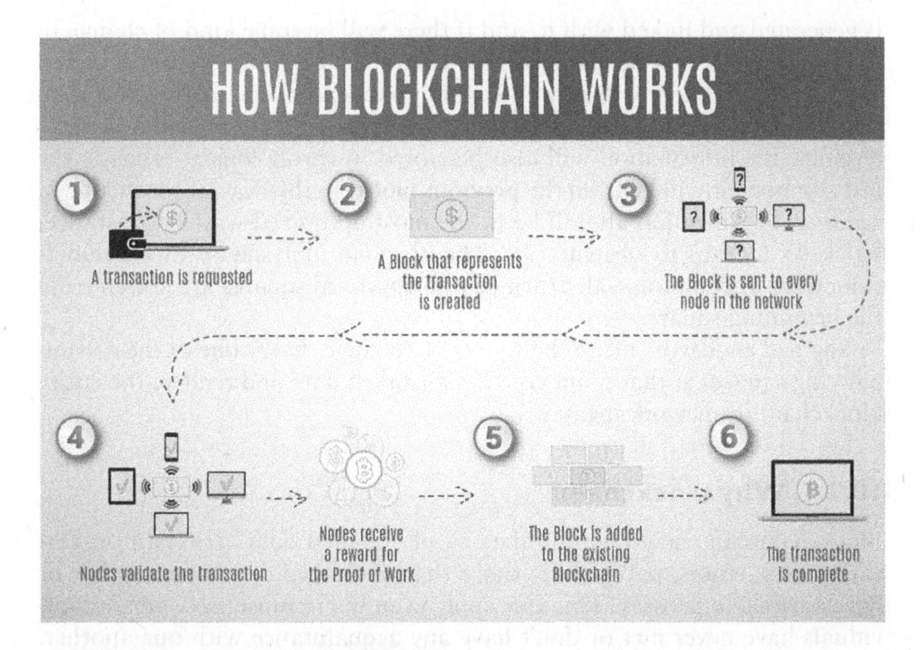

Figure 10.1 How blockchain works [7].

set appears as an ordered chain of interesting batches of data known as 'blocks', subsequently blockchain. These are safely connected via cryptography technology and the data that is contained inside each square could be subtleties of occasions, esteem exchanges, transactions, transport information, AI-based activities, or whatever other data, that would advance by the technology of the blockchain and so on, as before another square is added to the chain, individuals and data reside in the organization are needed to go to an agreement or understanding. On the other hand, the new square is added to the chain and duplicated across every one of the indistinguishable data sets.

A blockchain consists of the following:

- Set of data
- Hash functions

The information put away inside each square relies upon the kind of blockchain. For example, in the Bitcoin blockchain structure, the square keeps up information of the person of authority, sender–receiver, and amount of information. Because of highly secure encryption, conversion characteristics hash functions are the best available option for this particular technology because they can contain a large amount of data in chronological order and a very secure environment. These squares are created with the help of an SHA-256. While, the other square is developed, and its hash function is generated and linked with it, and if there will be some kind of change in the square there will be also a difference in the hash function that is linked with it. And if there will be any sort of change in the hash function then that will help to recognize the changes. Therefore, because of these characteristics, the information will also be stored in chronological order as the last component will be from the previous block. In this way, the robustness, security, organization all will be in the most optimized way. For instance, block 45 focuses to obstruct 46. The absolute first square in a chain is somewhat uncommon—all affirmed and approved squares are gotten from the beginning square.

Any bad endeavors incite the squares to change. Every one of the accompanying squares at that point conveys mistaken data and renders the entire blockchain framework invalid.

10.2.1 Why blockchain?

Blockchain can encourage the sharing of confided data across an organization of partners, particularly those that don't confide in one another or have various impetuses. This can apply even in circumstances where individuals have never met or don't have any acquaintance with one another. Shared (P2P), that is peer-to-peer, the move of significant worth can happen without the requirement for a confided-in outsider go-between. Blockchain spans the trust hole to empower a more prominent coordinated effort. Blockchains have five extraordinary qualities (Figure 10.2).

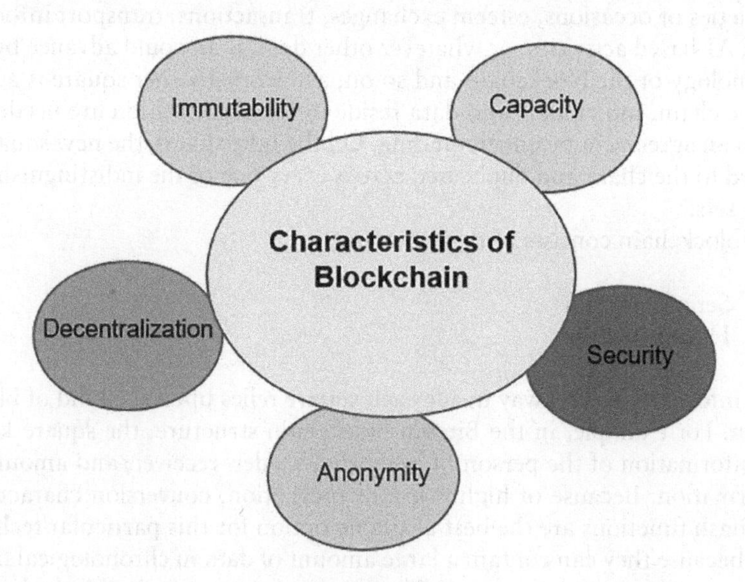

Figure 10.2 Characteristics of blockchain [8].

- Peer-to-peer communication without a central authority
- Transparent transaction processing with alternatively uncovered proprietorship
- Logic-based processing to trigger calculations and occasions
- **Decentralized information stockpiling and control:** The blocks are the exchanged data, and in decentralized style systems, these exchanges of blocks are taken care of different blockchain network stops. So, higher security is required
- **Data transparency and accuracy:** These characteristics give away the idea of, keeping a duplicate copy of exchanged blocks of data and that will be stored in squares as the records and will be available for all the companions to explore the history
- **Decentralized consensus:** Exactly when a concession to new trades is refined and set among the people in the blockchain and another square will be connected in the chain because every duplicate block of information is being recorded and that is very different and far more effective than the traditional way of storing the data
- **Secure encryption:** We have two separate sets of very special keys in this technology, public key and private key for high-level encryption and decryption process, and this way sender and receiver will able to have end to end secure communication

Blockchains can be understood as circulated information bases in which all part endorsed exchanges are forever put away, because of an agreement calculation. The exchanges are put away inside blocks, in sequential request. The quantity of squares is possibly limitless, and the inconceivability of changing them is acknowledged through the carried out calculations. The hash put away inside each square is evidence of the way that the squares are requested and unmodified. The agreement calculation is the interaction through which the circulated network agrees on the exchanges and squares' legitimacy. The plan of agreement calculations additionally considers the way that in the organization there are questionable hubs, cycles, or administrations that may get inaccessible, or correspondence might interfere. An arrangement can be reached with over half of the hubs inside the organization.

10.2.2 How blockchain is revolutionizing industries

In these past few years, blockchain is impacting the industries on a very high level, on a global level. Whatever industry you like to consider, whether there is healthcare facilities, banking, research and development departments, institutions, organizations and companies, large ones or small ones, transportation departments, railways, airlines, etc., all of these are utilizing this technology because of highly secure, amazingly backing up feature,

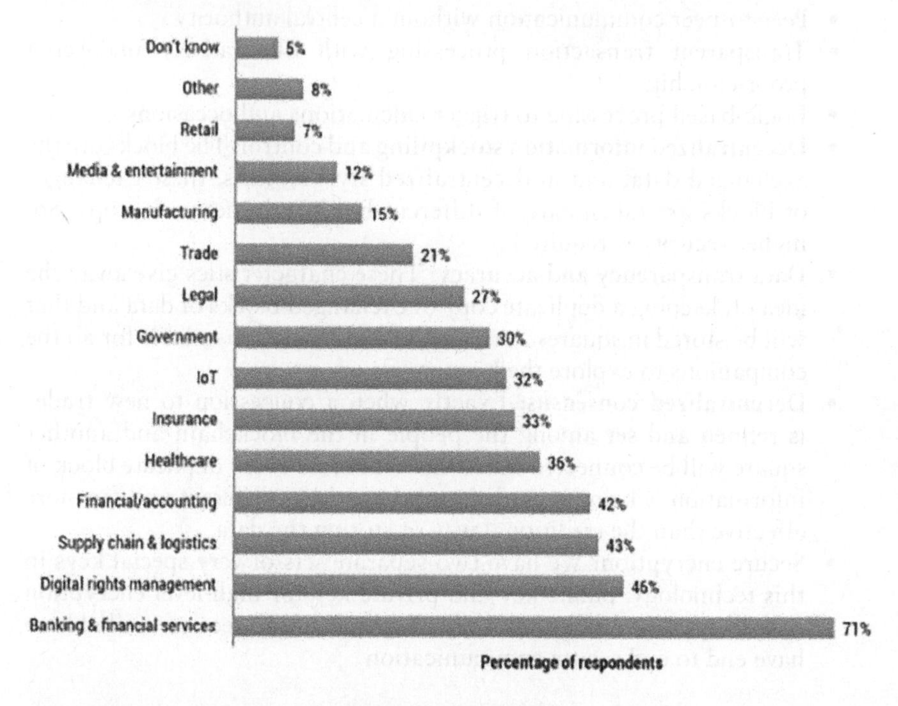

Figure 10.3 Blockchain impact on today's industries [9].

and also because of the chronological way of storing of the information. This technology is on the boom especially in the transportation systems in this modern era of time. Intelligent vehicular systems are being developed in a very secure and effective way because of blockchain technology (Figure 10.3).

Let's explore the ways blockchain technology is already improving our industry:

- New blockchain-based frameworks will permit simple coordination of records on a common disseminated record, making actual administrative work to a great extent superfluous.
- By utilizing smart agreements, endorsements and customs freedom can be snappier and more productive, decreasing handling times for merchandise at customs designated spots.
- Associations need refreshed, secure and credible information to decide. Blockchain guarantees dependable information across the transportation and coordination's biological system since the whole organization adds to information approval.
- With rising interest for same-day and 1-h conveyance administrations, customary following innovations won't scale. Blockchain innovation gives an adaptable, prompt answer for request following and authentication.

- With a blockchain, the production network for truck parts and utilized trucks could be followed on a computerized record, filling in as such a "CARFAX on steroids" for the business transportation industry.
- Associations like BiTA are dealing with making widespread principles for blockchain's mass appropriation in the transportation and coordination industry.
- Organizations in the business transportation industry do or die depending on their capacity to adjust to customer needs and execute new advances that help them increment proficiency and lower the expenses of delivery.
- In the previous year, blockchain innovation has arisen in the realm of shipping and coordination as an answer to a portion of the significant failures that have tormented the business for quite a long time.

In a blockchain, information is composed just a single time by various gatherings, and marks (check) by different accomplices take part in each of the squares as the development of each following square guarantees the incomprehensibility of data changing. As because of this feature a trust for being genuine/true data can be placed in blockchain technology. Blockchain plays an important part in different associations and organizations with the different domain as The record and verification of responsibility for physical (transfer) or narrative resource (information and protected innovation); permanence; agreement and concurrence on the legitimacy of information; a solitary rendition of truth, since the data set is imitated on the whole hubs of the organization; information straightforwardness can be modified, permitting limited information access as per who is perusing the information (i.e., who can peruse what information); and decentralization.

10.3 INTELLIGENT TRANSPORTATION SYSTEM

The proposed system for ITS consolidates the concentrated administration of SDN, calculation offloading utilizing the edge cloudlets, and the safe exchange of the executives utilizing the blockchain innovation. The vehicles produce the transportation information that is kept up utilizing the edge cloudlets put away utilizing the blockchain innovation which offers a safe stage for calculation offloading. In our proposed structure, we utilize private blockchain where the mining task is performed by the asset proficient edge workers when contrasted with asset restricted the Internet of Things (IoT) gadgets. We utilize this cycle as it is less asset serious when contrasted with the public blockchains. This gives edge workers the command over making and attaching the square to the blockchains to proficiently store the offloaded information from the ITS. In these conditions, the ITS foundation including savvy vehicles and administration framework goes about as the customers which can just demand the exchange administrations. The ITS uses incorporated organization the board utilizing SDN. SDN gives the adaptable organization the executives by utilizing q concentrated control technique over the organization. It empowers network programmability by deftly introducing stream rules for redid traffic sending. Organization gadgets and equipment act simply as sending stations where the mind is the regulator which effectively introduces sending stream rules on the relating equipment. SDN utilizes a proactive and responsive stream rule establishment component where the proactive philosophy is utilized to introduce stream rules on the SDN switches preceding the appearance of bundles. While in the responsive stream rule establishment, the streams are introduced when the bundles show up at the switch. In the majority of the conditions, half breed stream rule establishment methodology is utilized where some stream rules are proactively introduced while different guidelines are introduced when parcels show up at the switches. The security system involves the accompanying parts. By transportation, we expect the development of individuals or products starting with one spot then onto the next. In the accompanying, we will recognize explicitly all that relates to the development of merchandise and we will consider it as a feature of an examination bunch in inventory network and coordination. In ITS, it has to be taken into consideration, that: roads, traffic on it, the path and branches of the road, where the traffic is leaning more towards, as all these can be examined in a very optimum way in the blockchain technology-based systems (Figure 10.4).

Because of the blockchain, transportation can be controlled in a very unique and intelligent way, as we can have real-time processing reports, predictions, troubleshooting in the advance. These Frameworks work very closely with blockchain technology and with applications of AI. This work closes by affirming how the intermixing of emerging organizations and advances,

Figure 10.4 Traditional ITS [10].

similar to Mobility-as-a-Service, IoT, AI cognizance with significant learning, and 5G network, can result in a splendid way of reshaping the possible destiny of coordination and transportation. As in how to utilize, the BT in transportation is very logistic; to be sure, various papers have been presented recently. These are considered very effective but also very dangerous as well in which supply chains are supervised properly. As these are very competitive times for the industries and they need something, some frameworks on which they can rely and where their supply chain can be maintained in most optimum and in a most secure way. Also, in these times, cyberattacks are also one big factor in the competition and every industry wishes to keep their supply chain, stocks, their authentic information to be secure and safe, and with help of blockchain technology, it is highly possible for them. Because this technology provides great control over the information and historic data and great encryption techniques, thus creating greater trust among providers. Different significant issues can be tended to in inventory network the board by a blockchain, like expense and quality control, abstaining from falsifying, speed of the move, investigating halted products, hazard decrease, and adaptability. Some of the major organizations made an agreement on utilizing blockchain technologies and their applications in their associations for a more secure environment. Even different partners or colleagues can put their trust in the technology while their commination, exchange of some important information, etc. All of the exchanging information over servers via blockchain-based frameworks provide various features such, duplicate copies of all the information recorded in the most proper order, and cannot be manipulated by anyone, and even somehow if it's done, that information of breaching will be provided immediately. The breaching of information and manipulating with is it, is very difficult for the breachers or cybercriminals. The utilization of ensured keen objects of IoT could light up the fate of these sorts of the framework. Numerous organizations have put resources into store network uses of BT, similarly, Jio-coin, SKU-chain, and block verify, which contributed as for not to try to fake merchandise. So, many organizations and associations have chosen this type of technology.

10.3.1 Why intelligent transportation systems (ITS)

Wise transportation framework is the utilization of detecting, investigation, control, and interchanges innovations to ground transportation to improve well-being, versatility, and productivity. Canny transportation framework incorporates a wide scope of utilizations that interact and offer data to ease blockage, improve traffic the board, limit the ecological effect and increment the advantages of transportation to business clients and people as follows:

- Inadequate road advancement
- Low speed, increased accident rates

- It is unimaginable to expect to construct sufficient new streets or to satisfy the need
- Make transportation framework more productive, secure, and more secure user data, interchanges, and control advancements
- Improve the allure of public vehicles
- Tackle rising blockage which builds travel times and industry costs
- Decrease the ecological effects of transport
- Decrease in stops and postponements at crossing points
- Speed control and improvement
- Travel time improvement
- Capacity management
- Incident management

Energy stockpiling is fundamental to the keen city since it permits a neighborhood age module that provides direct feed to nearby structures, maintaining a strategic distance from the problem of going through the public lattice. In addition, building effectiveness centers on the digestion of a complete arrangement of cyber-physical programming, advancements, and parts inside the edge of an actual climate to influence structures' energy. The EU planned an objective of a decrease of energy interest at 25% by 2030. Building productivity doesn't simply save costs; it empowers authorities to introduce observing, control advances, exchanging energy utilization during seasons of pinnacle interest.

At last, insightful transportation frameworks center around programming, innovation, and actual foundation encouraging going around urban communities into a more proficient cycle, for instance, electric vehicles, electronic installment frameworks, and new travel plans of action, for example, vehicle sharing and carpooling. Canny vehicle frameworks represent a vital response to the attacking monetary effect of gridlock that is relied upon to punish the ECO.30, which is converted into a 63% expansion of the gross expense in 2016. Various insightful vehicle frameworks have been executed across shrewd leaving and cycling organizations.

10.3.2 Limitations of traditional ITS

Unequivocally, the primary motivation behind why AI instrument is excitedly cultivated during the coordinated concentrate, stack, and change information measure is not only that a more compelling framework could be set up but also that a self-ad lobbing and self-overseeing administration would be assembled dependent on such keeps learning capacity.

- The security worries of equipment organizations with a distributed control processor
- The correspondence issues on time postponement and information spillage

- The rising energy utilization of IoT gadgets
- The expanding correspondence requests on the network throughput
- The expansibility of the arrangement with a predefined existing blockchain administration framework

The dissected outcomes from gathered sensor information and the rich substance information from on-street modules will be ingested so clearly each time the amount investigation will have one stage toward exactness. This demonstrates that the on-street modules particularly the ISUs which are mostly answerable for concentrating rich configuration of information could change the assortment procedure by virtue of the predisposition factual investigation results. All in all, the ISUs, specifically, will in general effectively gather more helpful information and, therefore, such AI could be applied. Intermittently, the information assortment system of on-street ISUs will at that point keep on affecting the following degree of investigation of handled information and produce more substantial assessments to meet assumptions. In future works, it is important to consider the most proficient method to improve the effectiveness of working the blockchain put together framework with respect to the installed framework with low-energy utilization and the restriction of blockchain keen agreement that could upgrade the framework expandability. Moreover, it is relied upon to deal with the AI calculation which could recognize the auto collision and anticipate the traffic condition, and the agreement calculation of blockchain to upgrade the resilience and exchange pace of a constant framework.

10.4 VANET

The fast advancement of advances in the vehicular ad hoc network (VANET) field prompted the formation of various frameworks and improved investigates by various private, public areas and specialists all throughout the planet. These accomplishments give a solid and secure correspondence between vehicles themselves and vehicles with foundations. This chapter gives an outline of a large portion of the advances that have been presented in VANET-based ITS forward-thinking. We arranged them dependent on their functionalities, including blockage shirking, controlling the convergence, mishap evasion, and crisis the executives.

With the progressions of interchanges, detecting, and electronic frameworks, the regularly installed frameworks just as regulators are being supplanted by a high-level sort of framework. Therefore, this system implied as CPS is processing all the time while connecting to the server and keeping track and maintaining the processing between the digital and actual course (Figure 10.5).

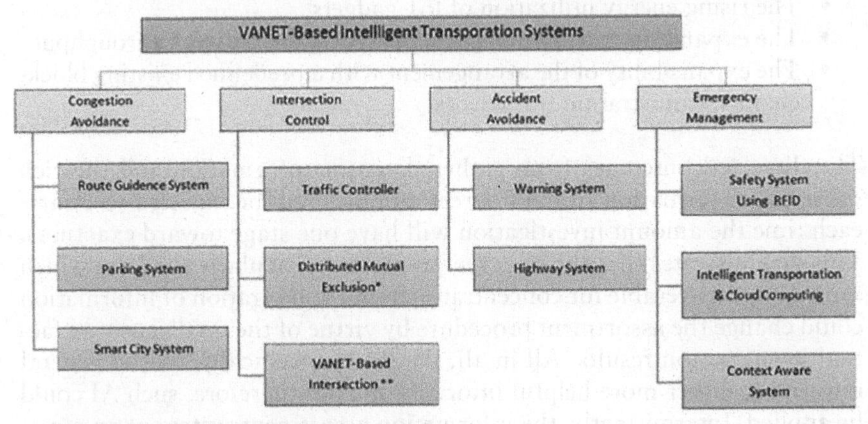

Figure 10.5 Overview of VANET [11].

Lately, this technology has been gaining more and more recognition and support from nontechnical individuals, common people, to large-level industries all over the globe. One of its instances is ITS, which is being considered as the technology of the future and for the future and its main aim is to creative runtime, intelligent processing and reasoning, more secure transportations systems.

Simultaneously, In this field of vehicular engineering, there is a vast level of advancement in the auto-motors sciences and technologies. One of the advanced ITS models is shown in Figure 10.6a. As control units like ECUs and OBUs, programming and firmware frameworks, various types of sensors, and many various kinds of tools with different devices. To proceed further the system uses the processed information from sensors, devices, derived information from algorithms, and each of these segments is based on various sorts of wired (CAN transport and LIN transport) and remote (Bluetooth) innovations. These trend-setting innovations may fuse into conventional vehicles soon. Additionally, with these cutting-edge innovations, the vehicles are becoming more intelligent and self-governing and with learning features continuously which can make the system unstable at some point.

Another worldview called the internet of vehicles (IoV) is presented inside the ITS which is driven by the keen vehicles, IoT, and AI methods. In this worldview, the vehicles are associated with one another, individuals, and frameworks through correspondence advances so the vehicles can drive securely and cleverly through observing and detecting the adjoining conditions (Figure 10.6a). Henceforth, the IoV biological system is viewed these days as an all-inclusive part of CPS.

Moreover, ITS utilizes IoV environment including, RSU, that is, street-side units, and some deeply well-separated levels of cloud. Via network

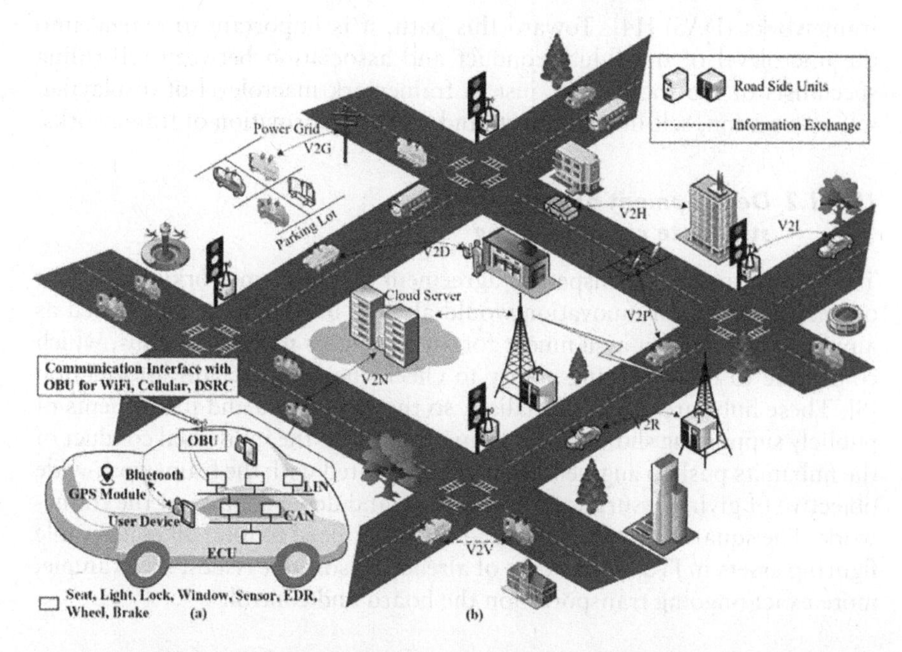

Figure 10.6 (a) Communication design of an intelligent vehicle [12]; (b) Communication design of vehicles on streets and interacting with smart devices, people, buildings, and grids. [12].

utilizing utility such as Hubs V2W controls over the intelligent monitoring vehicles, and RSUs generally go by the passengers and base station to help VCS (Figure 10.6b). While figuring hubs on the off chance that they have computational and capacity abilities.

10.4.1 Why blockchain technology for ITS

10.4.1.1 Decentralized autonomous transport systems

Decentralized self-sufficient vehicle frameworks incorporate the primary segments of the square. P2P organizations, in view of agreement circulated coordination and monetary advantages, are a characteristic method to show an unpredictable vehicle framework. Each registering hub (for instance, IoT gadgets, vehicles, or different articles with processing power) can be considered as a self-ruling specialist in this framework [13].

An enormous number of such hubs can be associated with a typical arrange and speak with one another through different kinds of square-based decentralized applications, bringing about a decentralized self-sufficient association (DAOs), subordinate to explicit necessities and errands. Then, while shaping the large-scale level, we will go to the production of a decentralized self-governing framework and surprisingly a local area of frameworks (DAS) [14]. Toward this path, it is important to extend into the microlevel of individual conduct and association between self-ruling specialists of the framework, just as framework macrolevel of displaying, self-association, self-improvement, and self-transformation of frameworks.

10.4.1.2 Development of mechanism to stimulate crowdsourcing

The opposition of the dispersed agreement in the frameworks dependent on the square chain innovation would already be able to be considered as a publicly supporting assignment for an enormous number of hubs, which contribute to their figuring ability to check the square chain information [8]. These hubs are single specialists, so the motivators and instruments of publicly supporting should serve to guarantee that the individual conduct of the hub in its push to augment its pay is facilitated with the framework-wide objective of giving assurances of insurance and dependability of the framework. The square innovation utilized can be utilized to total all conceivable figuring assets in ITS to take care of already unsolvable issues, for example, more exact ongoing transportation the board and control.

10.4.1.3 Software that determines the trust in its systems

Certainty dependent on a chain of squares in a framework assumes a significant part in the formation of decentralized ITS, which will permit to apply this innovation to numerous assignments like P2P trade, installment, and correspondence [9]. This sort of trust is ensured by the code and check of most cycle members. The innovation can possibly fundamentally decrease primarily complex frameworks, which will lessen social issues. This will permit money and resources to move unreservedly between lawful elements and people. For instance, based on P2P trust, the vehicles utilized can be exchanged and enrolled straightforwardly through the application blockchain rather than incorporated bodies or stages.

Toward this path, gives anticipate further exploration are the crucial reasoning for building trust and trust the board.

10.4.1.4 Intellectual contracts based on intelligent transport

Savvy contract fills in as an "activator" of the blockchain, giving static information, utilizing an assortment of calculations (for instance, AI, examination of huge information, and so on) and high-level projects of rationale, to assemble the product environment of ITS and improve the knowledge of its applications [10]. A self-executing canny agreement essentially lessens social intricacy by diminishing the significance of the human factor and can go about as programming specialists in the interest of their maker or even themselves. Thus, there is a dire need to contemplate the turn of events and execution of explicit shrewd agreements and to oversee and control ITS dependent on them.

10.4.1.5 Data security and privacy protection

Even though blockchain has shown high unwavering quality and security, the encryption structure should be additionally fortified in ITS with an enormous number of gadgets to ensure against assaults.

Various scientists [11] proposed the possibility of PTMS (Parallel Transportation Management and Control System), which advances the genuine vehicle framework by simultaneously communicating with its separate fake or virtual partners. Blockchain is quite possibly the most secure and solid structure for PTMS, and hence, blockchain can be viewed as a significant advance towards PTMS. One of the potential uses of blockchain innovation in PTMS is appeared (Figure 10.7).

A blockchain-based PTMS will remember that all articles for actual space, including IoT gadgets, vehicles, and resources, can be effortlessly digitized through the "blockchain of things" and enrolled in the blockchain on the web. The transmission of huge information on the internet can likewise be incorporated into the blockchain. Likewise, it is feasible to make at least one fake vehicle framework in the code space of savvy contracts utilizing the Ethereum stage, which offers programmable contents to help complex displaying and calculation [16].

In view of these together grown genuine and counterfeit vehicle frameworks, we can plan and lead multidirectional computational tests to assess and approve explicit conduct, instruments, and procedures in ITS (e.g., to assess traffic conditions). These tests can be planned as "Consider the possibility that" kind of situation yield and displaying, in light of predefined

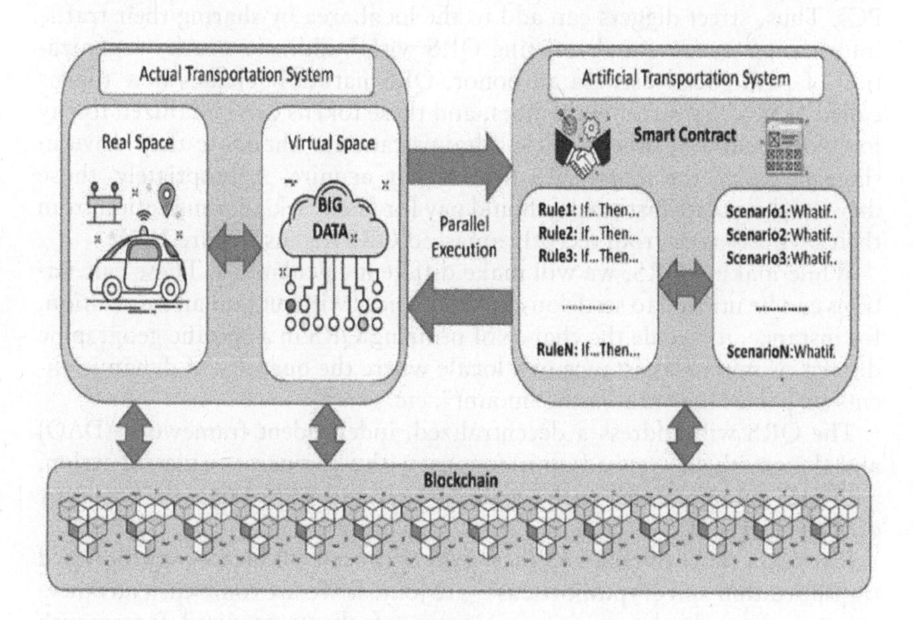

Figure 10.7 Parallel transport management [15].

"Assuming Then" rules. The ideal arrangement will be created in countless computational investigations and will get back to genuine vehicle frameworks. This cycle is rehashed interminably driving the real vehicle framework to the last methodology of its ideal fake analogs [17].

10.4.1.6 Quick road system

One of the potential situations for the utilization of block checks in the ITS might be assistance to get an unobstructed entry. Allow us to call it Quick Road System (QRS). This help is pointed toward making a decentralized organization of street paths partaking continuously. On the off chance that a driver needs to speed up for some reason and he turns to heavy traffic speed path and behind that vehicle, another vehicle is coming in speed too. In such scenarios we can avoid life-threatening accidents and furthermore, we can even provide a more accurate path and routes for drivers according to their needs.

A unique application introduced on the driver's cell phone or coordinated into the vehicle programming can be enlisted as one of the QRS processing hubs called "Street excavators". Real-time information is checked and put away on a local area upheld P2P network through which all path and installment sharing conduct are composed and performed. The street excavators are associated with the P2P network with no focal position.

As an agreement calculation, an imaginative agreement calculation can be proposed, called "evidence of traffic", which urges street excavators to drive with the QRS application running on their cell phones or installed PCs. Thus, street diggers can add to the local area by sharing their traffic information in route and assisting QRS with building a nearby organization of path social use. As an honor, QRS naturally creates new tokens called "QRS" for street excavators, and these tokens can be utilized to pay for movement and other vehicle administrations. The more they drive in sluggish paths, the more QRS tokens they acquire. Appropriately, those drivers who utilize fast tracks should pay for fast-track administration from their own assets or from recently amassed QRS tokens (Figure 10.8).

While making QRS, we will make different calculations. These calculations can be utilized to settle on explicit choices without human association, for instance, to decide the chance of utilizing QRS in a specific geographic district or initiate assistance in a locale where the quantity of dynamic clients surpasses the "minimum amount", etc.

The QRS will address a decentralized, independent framework (DAO) and along with different administrations, with a comparative plan of action, address a future pattern of the social vehicle and will change the economy of sharing [18–21].

It is just a short time before these two advancement ideas—appropriated transportation and cryptoporticus—are joined. We use tomb cash advancements to acquire the minimum amount of clients required for smooth

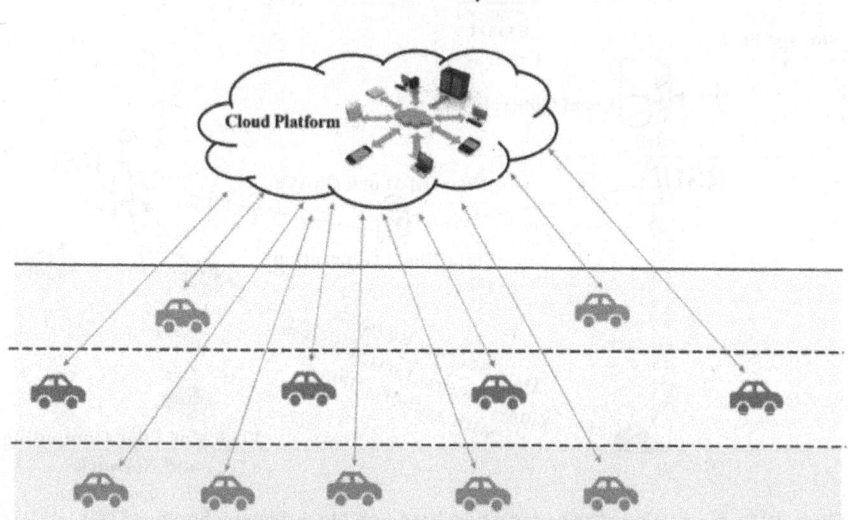

Figure 10.8 Maintaining service for a free passage [15].

activity of the circulated transport organization, just as to make a really dispersed vehicle arrangement [22,23].

10.4.2 Some proposed models

As of late, blockchain innovation has been embraced in vehicular information the board situations to address the security and protection concerns, yet in addition to set up a trustworthy relationship between hubs and frameworks to every aspect of the system. As one of the great examples can be a consortium blockchain and it provides security and appropriated information the executive's framework inside the vehicular edge processing organizations [24]. The benefits from the use of savvy contracts in this proposed framework are twofold. If we talk about the intelligent information and agreements sharing from person to person in vehicles and vehicles to stations, for instance, RSU (Figure 10.9). Besides, the splendid feature of blockchain technology ensures that the data can't be shared without approval and cannot be tampered with. In the proposed structure with the objective that the vehicles can pick the ideal and more strong data source having astounding controlled and secure data. To manage the remaining vehicles appropriately and conclusively, a three-weight passionate reasoning model is used by thinking about coordinated effort repeat, event advantageousness, and course closeness. With this standing arrangement, the proposed system improves the disclosure speed of dangerous and bizarre

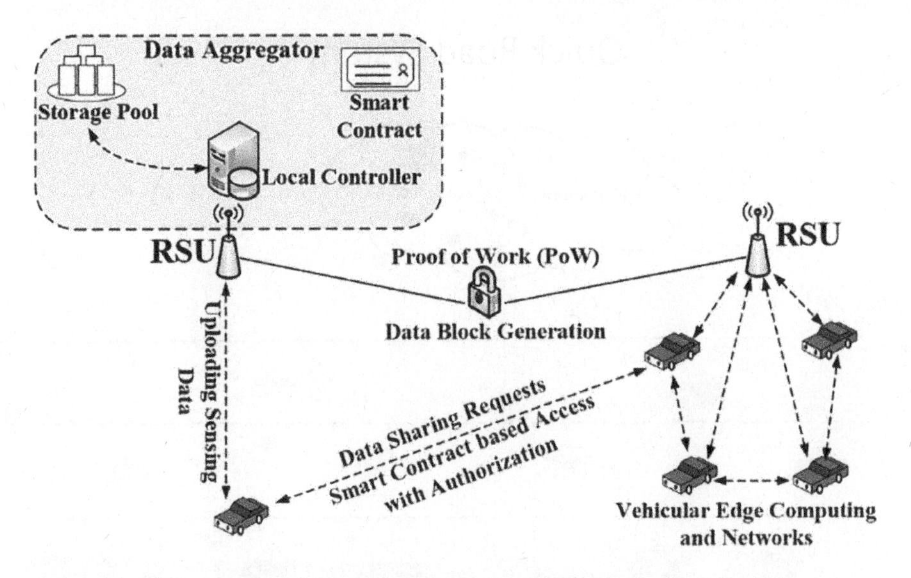

Figure 10.9 Information sharing structure based on blockchain technology [24].

vehicles over customary standing plans. In a similar way [68], engineers gave an astonishing model DrivMan based on blockchain technology, in which, the vehicle and station can build trust the board, information provenance, and protection through keen agreement, physical unclonable function, and public.

With the help of blockchain technology, we can use key function schemes in DrivMan Model, where trust can be built over a network. A network is not so easy to trust in an ideal world, yes, but this idea can fill in that trust. Furthermore, DrivMan accomplishes information provenance since physical unclonable function assists with allotting a one-of-a-kind crypto unique finger impression to each vehicle. Key function mechanism, that is, PCI is used for assigning the various corresponding keys to their respected vehicles by presentation authority (CA). Because this can prevent cyberattacks or cybercriminals in a very effective way and can deny permission from any wrong party. By use of PKI and CA, we can ensure real-time intelligent security.

Blockchain is very beneficial for vehicular information partaking in ITS. For instance, a sight and sound information-sharing methodology is dependent on blockchain technology and within cryptographic (encryption and decryption) methodologies [25]. With the help of blockchain technology along ITS can be used to keep the multimedia and mix media data safe and secure along with sharing multimedia and mixed media data with vehicles in a very effective way. Mainly it can really be helpful for preventing the intrusion of cyberattacks and cybercrimes and keeping the customer and server very safe. While utilizing these techniques we can get the assurance spillages of customers, vehicles, and RSUs from cybercriminals and cyberattacks

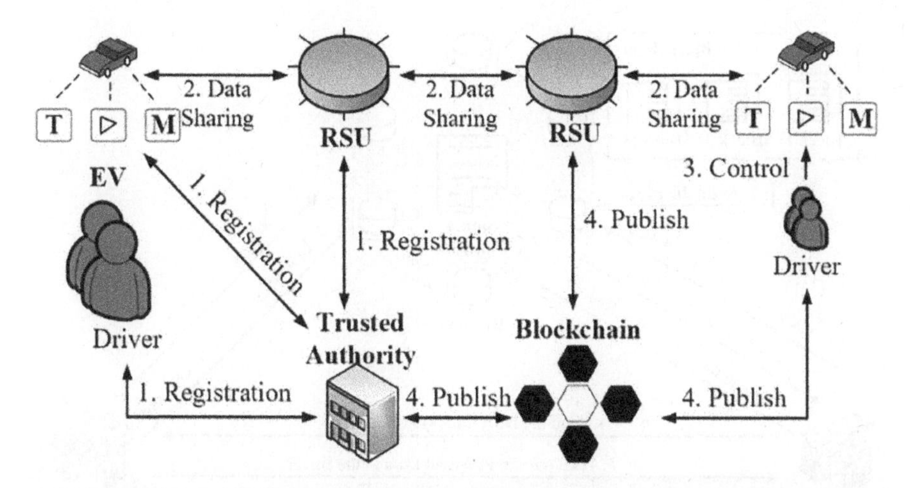

Figure 10.10 Security and protection with multimedia data blockchain-based model [25].

during the data sharing. Security is the main charm and advantage of these technologies for these models. addresses the framework model of this methodology. In the given model, an effective and secure way of controlling the sharing data and maintaining the communication through a server is the key feature (Figure 10.10) [26]. One of the main key features of this model is that security is really high because of the utilization of the key function schemes here. This can ease complexities, such as shared information security, and effective monitoring. This model provides the feature of key function mechanism and because of this great feature, it is very easy to access and control areas, easy monitoring, and very effective handling of the server along with the great security and robustness. This even gives the time-stamp feature for defending any cybercrime and can stop cybercriminals and hackers. The main advantage could be real-time intrusion detection and prevention.

10.4.2.1 Data and resource sharing

In any case, information and asset exchanging may prompt security issues, for example, access control implementation and protection of clients. Besides, thinking about the versatility of the vehicles and remote connection, the exchanging situations could be defenseless against denial-of-service attack and sticking like malignant assaults. Truth be told, such assaults could be accidental too. Another worry is to dispose of any conceivable disturbance because of malignant assaults and furthermore, guarantee irregular just as reliable exchanging in between the sharing parties.

In particular, blockchain could offer secure, shared, and decentralized answers for information and asset exchanging among various substances in the IoV setting.

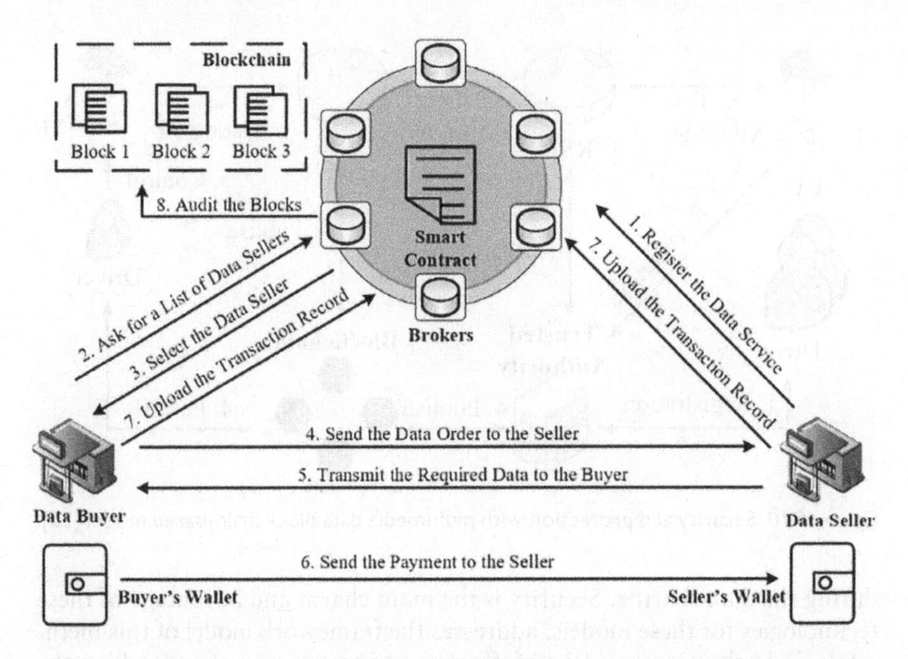

Figure 10.11 Peer-to-peer architecture for information sharing by utilizing consortium blockchain technology [27].

Specifically, blockchain innovation has been incorporated with vehicular information exchanging to encourage shared exchanging. Using blockchain technology we can achieve a very secure and robust IoV-based system [27]. Here, blockchain technology is utilized for securing the server and the system from external attacks, hackers, unwanted data, and traffic, and keep the flow smooth and in proper working condition. Consortium-based blockchain utilizing systems is used mainly for exchanges reviewing and confirmation (Figure 10.11). Furthermore, an iterative twofold closeout system is received to enhance the cost of information, amplify social government assistance while saving the protection of both the purchasers and merchants with the goal that more clients will be urged to partake in information exchange. Besides, the information transmission cost is likewise considered to upgrade the framework strength.

What's more, blockchain innovation has been fused with vehicular asset sharing. For instance, a shared framework utilizes blockchain to give a registering asset exchanging stage (Figure 10.12) [28]. The point of this proposition is to address the asset exchanging issues for edge-cloud-based frameworks, for example, guaranteeing an honest bid and permitting both the vendor and purchaser to take part in the exchange. In this vast era of advancement, it became very difficult to keep standing in the market, and in this proposition, there is a way to do that. clients and

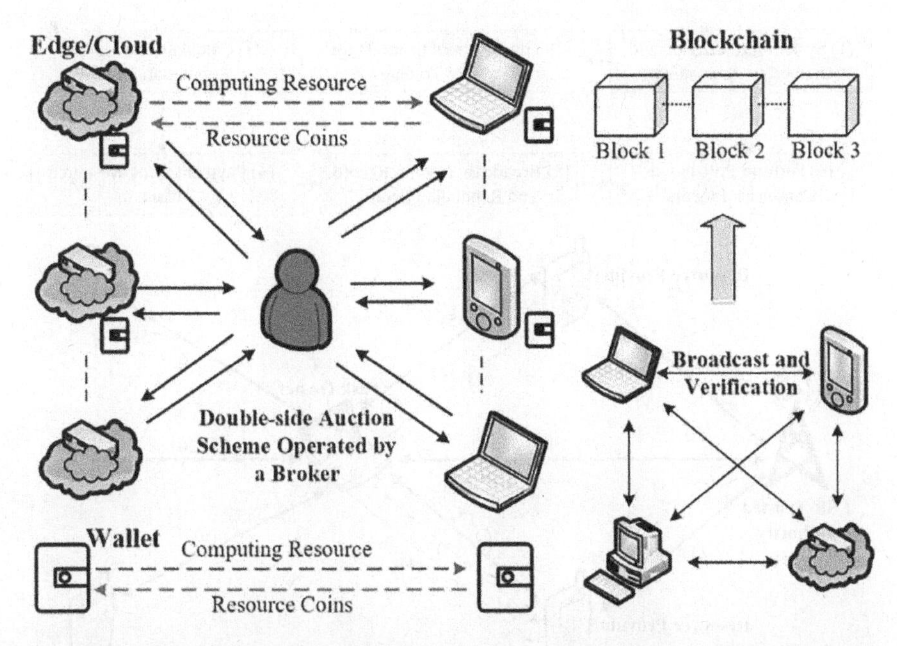

Figure 10.12 The working flow procedures of the proposed blockchain-based processing resource trading platform [28].

customers can exchange assets and show a very specific rule to support brokers with the goal that they offer honest offers. To do this, this proposition utilizes an iterative twofold-sided closeout component. Accordingly, it additionally guarantees the greatest social government assistance and empowers singular reasonableness just as adjusted spending plan. Figure 10.13 shows the nitty-gritty interaction of the processing asset exchanging dependent on blockchain. Also, a decentralized structure alludes to D2D-ECN for IoT applications [29]. In this approach, we use fair prioritizing of resources, stamp-time, and fair deals and agreements and it also looks after the situations like overflooding and computational errors, etc. Also, it sets up the reliability among the asset specialist organizations and undertaking holders while tending to the productivity issue of asset the executives. Moreover, in this agreement system, elements having higher priority can share resources and exchange data on the basis of their priorities, and the duplicate information of that information sharing and exchanging gets stored as a duplicate copy but is a permanent record with no tempered possibility. In this way, the system can process in real-time by going through all the previous data and as well as present data and can run in real-time and even can predict the future situations and routes. Still to give compensations to clients and customers hypothesis-based function is created.

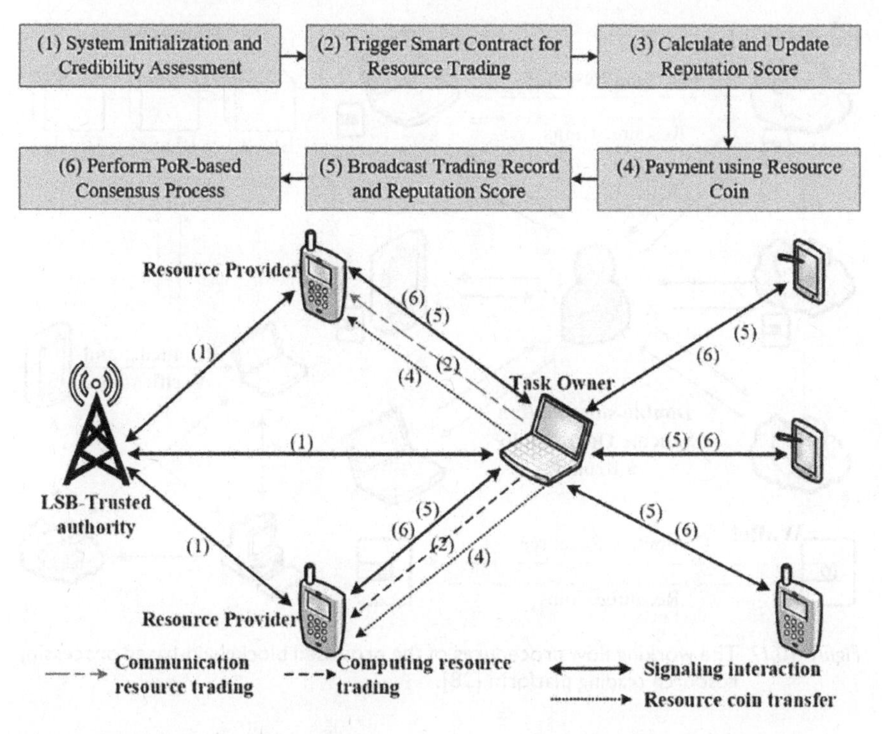

Figure 10.13 The working procedures of the resource exchanging arrangement dependent on blockchain [29].

10.4.3 Security concerns

The blockchain-based ITS proposed in this article outlines a thorough arrangement as follows: (a) IoT gadgets incorporated by IoV hubs, conveyed processor hubs, and correspondence intermedia; (b) blockchain-based framework sent on BaaS to serve the information and installment exchanges; and (c) ITS itself as a product to help the entire arrangement. Consequently, in this part, the perspectives on these three viewpoints to feature the issues hidden in this proposed framework are clarified. Moreover, it likewise put eyes on information security and protection as a key necessity for an online framework.

10.4.3.1 Infrastructure concerns

As the frameworks proposed above, it is referenced that the potential issues could occur on the IoT framework (Figure 10.14).

Equipment obliteration. RSUs and ISUs are sent out and about, which implies it causes a potential danger that hubs could be annihilated or hacked.

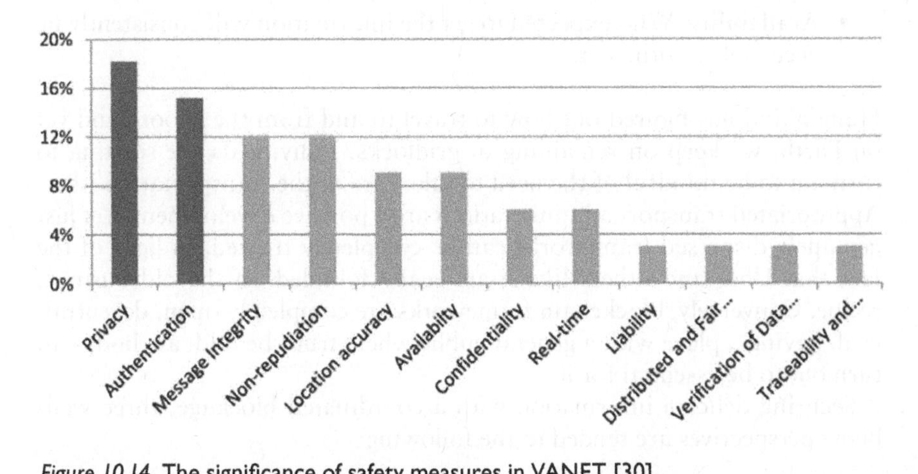

Figure 10.14 The significance of safety measures in VANET [30].

In the design proposed over, the component to evade such equipment annihilation is to multi-send such spotted hubs as an appropriated network with the goal that the information through the framework could be adjusted in agreement technique which diminishes the chance of single-point assault issue. Organization assault. There are the following two normal organization assault conditions:

- **External network attack:** A distributed denial-of-service attack, which endeavors to send malevolent solicitations that began from an assortment of endpoints to upset ordinary help organization, is one of the common assaults that could typically occur in a conventional concentrated framework [31,32].
- **Internal data flood:** As a real-time framework hyper coordinated over the traffic situation, it is profoundly conceivable to confront the tension on the exchanges between hubs through the framework inside, which is considered as an inward information flood issue [33].

10.4.3.2 Sensitive data protection

In a keen rush hour gridlock setting, information is touchy, even the speed of vehicles, either by and large planning the traffic condition in a full-scale view or giving responses from the explicit significant vehicle in a miniature view [34]. With blockchain coordinated, it might raise three parts of safety as follows [32]:

- **Confidentiality:** Information to be moved must be perused by explicit clients, relegated by keen agreement.
- **Integrity:** Any information exchange will be recorded in the record so the individuals who get and uncover the information will be recorded.

- **Availability:** Who expected to get the information will consistently be accessible to utilize it.

Humankind has figured out how to travel to and from the Moon, and yet on Earth, we keep on remaining in gridlocks. Individuals are starting to turn out to be mindful of the need to take care of the transportation issue. Appropriated transport administrations are a positive development. Yet just genuinely dispersed frameworks can be completely trusted, in light of the fact that Uber and others like it are really founded on the old business recipe. Conversely, blockchain frameworks are completely open, decentralized, having a place with a general public where truth be told, anybody can turn out to be essential for it.

Securing delicate information with a coordinated blockage, three well-being perspectives are tended to the following:

- **Confidentiality:** The information to be moved must be perused by specific clients assigned by a "savvy contract".
- **Honesty:** Any information exchange will be recorded in a register, which gets and uncovers the information recorded in it.

10.5 CONCLUSION

In this chapter, we focused on the well-recognized proposed works and models, as well as research by various researchers, engineers, and scientists on ITS, mainly based on combining it with the blockchain technology to see what are the applications, characteristics, advantages, disadvantages, and security concerns raised by these well-respected people.

Most of the research papers and works we have gone through are practical proposed models with their features, applications, security concerns but still, they are not completely working models, an extension of their work is to be picked up by us. We understood that blockchain technology is much the importance is because it's an amazing chronological ordering of data, duplicate permanent records of actual data in a proper format, and its highly secure nature over the sever. As one of the greatest advantages of blockchain technology is being able to create a chronological order of the stored information and moreover this arrangement is permanent forever, and with its secure properties including, all of these characteristics make this technology so unique and so rich. Blockchain technologies utilize one of the very versatile technologies, IoT, and its tools and devices for its constructions and development. Therefore, this will enable us to achieve fully IoT-based ITS utilizing blockchain, which will result in highly advanced ITS like VANET.

Blockchain technology tied with the transportation area are in a beginning stage of improvement and keeping in mind that numerous creators advocate the utilization of this innovation, there are additionally a few worries about practical models that are being proposed is that they are not entirely functioning and results of these projects haven't been up to the mark as well. Until we will have some really accurate and properly functioning model with amazing results, much work is yet to be done and research should be concluded. As we have concluded, right now there is a big passion among organizations and industries for creating ITS all over the globe. In any case, much exertion is as yet required, primarily with respect to transportation designing analysts, so this promising innovation can arrive at its last level of development.

Later on, with the fast advancement of man-made reasoning, and the coming development of remote correspondence on 5G, the savvy traffic is forming into a thorough arrangement with huge size of IoT gadgets being associated and spoken with one another that the traffic condition could be anticipated and front led. Moreover, a more clever framework will be presented as the upheaval of things to come to the smart world.

REFERENCES

1. S. Nakamoto Bitcoin: A peer-to-peer electronic cash system. Available at: https://bitcoin.org/, (2008).
2. M. Friedlmaier, A. Tumasjan, I.M. Welpe. Disrupting industries with blockchain: The industry, venture capital funding, and regional distribution of blockchain ventures. In *Proceedings of the 51st Hawaii International Conference on System Sciences*, Waikoloa Village, HI, (3–6 January 2018).
3. N. Hackius, M. Petersen, Blockchain in logistics and supply chain: Trick or treat? In *Proceedings of the Hamburg International Conference of Logistics (HICL)*, Hamburg, Germany, pp. 3–18 (12–13 October 2017).
4. B. Dickson, Blockchain has the potential to revolutionize the supply chain. Available at: https:// techcrunch.com/2016/11/24/blockchain-has-the-potential-to-revolutionize-the-supply-chain (accessed on 28 November 2019).
5. M.J. Casey, P. Wong, Global supply chains are about to get better, thanks to blockchain. Available at: https://hbr.org/2017/03/global-supply-chains-are-about-to-get-better-thanks-to-blockchain (accessed on 2 December 2019).
6. C. Carter, L. Koh, *Blockchain Disruption in Transport: Are You Decentralized Yet?* Catapult Transport, Systems: Milton Keynes, UK, (2018) Available at: https://trid.trb.org/view/1527923.
7. ZigNuts, How blockchain architecture works? Basic understanding of blockchain and its architecture. Blogs, Cryptocurrency, Blockchain, Web Development, Available at: https://www.zignuts.com/blogs/how-blockchain-architecture-works-basic-understanding-of-blockchain-and-its-architecture/ (18 March 2021).
8. H. Atlam, A. Alenezi, M. Alassafi, G. Wills, Blockchain with internet of things: Benefits, challenges and future directions. *International Journal of Intelligent Systems and Applications*, (2018) Doi: 10. 10.5815/ijisa.2018.06.05.

9. A. Mullaney, Industries where blockchain is having an impact. Blog Available at: https://blog.cutter.com/2017/10/31/industries-where-blockchain-is-having-an-impact/ (31 October 2017).
10. Y. Pei, S. Biswas, D. Fussell, K. Pingali, Kalman filtering without Bayesians and Gaussians. Available at: https://www.sigarch.org/kalman-filtering-without-bayesians-and-gaussians/ (12 December 2017).
11. S. Baras, I. Saeed, H.A. Tabaza, M. Elhadef, VANETs-based intelligent transportation systems: An overview. *Advances in Computer Science and Ubiquitous Computing*, 474, ISBN :978-981-10-7604-6 (2018).
12. M.B. Mollah, J. Zhao, D. Niyato, Y.L. Guan, C. Yuen, S. Sun, K.-Y. Lam, L.H. Koh, Blockchain for the internet of vehicles towards intelligent transportation systems: A survey. *IEEE Internet of Things*, (2020) Doi: 10.1109/JIOT.2020.3028368.
13. Y. Yuan, F.-Y. Wang, Towards Blockchain-based Intelligent Transportation Systems. In *19th IEEE International Conference on Intelligent Transportation Systems (ITSC2016)*, (2016) Doi: 10.1109/ITSC.2016.7795984.
14. Q.-J. Kong, L.-F. Li, B. Yan, S. Lin, F.-H. Zhu, G. Xiong, UTN-model-based traffic flow prediction for parallel-transportation management systems. *IEEE Intelligent Systems*, (2013).
15. L. Eremina, A. Mamoiko, L. Bingzhang, Use of blockchain technology in planning and management of transport systems. In *E3S Web of Conferences 157*, 04014, (2020) Doi: 10.1051/e3sconf/202015704014.
16. T.V. Lakshman, A.K. Agrawala, IEEE transactions on software engineering, SE-12 600-607 (1986).
17. F. Kitahara, K. Kera, K. Bekki, Autonomous decentralized traffic management system. In *Proceedings of International Workshop on Autonomous Decentralized Systems*, (2000) Doi:10.1109/IWADS.2000.880891.
18. K. Mori, Autonomous decentralized systems technologies and their application to a train transport operation system. In: Winter V.L., Bhattacharya S. (eds) *High Integrity Software. The Kluwer International Series in Engineering and Computer Science*, vol 577. Springer, Boston, MA, (2001). Doi: 10.1007/978-1-4615-1391-9_5.
19. F.-Y. Wang, Artificial societies, computational experiments, and parallel systems: A discussion on computational theory of complex social-economic systems. *Complex Systems and Complexity Science*, 1(4), 25–35 (2004).
20. Y.-S Lv, Y. Ou, S.-M. Tang, F.-H. Zhu, H.-X. Zhao, *Journal of Jilin University* (Engineering and Technology Edition), 39, 87–90 (2009).
21. H.M. Kim, M. Laskowski, Toward an ontology-driven blockchain design for supply-chain provenance. *Intelligent Systems in Accounting, Finance and Management*, 25, 18–27, (2018) Doi: 10.1002/isaf.1424.
22. H.R. Morley, Industry skeptical of pace of logistics tech adoption. *International Logistics*, (2017) Available at: https://www.joc.com/international-logistics/logistics-providers/industry-skeptical-pace-logistics-tech-adoption_20170620.html.
23. W. Lehmacher *Why Blockchain Should be Global Trade's Next Port of Call*, World Economic Forum: Geneva, (2017) Available at: https://www.weforum.org/agenda/2017/05/blockchain-ports-global-trades/.

24. J. Kang, R. Yu, X. Huang, M. Wu, S. Maharjan, S. Xie, Y. Zhang, Blockchain for secure and efficient data sharing in vehicular edge computing and networks. *IEEE Internet of Things Journal*, (2018) Doi: 10.1109/JIOT.2018.2875542.
25. K. Shi, L. Zhu, C. Zhang, L. Xu, F. Gao, Blockchain-based multimedia sharing in vehicular social networks with privacy protection. *Multimedia Tools and Applications*, 1–21, (2020).
26. Y. Chen, X. Hao, W. Ren, Y. Ren, Traceable and authenticated key negotiations via blockchain for vehicular communications. *Mobile Information Systems*, (2019). Doi: 10.1155/2019/5627497.
27. C. Chen, J. Wu, H. Lin, W. Chen, Z. Zheng, A secure and efficient blockchain-based data trading approach for internet of vehicles. *IEEE Transactions on Vehicular Technology*, 68(9), 9110–9121, (2019).
28. Z. Li, Z. Yang, S. Xie, Computing resource trading for edge cloud-assisted internet of things. *IEEE Transactions on Industrial Informatics*, (2019) Doi: 10.1109/tii.2019.2897364.
29. G. Qiao, S. Leng, H. Chai, A. Asadi, Y. Zhang, Blockchain empowered resource trading in mobile edge computing and networks. In *ICC 2019–2019 IEEE International Conference on Communications (ICC)*, IEEE, pp. 1–6 (2019).
30. S. Goudarzi, H. Abdullah, M. Satria, M. Baee, H. Anisi, A. Muhammad A systematic review of security in vehicular ad hoc network. In *2nd Symposium on Wireless Sensors and Cellular Networks (WSCN'13)* (2013).
31. M. Ali, J. Nelson, R. Shea, Blockstack: A global naming and storage system secured by blockchains. In: *USENIX Annual Technical Conference*, Denver, CO, Berkeley, CA: USENIX Association (22–24 June 2016).
32. A. Dorri, S.S. Kanhere, R. Jurdak, Blockchain for IoT security and privacy: The case study of a smart home. In: *Proceedings of the 2017 IEEE International Conference on Pervasive Computing and Communications Workshops (PerCom Workshops)*, Kona, HI, New York: IEEE (13–17 March 2017).
33. Q. Ren, K.L. Man, M. Li, Using blockchain to enhance and optimize IoT-based intelligent traffic system. In: *Proceedings of the 2019 International Conference on Platform Technology and Service (Platcon)*, Jeju, South Korea, New York: IEEE (28–30 January 2019).
34. G. Zyskind, and O. Nathan, Decentralizing privacy: Using blockchain to protect personal data. In: *Proceedings of the 2015 IEEE Security and Privacy Workshops (SPW)*, San Jose, CA, New York: IEEE (21–22 May 2015).

Chapter 11

Optical fiber communication and sensitivity

Salma Masuda Binta
Bangladesh Army University of Science and Technology

Mohammed Hossam-E Haider
Military Institute of Science and Technology

Mohammad Farhan Ferdous
Japan–Bangladesh Robotics and Advance Technology Research Center (JBRATRC)

CONTENTS

DOI: 10.1201/9781003156789-11

NOMENCLATURE

PCF	Photonic crystal fiber
SPR	Surface plasmon resonance
FBG	Fiber Bragg Grating
MZI	Mach–Zehnder Interferometer
TIR	Total internal reflection
ARROW	Anti-resonant Reflecting Optical Waveguide
RI	Refractive index
PBG	Photonic bandgap
MOFs	Metal–organic Frameworks
PXRD	Powder X-ray Diffraction
TGA	Thermogravimetric Analysis
FT-IR	Fourier Transform Infrared
PDE	Partial differential equation
PDEs	Partial Differential Equations
CPM	Cross-Phase Modulation
FWM	Four-Wave Mixing
GaAS	Gallium Arsenide
InGaAs	Indium Gallium Arsenide
SBS	Stimulated Brillouin-Scattering
SRS	Stimulated Raman-Scattering
FEM	Finite element method
FEA	Finite element analysis
TE	Transverse electric

11.1 PART I: INTRODUCTION

11.1.1 Introduction

Photonic crystal fibers (PCFs) with a periodic arrangement of microscopic air holes are used in sensing physical phenomena such as temperature, strain, refractive index (RI), pressure, and so on [1]. Over the last few years, these PCF sensors have drawn significant interest and attention due to their high sensitivity, lightweight, isolation with an electrical signal, and great design flexibility compared with conventional optical fiber. Because of the flexible design, PCF is filled with temperature-sensitive to realize the PCF temperature sensor. Many researchers have worked on the design or fabrication of PCF temperature sensors. In recent years, the sensitivity of the sensor is improved by combining PCF with surface plasmon resonance (SPR) technology. SPR is a phenomenon where an electromagnetic mode is generated on the metal surface due to the interaction between free electrons of metal and optical electromagnetic fields. Plasmon has a significant role in designing PCF-based temperature sensors. Usually, silver, gold, graphite, aluminum, and copper are used as plasmonic materials. However, silver as a plasmonic material has an oxidization problem that reduces the sensing range and performance. On the other hand, gold is chemically stable. The use of gold as a plasmonic material for designing temperature sensors is preferable for better performance. Although there are different types of optical fiber sensors such as the fiber Bragg grating and Mach–Zehnder interferometer, the recent trend indicates that researchers have shown significant interest to design PCF temperature sensors based on SPR.

11.1.2 Nanofibers

Optical nanofibers, conjointly referred to as photonic nanowires, area unit optical fibers with diameters within the variation from tens to a number of nanometers. This implies that the diameter is usually well below the optical wavelength. The choice term sub-wavelength fibers emphasize this necessary fact. Such nanowires will have peculiar mechanical and optical properties. Due to the massive ratio distinction between fiber and air, the numerical aperture is extremely high, and therefore the effective mode space is extremely tiny. For precise calculations of the mode properties, a full vectorial model area unit is needed, because the paraxial approximation isn't consummated. Silica nanowires have associated exceptional mechanical strength, giving bending with radii of a number of micrometers. The high numerical aperture keeps the bend losses low even for such tight bending. Tightly wound fibers are often used for miniature fiber resonators. The light that is guided in nanofibers will expertise robust nonlinearities because of the little effective mode space and is related to important impermanent fields simply outside the fiber surface. For fiber diameters below $\approx 0.6\,\mu m$ (in the case of silicon dioxide fibers), the mode radius of guided lightweight will increase because the fiber diameter is an additional weak-end, which is a result that the "guiding power" of an agent fiber becomes weaker. Most of the optical power then propagates within the impermanent field outside the fiber.11.

11.1.3 Silicon nanowire embedded circular PCFs for temperature sensing

Using the finite element method (FEM), a chemical element nanowire embedded circular PCFs. Silicon nanowires are often used as optical sensing elements moreover as a temperature sensor. The application of chemical element nanowire (SiNW) as a sensing nanomaterial for the detection of biological and chemical species has gained attention thanks to its distinctive properties [2]. With the quick growth and development of advanced engineering, several sensing nanomaterials with distinctive properties, desired size, and chemical compositions are invented to be incorporated among the electrical device. One of them is that the application of one-dimensional (1D) properties, desired size, and chemical compositions are invented to be incorporated among the electrical device. One in all them is that the application of 1D nanostructures (nanotubes, nanowires, nanorods, nanobelts, and heteronanowires) among the transducers in previous studies which will enhance the sensing element performance, as an example, nanowires [3], carbon nanotubes [4], CuS nanowires [5], NiO-Au nanobelts [6], CuS nanotubes and graphene oxide-modified metal nanoribbons [7]. Silicon nanowire is one in all the 1D nanostructures and has emerged because of the promising sensing nanomaterial upon its distinctive mechanical, electrical, and optical properties. The most reason why SiNWs have attracted attention within the development of ultrasensitive sensors is thanks to their high surface to volume ratios therefore greatly enhancing the detection limit to concentration and giving high sensitivity. Additionally, the dimension of SiNW is within the vary of 1 100 nm, therefore creating it comparable and compatible to the dimensional scale of biological and chemical species. Having the smallest dimension, SiNWs exhibited sensible lepton transfer in detection as a result of the buildup of charge in SiNWs directly happens among the majority of fabric leading to the quick response of detection.

11.1.4 Photonic crystal fibers

Based on the properties of photonic crystals, PCF is a class of optical fibers. PCFs have claddings that contain tiny air holes in a pure silica background. Such a cladding concept provides flexibility to design a wide range of index contrast between the core and the cladding. In order to study the coupling between guided modes in multicore PCFs and verify the predictions of the effective index model for PCF, we have fabricated a PCF with two closely coupled cores. Dual-core PCF temperature sensor in terms of wavelength sensitivity. The high-temperature coefficient liquid and plasmonic material are deposited in the outer layer of the PCF to make the fabrication easier than existed optical sensor. PCFs with a periodic arrangement of microscopic air holes are used in sensing physical phenomena such as temperature, strain, RI, pressure, and so on. There is an increasing interest in studying such PCFs, and many research groups around the world are demonstrating various kinds of PCFs with regular or irregular structures, with solid or hollow cores using various materials such as polymer or silicon. The PCFs show very interesting characters such as a wide wavelength range which can be utilized to transmit ultrashort pulses, zero-dispersion wavelength in the 800 nm region suitable for nonlinear applications. For example, supercontinuum generation and extraordinary dispersion properties at visible and near-infrared wavelengths. An additional important characteristic of PCFs is their strong birefringence induced by the size and geometrical arrangement of air holes. As a result of useful properties, growing interest is being shown in PCFs for a range of applications in optical signal processing, sensing, and optical communication systems. Traditional optical fibers, whose RI of the core is higher than that of the cladding, confine the light field by the total internal reflection (TIR) (Figures 11.1 and 11.2).

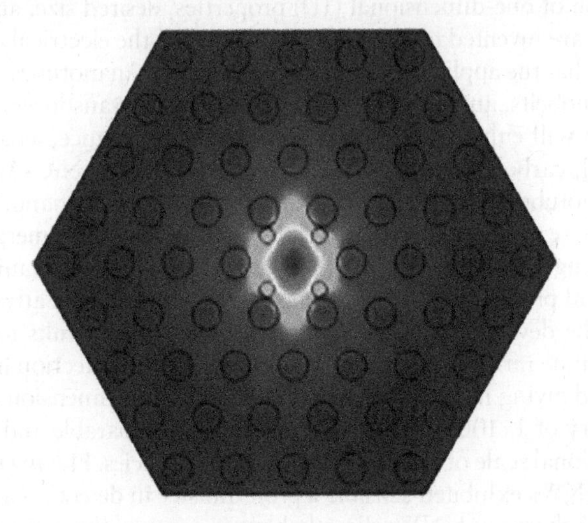

Figure 11.1 Hexagonal PCF.

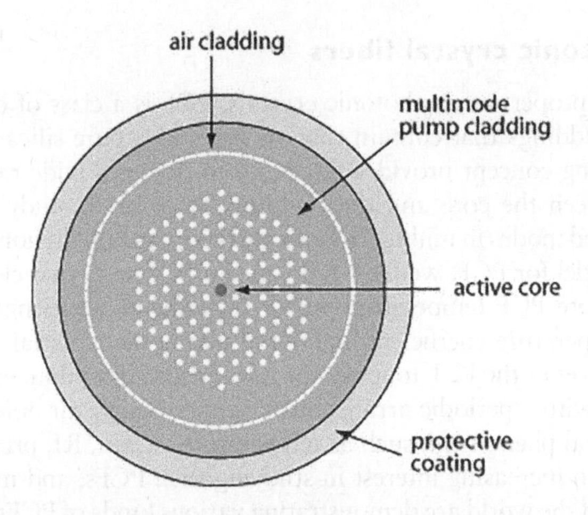

Figure 11.2 **Structure of a PCF with an air cladding.**

PCFs or holey fibers guide the light via one of the two mechanisms: effective-index guidance and photonic bandgap (PBG) guidance. In the effective-index PCFs, the effective RI of the composite material is lower than RI of the higher-index material, which makes it possible to trap the light in a core region formed from the higher-index material by a form of TIR. In other words, in these PCFs, the light is guided based on TIR between the core and the cladding region 6. For example, an air-silica micro structured material has an effective RI lower than that of pure silica. Therefore, propagation of the light in the silica core region is internally reflected from the photonic crystal cladding region.

11.1.5 Optical properties of PCFs

Since the first experimental demonstration of a PCF in 1996 by Knight et al., the optical properties and the fabrication of such fibers have attracted significant attention. The fiber structure with a lattice of air holes running along the length of the fiber provides a large variety of novel optical properties and improvements compared to standard optical fibers. PCF consists of air hole rings encompassing the core and these air hole rings running throughout the length of the fiber. It has several advantages over common step-index fiber; usually, it consists of coalesced silicon oxide glass, and therefore there is no complexity in fabric; it is also resistant against high electrical and force fields, and dangerous atmosphere. Also, this has some excellent optical properties such as high birefringence, low confinement loss, endlessly single mode, and flat dispersion nature. PCF is named either by its cladding air hole structure such as square lattice PCF, hexagonal PCF, octagonal PCF, and decagonal PCF or by its core nature such as solid-core PCF and hollow-core PCF. Highly birefringent DPCF can be used in telecommunication, polarization-maintaining fiber, sensor, etc. In all-solid PCFs consisting of high-index rods in a low-index background, the PBG effect can be more easily understood from the anti-resonant reflecting optical waveguide model. Most PCFs, however, are based on modified TIR, similar to that in conventional optical fibers, because RI of the core is higher than the effective RI of the cladding. More recently, inhibited coupling between the core and cladding modes has been identified as the principle of Kagome-lattice hollow-core PCFs. Second, the top of the LP 01 band falls slightly below the low-index material line in the very low-frequency region. This region is commonly referred to as the TIR-guiding region because a defect composed of the same material as the background would have a higher RI than that of the cladding. The optical properties of PCFs fabricated from pure silica are determined by the position, geometry, and size of the air holes. A commonly accepted classification of PCFs divides the fibers into two main classes: (a) Index-guiding PCFs and (b) PBG guiding fibers. In an index-guiding PCF RI of the core is higher than the effective RI of the cladding and the fiber operates by the principle of TIR. PBG fibers require a lattice of holes with a PBG at the frequency, at which the fiber is intended to transmit light. Due to the different nature of the guiding mechanisms, the two classes of PCFs are treated in individual sections.

11.1.6 Advantages of PCFs

- Many physical properties can be engineered (power fraction, birefringence, chromatic dispersion).
- The waveguide dispersion can be engineered to have the zero-dispersion wavelength at any desired wavelength. This is useful for nonlinear applications, where normal dispersion is a limiting factor.

- By changing the core diameter of the fiber, the zero-dispersion wavelength can be shifted to the visible range.
- PCF can be filled with gases or liquids for sensing.

11.1.7 Application of PCFs

Their special properties make PCFs very attractive for a very wide range of applications. Some examples are as follows:

- Fiber lasers and amplifiers, including high-power devices, mode-locked fiber lasers, etc.
- Nonlinear devices for supercontinuum generation, Raman conversion, parametric amplification, or pulse compression
- Telecom components for dispersion control, filtering, or switching
- Fiber-optic sensors of various kinds
- Quantum optics, e.g., generation of correlated photon pairs, electromagnetically induced transparency, or guidance of cold atoms

Even though PCFs have been around for several years, the huge range of possible applications is far from being fully explored. It is to be expected that this field will stay very lively for many years and many opportunities for further creative work, concerning both fiber designs and applications. Optical fibers with high birefringence can find important applications in optical fiber communications, fiber filters, fiber sensors, fiber lasers, and so on (Figure 11.3).

Photonic crystals can reflect light very efficiently.

- Creates an allowed photon state in PBG.
- Can be used as a cavity in lasers.

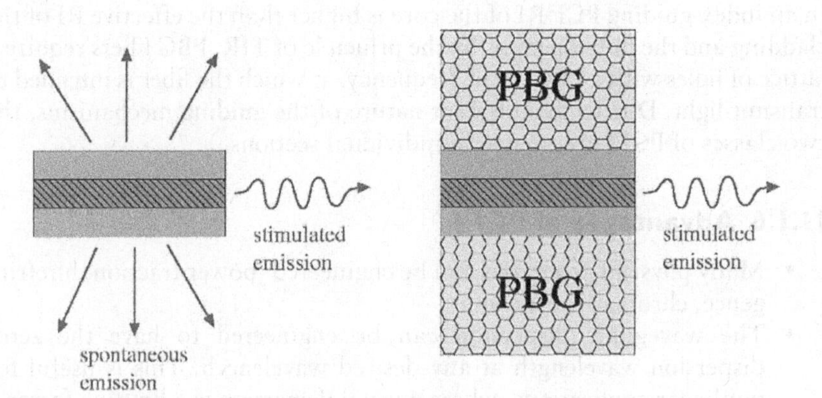

Figure 11.3 Suppression of spontaneous emission.

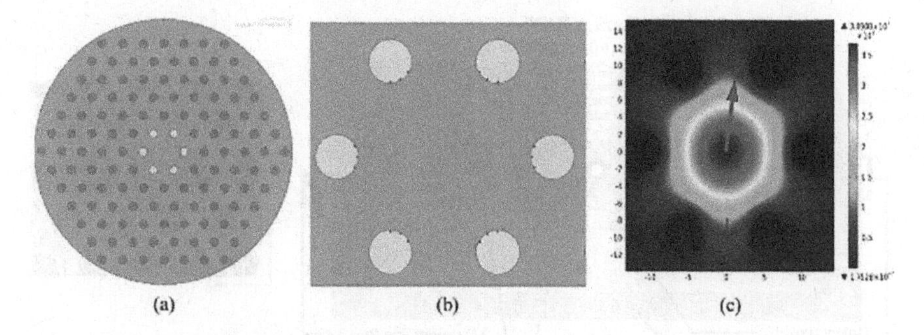

Figure 11.4 (a) Cross-section of the produced PCF-based SPR temperature sensor. (b) Cross-section of analyte channel filled with nanowires. (c) Electric field distribution of the fundamental mode.

11.1.7.1 PCFs as temperature sensor

Sensors-supported PCFs with distinctive properties and style flexibility are expected to beat the difficulties and bit by bit turning into a brand-new analysis focus. PCF-based SPR sensors especially gather continuous analysis interests in medical medicine, environmental management, drug management, and food safety management that is due to the characteristics of low-price fabrication, straightforward mensuration system, and capability of remote sensing. PCF-based SPR sensing will be accomplished once the part matching condition is met between the exciting lightweight and also the surface plasmons. Within the past years, PCF-based SPR sensors for exploitation in an index of refraction sensing of liquid settings have promoted the development of fiber optic sensors (Figure 11.4).

With the development of the sensors, the propagation modes of PCF are highly temperature-dependent. PCF-based SPR sensing has been involved in temperature sensors. The sensors can realize temperature sensing by adding some liquid with a high thermo-optic coefficient into the air holes.

11.1.7.2 Photonic crystal lasers

Incorporation of 2D photonic crystal with light-emitting semiconductor quantum well provides confinement and gain necessary for lasing (Figure 11.5).

11.1.7.3 Superprism effect

The light path shows an extremely wide swing with a slight change of incident light angle (Figures 11.6 and 11.7).

- Based on highly anisotropic dispersion by a photonic band (negative refraction)
- Wavelength demultiplexer

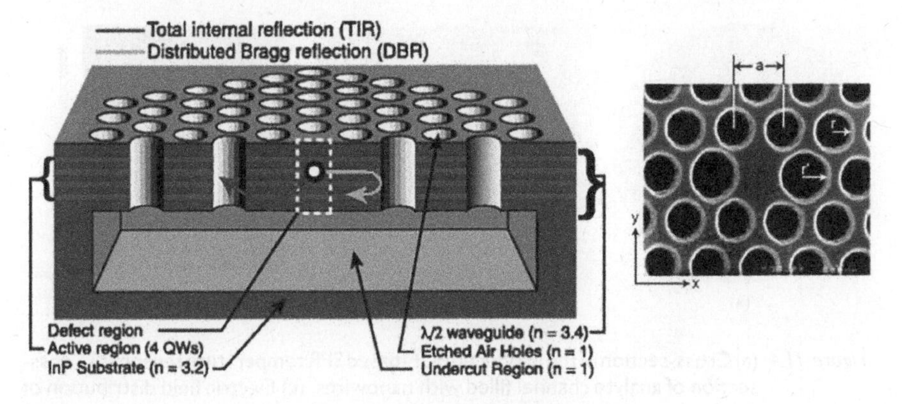

Figure 11.5 2D photonic crystal with light-emitting semiconductor.

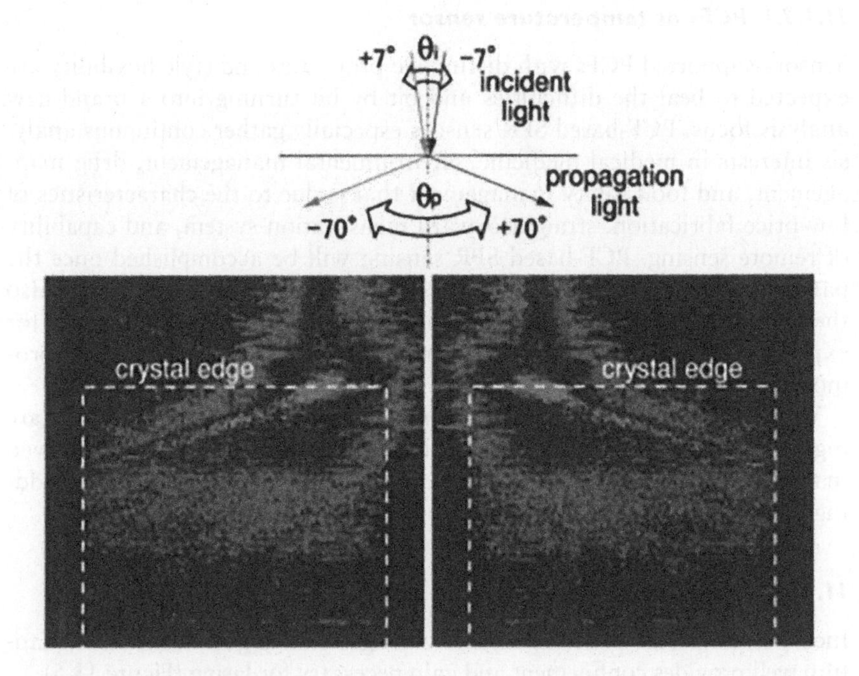

Figure 11.6 Superprism effect.

11.1.7.4 Artificial opals

- Chemical synthesis uses chemical vapor deposition and wet-etching to form air spheres surrounded by silicon shells
- Easier to achieve smaller dimensions (Figure 11.8)

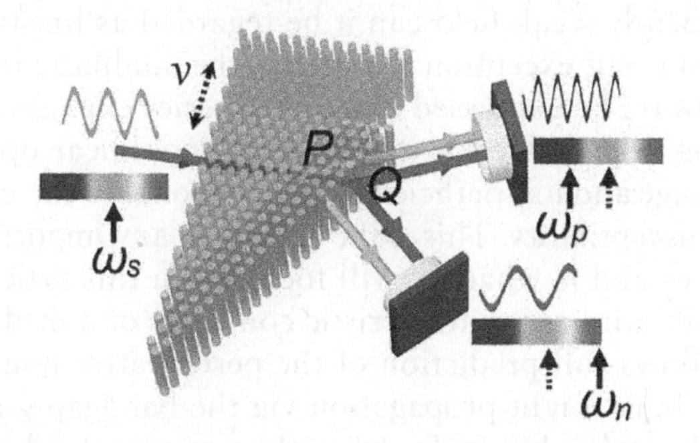

Figure 11.7 Sketch map of light traveling through the moving PhC prism. The detecting frequencies of outgoing lights vary in two different ways.

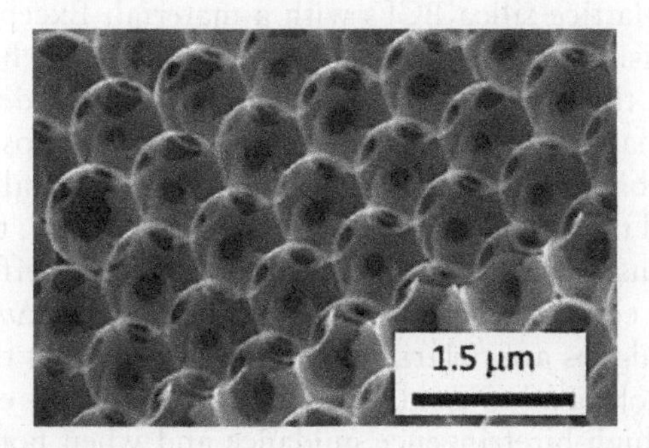

Figure 11.8 Artificial opals.

11.1.8 Propagation of light along PCFs

Using FEM, the propagation of light along the hollow-core PCFs filled with low-temperature glow discharge plasma. RI of glow discharge plasmas is a function of plasma parameters and incident wavelength. In weakly ionized helium, neon and He–Ne plasmas with low electron density and low electron temperature, the rise in the electron density will increase the plasma oscillation frequency, while the rise in the electron temperature and helium content will increase the collision frequency. In the end, the increase of the wavelength, the plasma oscillation, and the collision frequency will cause an increase in the imaginary part of RI. With the air holes occupied by plasmas, the optical field distribution is simulated in PBG fibers for two different wavelengths. Light propagation in a medium induces a medium response via the interaction of light with electrons and nuclei in the medium. The medium response, in turn, affects the light propagation and brings about various optical effects. Generally, the response of a medium to an applied field is nonlinear. Only in the case of a sufficiently weak field can it be regarded as linear. However, if the applied field is not exceptionally strong, the nonlinear response of the medium can always be expressed as a power series expansion of the field. The *n*th term is responsible for the *n*th-order nonlinear optical effect in the usual language and its coefficient is proportional to the *n*th-order nonlinear optical susceptibility. This is the case for many important nonlinear optical processes and is what we will focus on in this article. Nonlinear optical susceptibilities are characteristic constants of a medium and their specification allows full prediction of the perturbative nonlinear optical effects in a medium. Light propagation via the bandgap-guiding mechanism can be achieved in LC-PCFs, when the periodic cladding leads to the formation of bandgap windows, within which light can be confined in the central defect core. The most direct approach which has been extensively studied both theoretically and experimentally in the literature is to infiltrate triangular-lattice silica PCFs with a material. Except for rare cases regarding the visible spectrum (Wolinski et al. 2007), both LC indices are higher than that of silica. The presence of high-index inclusions in the silica PCF matrix leads to the formation of bandgaps. In the case of LC-PCFs, these bandgaps depend not only on the material but also on the polarization and the LC orientation. In the extreme case, the two orthogonal polarizations sense a high-index cladding with an effective capillary index difference equal to the LC material birefringence $\Delta n = n_e - n_0$.

Thus, the bandgaps are polarization-dependent so that the LC-PCF can support single-polarization guidance when the bandgap exists for a sole polarization or high-birefringence guidance and when both polarizations are guided within their corresponding bandgaps (Ren, Shum, Hu, Yu, & Gong, 2008).

Figure 11.9a shows the effective modal indices for the two orthogonal polarizations and the corresponding modal birefringence Δn_{eff}, assuming a perfect alignment of the LC molecules along the *e*-axis. The bandgaps for each polarization define the spectral windows of high transmittance. When these overlaps, both polarizations are guided, exhibiting very high values of Δn_{eff}. As one approaches the bandgap edges, the confinement losses and the effective modal area become higher, as shown in Figure 11.9b. It is stressed that the results presented in Figure 11.9 are calculated for the extreme case of perfect alignment of the LC molecules along with one of the orthogonal axes, that is, in the limit of infinite voltage. By adjusting the control voltage, it is possible to shift the bandgaps in an intermediate state for *e*-polarized light and thus tune both the value of Δn_{eff} and the propagation losses.

11.1.9 Spectral sensitivity

Spectral sensitivity is the relative efficiency of detection of light or another signal as a function of the frequency or wavelength of the signal.

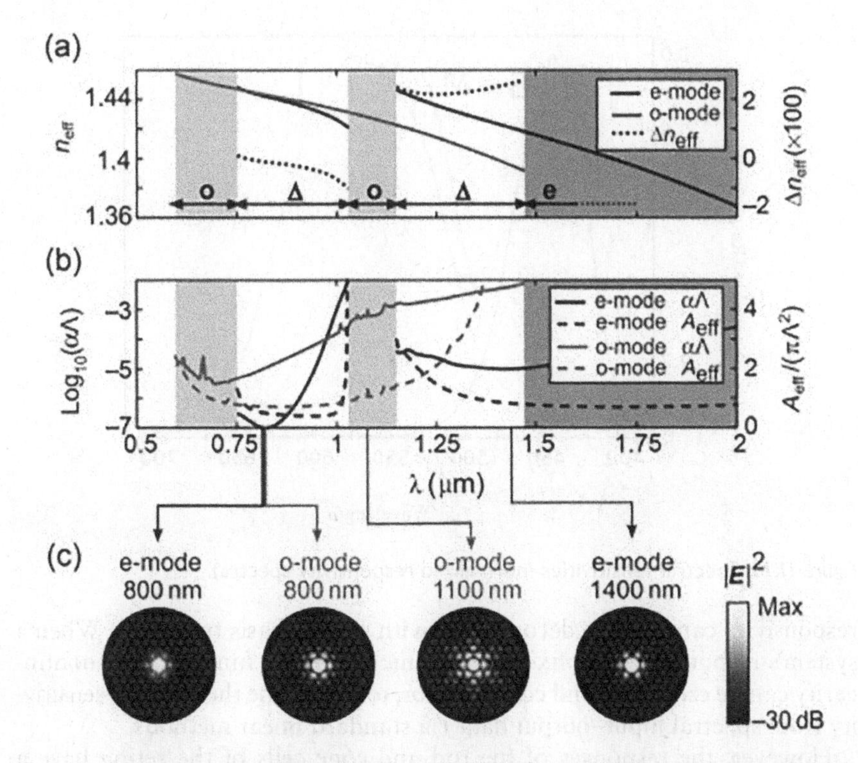

Figure 11.9 (a) Modal effective indices and birefringence for the bandgap-guiding SiO_2/E7 LC-PCF studied. Shaded regions indicate single-polarization spectral windows. (b) Normalized confinement losses and modal effective area. (c) Optical power profiles for guided modes of *o*- and e-polarization calculated at different operational wavelengths.

In visual neuroscience, spectral sensitivity is used to describe the different characteristics of the photopigments in the rod cells and cone cells in the retina of the eye. It is known that the rod cells are more suited to scotopic vision and cone cells to photopic vision and that they differ in their sensitivity to different wavelengths of light. It has been established that the maximum spectral sensitivity of the human eye under daylight conditions is at a wavelength of 555 nm, while at night the peak shifts to 507 nm.

In photography, film and sensors are often described in terms of their spectral sensitivity, to supplement their characteristic curves that describe their responsivity. A database of camera spectral sensitivity is created and its space analyzed. For X-ray films, the spectral sensitivity is chosen to be appropriate to the phosphors that respond to X-rays, rather than being related to human vision.

In sensor systems, where the output is easily quantified, the responsivity can be extended to be wavelength dependent, incorporating spectral sensitivity. When the sensor system is linear, its spectral sensitivity and spectral

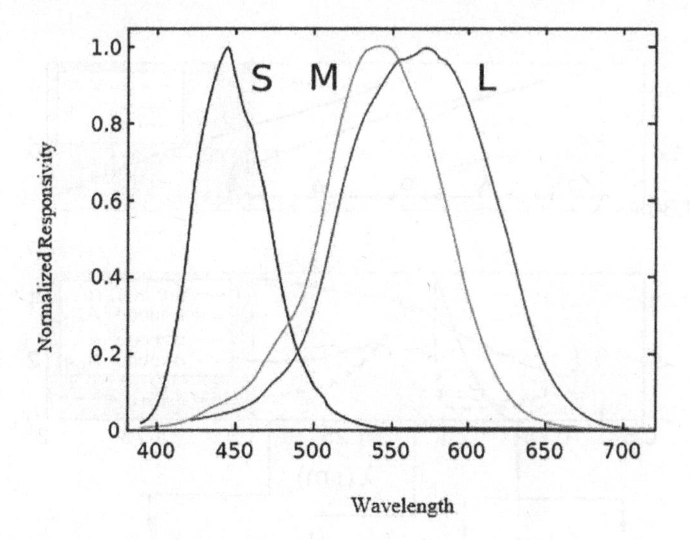

Figure 11.10 Spectral sensitivities (normalized responsivity spectra).

responsivity can both be decomposed with similar basis functions. When a system's responsivity is a fixed monotonic nonlinear function that nonlinearity can be estimated and corrected for, to determine the spectral sensitivity from spectral input–output data via standard linear methods.

However, the responses of the rod and cone cells of the retina have a very context-dependent (coupled) nonlinear response, which complicates the analysis of their spectral sensitivities from experimental data. However, despite these complexities the conversion of light energy spectra to the effective stimulus, the excitation of the photopigment, is quite linear, and linear characterizations such as spectral sensitivity are therefore quite useful in describing many properties of color vision.

Spectral sensitivity is sometimes expressed as quantum efficiency, that is, as the probability of getting a quantum reaction, such as a captured electron, to a quantum of light, as a function of wavelength. In other contexts, the spectral sensitivity is expressed as the relative response per light energy, rather than per quantum, normalized to a peak value of 1, and quantum efficiency is used to calibrate the sensitivity at that peak wavelength. In some linear applications, the spectral sensitivity may be expressed as a spectral responsivity, with units such as amperes per watt (Figure 11.10).

11.1.10 Physiological temperature sensing

Thermal sensing and imaging in the physiological temperature range are of great importance for studying physiological processes and treating diseases. Metal–organic frameworks exhibit great promise for developing luminescent thermometers due to their remarkable structural diversities and tunable luminescence properties. Here, we synthesized a series of luminescent mixed-lanthanide metal–organic frameworks, Eu*x*Tb1-*x*BPT (*x*=0.019, 0.058, 0.106; H3BPT=biphenyl-3,4,5-tricarboxylate acid) and adopted powder X-ray diffraction, thermogravimetric analysis, and Fourier transform infrared to characterize the resulting products. The temperature-dependent photoluminescence mission spectra were recorded to investigate their potential applications in physiological temperature readout. It is found that the intensity ratio of Tb3+ to Eu3+ is linearly correlated with temperature and the relative sensitivity is higher than 1.5%/°C over the entire physiological temperature range. Furthermore, the temperature-dependent luminescence color emission allows for visual colorimetric temperature measurements. Luminescence lifetime testing and triplet energy level measurement were further conducted to study the mechanism.

11.1.11 Surface plasmon resonance (SPR)

SPR is a coupled state of photons and collective oscillating electrons which is highly sensitive to variations in RI of the surrounding dielectrics. It offers miniaturization, a high degree of integration, and remote sensing capabilities. Based on those features, a chemical sensor based on SPR within optical fiber was first demonstrated in 1993. Since that time, driven by the need for further miniaturization of SPR-based sensors, various optical fiber support SPR sensors have been investigated. Sharma and Gupta designed a fiber-optic temperature sensor based on SPR with gold as the metallic layer and TiO_2 as the sensing layer. The dependence of the performances of the sensor on different properties of the optical fiber metallic layers, sensing regions, and the incident waves are used in the fabrication of metal-covered hollow waveguides.

11.1.12 Literature review

Over the last few years, these PCF sensors have drawn significant interest and attention due to their high sensitivity, lightweight, isolation with an electrical signal, and great design flexibility compared with conventional optical fiber. Because of the flexible design, PCF is filled with temperature-sensitive to realize the PCF temperature sensor. Many researchers have researched the design or fabrication of PCF temperature sensors [8].

In recent years, the sensitivity of the sensor is improved by combining PCF with SPR technology. SPR is a phenomenon where an electromagnetic mode is generated on the metal surface due to the interaction between free electrons of metal and optical electromagnetic fields. Plasmonics has a significant role in designing PCF-based temperature sensors. Usually, silver, gold, graphite, aluminum, and copper are used as plasmonic materials. However, silver as a plasmonic material has an oxidization problem that reduces the sensing range and performance. On the other hand, gold is chemically stable. The use of gold as a plasmonic material for designing temperature sensors is preferable for better performance. Although there are different types of optical fiber sensors such as the fiber Bragg grating and Mach–Zehnder interferometer the recent trend indicates that researchers have shown significant interest to design PCF temperature sensors based on SPR.

Liu [9] has presented a PCF temperature sensor based on coupling between the liquid core mode and defect mode with high sensitivity of 1.85 nm/°C. Vera [10] considered a temperature sensor using a sagnac-loop interferometer based on a side-hole PCF filled with metal, the high-temperature sensitivity of 9.0 nm/°C could be achieved. Li [11] suggested a high sensitivity fiber temperature sensor using a selective ethanol-filled PCF, the temperature sensitivity of the sensor is 1.65 nm/°C in the range of 25°C–33°C. Zhao et al. have proposed an SPR-based temperature sensor that has a sensitivity of 1.575 nm/°C [12]. However, they used silver as plasmonic material that has an oxidization problem. Chen et al. reported a liquid sealed temperature sensor with low sensitivity about –166 pm/°C [13].

11.1.13 Objective of the chapter

- To study the different properties of silicon PCFs as temperature sensors
- To study the different characteristics of silica PCFs
- To find out the different features and characteristics of an optical sensor
- Design and computational analysis of dual-core PCF temperature sensor based on SPR

11.1.14 Chapter outline

This Chapter is arranged into seven Sections. All the Sections share the same reference list at the end of the Chapter.

Section 11.1 consists of the introduction of nanowire and the applications of nanowire as PCFs. This chapter also discusses the use of nanowire as a temperature sensor along with nanowire features of the general waveguide. A brief description of the previous work has also been stated.

Section 11.2 discusses the analysis technique of silicon nanowire. A rigorous full-vectorial FEM has been used to characterize such nanowires. Starting from the brief history of FEM, its solution process, principle, necessity, and application are described. Then I represented COMSOL Multiphysics which is the simulation software that has been used to analyze. The advantages and preferences for COMSOL Multiphysics are also discussed here.

Section 11.3 discusses the description of the characteristics of the nanowire. Effective index, effective area, confinement factor, hybridness, dispersion, nonlinearity are discussed among various characteristics of Silica nanowire. Corresponding equations and their evaluation are also given. The purpose of this discussion is to give a basic understanding of the properties we analyzed.

Section 11.4 shows the results and discussions of different properties of silicon nanowire and silica nanowire. Different modes of nanowires have been studied. Different properties are explained with the help of graphs and necessary figures. The equations used for analysis have already been described in Section 11.3. This chapter shows the characteristics of Silica nanowire as an optical sensor and also presents the sensitivity and sensitive area of the nanowire.

Section 11.5 shows the geometry of the temperature sensor.

Section 11.6 presents a performance analysis of the temperature sensor.

Section 11.7 has the conclusion part of the thesis work. Some of the opportunities for future works are also mentioned which could help us discover more fascinating optical properties of nanowire as PCF temperature sensors.

11.2 PART 2: ANALYSIS TECHNIQUE

11.2.1 Finite element method

11.2.1.1 Introduction

The basic idea of FEM is to discretize the domain of interest where the partial differential equation (PDE) is defined in order to obtain an approximate solution of the PDE by a linear combination of basic functions defined within each subdomain. Then the assembly of subdomains, which is based on the process of within each subdomain. Then the assembly of subdomains, which is based on the process of putting the finite elements back into their original positions, results in a discrete set of equations which are analogous to the original mathematical problem.

The entire domain under investigation is approximated as an assembly of discrete elements, so-called finite elements, interconnected at points common to two or more elements, so-called nodes, as illustrated in Figure 11.11. Each finite element is an independent geometric region of the domain over which equations with unknown variables of the given problem are defined using the governing equations of the mathematical model of interest.

In each finite element, these equations are solved by assuming basis functions that interpolate the unknown variables over the finite element, in order to approximate the solution of the problem within the element. The basis function is defined within the finite element using the values of the unknown variables at the nodes. The approximate solution to the problem within the element is obtained as a linear combination of nodal values of the variables

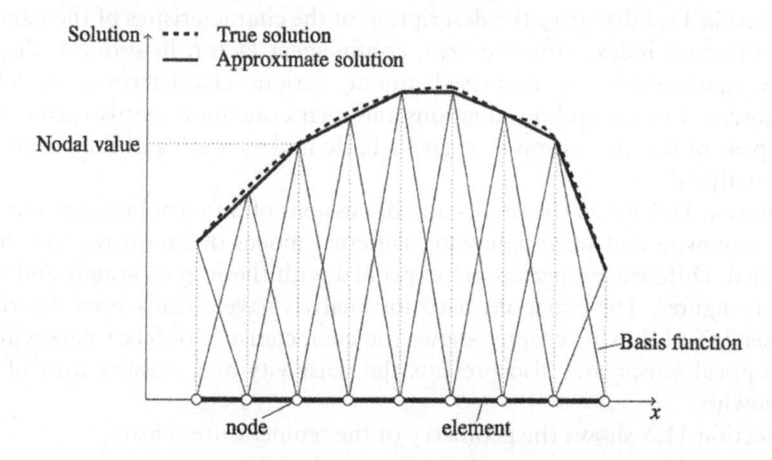

Figure 11.11 Schematic view of the linear discretization of the domain in elements and nodes. The true solution is represented as a continuous function (dotted line) and the approximate solution is described as a piecewise polynomial (solid line).

and the basic functions for the element. The equations for the finite element relate the nodal values of the variables to other parameters. The formulation of the finite element analog of a model equation follows two main approaches—weighted residual and weak formulation. Then with appropriate loadings, boundaries, and initial conditions applied to the elements/nodes, the local element equations for all the finite elements are assembled and solved simultaneously to obtain a continuous solution in terms of its values at the nodes.

During the finite element analysis (FEA), errors can be introduced due to the approximations of the domain discretization, the solutions of the element equations, and the solution of the assembled system of equations. The estimation of these errors is not simple, and therefore the exact solution of the problem cannot be obtained in most of the cases.

FEM is a numerical technique for solving physical problems which are described by PDEs or can be formulated as functional minimization. It is necessary to use mathematics to comprehensively understand and quantify any physical phenomena, such as structural or fluid behavior, thermal transport, wave propagation, and the growth of biological cells. Most of these processes are described using PDEs. However, for a computer to solve these PDEs, numerical techniques have been developed over the last few decades and one of the most prominent today is FEM. FEM was originally developed to study stresses in complex aircraft structures; it has since been extended and applied to the broad field of continuum mechanics, including fluid mechanics and heat transfer. Because of its capability to handle complex problems and its flexibility as an analysis tool, FEM has gained a prominent role in engineering analysis and design.

FEM mainly cuts a structure into several elements (pieces of the structure) which is known as "discretization" and then reconnects elements at "nodes" as if nodes were pins that hold elements together. This process

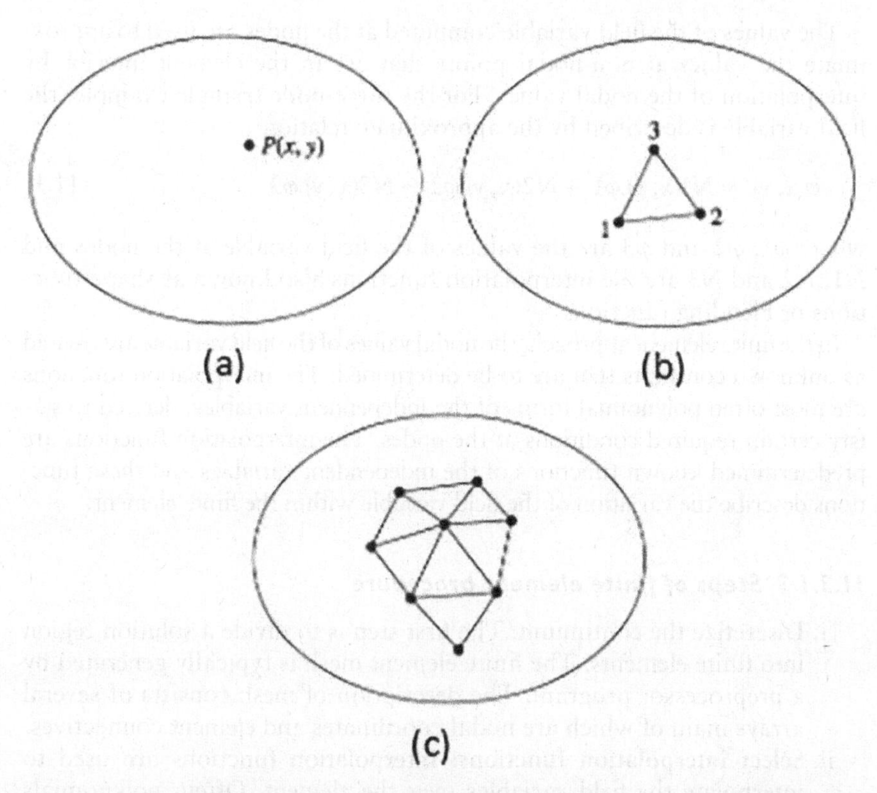

Figure 11.12 (a) A general 2D domain of field variable φ (x, y). (b) A three-node finite element is defined in that domain. (c) Additional elements show a partial element mesh of the domain.

results in a set of simultaneous algebraic equations [39]. By increasing the number of elements, we can achieve any precision.

11.2.1.2 Principle of FEM

FEM is a computational technique used to obtain approximate solutions to boundary value problems in engineering. Boundary value problems are also called field problems. The field is the domain of interest which is most often a physical structure represented as an assembly of "finite elements". The field variables are the dependent variables of interest governed by the differential equation. The boundary conditions are the specified values of the field variables or related variables such as derivatives on the boundaries of the field. For simplicity we assume a 2D case with a single field variable φ (x, y) to be determined at every point $P(x, y)$ such that a known governing equation is satisfied exactly at every such point. A node is a specific point in the finite element at which the value of the field variable is to be explicitly calculated [14] (Figure 11.12).

The values of the field variable computed at the nodes are used to approximate the values at non-nodal points that are in the element interior by interpolation of the nodal values. For the three-node triangle example, the field variable is described by the approximate relation

$$\varphi(x,y) = N1(x,y).\varphi 1 + N2(x,y).\varphi 2 + N3(x,y).\varphi 3 \tag{11.1}$$

where $\varphi 1$, $\varphi 2$ and $\varphi 3$ are the values of the field variable at the nodes and $N1$, $N2$ and $N3$ are the interpolation functions also known as shape functions or blending functions.

In the finite element approach, the nodal values of the field variable are treated as unknown constants that are to be determined. The interpolation functions are most often polynomial forms of the independent variables, derived to satisfy certain required conditions at the nodes. The interpolation functions are predetermined known functions of the independent variables and these functions describe the variation of the field variable within the finite element.

11.2.1.3 Steps of finite element procedure

i. **Discretize the continuum:** The first step is to divide a solution region into finite elements. The finite element mesh is typically generated by a preprocessor program. The description of mesh consists of several arrays main of which are nodal coordinates and element connectives.
ii. **Select interpolation functions:** Interpolation functions are used to interpolate the field variables over the element. Often, polynomials are selected as interpolation functions. The degree of the polynomial depends on the number of nodes assigned to the element.
iii. **Find the element properties:** The matrix equation for the finite element should be established which relates the nodal values of the unknown function to other parameters. For this task different approaches can be used; the most convenient are: the variational approach and the Galerkin method [15].
iv. **Assemble the element equations:** To find the global equation system for the whole solution region we must assemble all the element equations. In other words, we must combine local element equations for all elements used for discretization. Element connectivities are used for the assembly process. Before the solution, boundary conditions and loads should be imposed.
v. **Solve the global equation system:** The finite element global equation system is typically sparse, symmetric, and positive definite. Direct and iterative methods can be used for the solution. The nodal values of the sought function are produced as a result of the solution.
vi. **Compute additional results:** In many cases, we need to calculate additional parameters. For example, in mechanical problems, strains and stresses are of interest in addition to displacements which are obtained after a solution of the global equation system.

vii. **Accuracy:** In FEA only, the model of the real problem is solved. A good finite element model, once set up, is about a 95% accurate solution of the field equations which themselves are based on a theoretical model which is idealized from reality. So, it is better to refine the mesh in areas of high stress, repeat the procedure two or three times, and check the iteration effects (see Figure 11.14).

11.2.1.4 Applications

FEM can handle problems possessing any or all of the following characteristics:

- Differential, integral, integrodifferential, or variational equations.
- Boundary value problems are also called equilibrium or steady-state problems, Eigen Any mathematical or physical problem described by the equations of calculus, e.g., problems for resonance and stability phenomena and initial value problems such as diffusion, vibration, and wave propagation.
- The domain of the problem may be any geometric shape, in any number of dimensions such as 1D, 2D, or 3D.
- Physical properties, e.g., density, stiffness, permeability, and conductivity may also vary throughout the system.
- The external influences, generally referred to as loads or loading conditions, may be in any physically meaningful form, e.g., pressure, thermal, inertial forces, etc. The loads are typically applied to the boundary of the system (boundary conditions), to the interior of the system (interior loads), or at the beginning of time (initial conditions).
- Problems may be linear or nonlinear.

11.2.1.5 Advantages of FEM

FEM can handle a wide variety of engineering problems, such as solid mechanics, dynamics, heat problems, fluids, and electrostatic problems

- Complex restraints and indeterminate structures can be solved
- Optimizing product performance and cost
- Reduction of development time
- Elimination or reduction of testing
- First-time achievement of the required quality
- Improved safety
- Satisfaction of design codes
- Improved information for engineering decision-making
- Understanding components allows for more rational design
- Satisfaction with legal and contractual requirements

11.2.2 COMSOL multiphysics

FEM is extensively used in science and engineering applications based on approximations of PDEs. Numerical analysis tools make the solutions of coupled physics, mechanics, chemistry, and even biology accessible to the novice modeler. COMSOL is a very notable software for FEA. It was started in July 1986 by Svante Littmarck and Farhad Saeidi at the Royal Institute of Technology (KTH) in Stockholm, Sweden. COMSOL Multiphysics (an acronym for Common Solution) is a cross-platform FEA, solver, and 25 Multiphysics simulation software [16]. Preprocessing and post-processing was the main concern while designing the COMSOL Multiphysics software. FEA software package is a variety of physics and engineering applications, especially coupled phenomena, or 25 Multiphysics. COMSOL Multiphysics also offers a widespread interface to MATLAB® and its toolboxes for a large variety of programming, preprocessing, and post-processing possibilities. The packages are cross-platform. COMSOL Multiphysics also permits us for entering coupled systems of PDEs.

The uniqueness of COMSOL Multiphysics over other FEA software packages is its ability to solve coupled phenomena. These are phenomena for which the solution of each physical problem cannot be obtained independently of the others. Examples of these include combined fluid dynamics/chemical engineering or heat transfer/mechanical engineering problems.

11.2.2.1 Features of COMSOL multiphysics

COMSOL Multiphysics offers a unique user-friendly working environment and also provides a wide range of tools for fast and effective analysis. There are numerous important features of the software as follows:

- Graphical user interface
- Tools for creating the geometry and internal boundaries and domains
- Automatic mesh generation and refinement
- The ability to solve different equations on different meshes
- The Multiphysics capability permits the addition of equations to represent additional phenomena
- The ability easily to make parameters depend upon the solution
- The parametric solver
- The post-processing graphical features

11.2.2.2 Modeling environment

COMSOL Multiphysics is a powerful interactive environment for modeling and solving all types of scientific and engineering problems which are based on PDEs. With this software, one can easily extend conventional models for one type of physics into 26 Multiphysics models that solve coupled physics phenomena and do so simultaneously. Accessing this power does not need any depth knowledge of mathematics or numerical analysis. To built-in physics modes, it is possible to build models by defining the relevant physical quantities such as material properties, loads, sources, fluxes, and constraints rather than by defining the equations. COMSOL Multiphysics then internally compiles PDEs set that represents the entire model. COMSOL Multiphysics has unique features in representing multiple linked domains with complex geometry, highly coupled and nonlinear equation systems, and arbitrarily complicated boundary, auxiliary and initial conditions. But with this modeling power come great opportunities and perils.

11.2.2.3 COMSOL multiphysics and its importance

For long scientists and engineers had to make assumptions in order to be capable of realizing their design ideas. With the progression of time, these assumptions are being refined and, in some cases, even eliminated to get more accurate results. Multiphysics is one major enabler to eliminate assumptions by coupling related physical applications together to include all the necessary factors for a complete model. COMSOL Multiphysics is a simulation software designed to provide the most accurate results by minimizing the assumptions its users must make. COMSOL Multiphysics users are free from the restrictive nature which is generally associated with simulation software. The COMSOL user has full control of their model. The ability to couple any number of physics together and input user-defined physics and expressions directly into a model allows COMSOL users to be creative in a way which is not possible with other simulation software. So, the features of COMSOL have a great deal of flexibility when dealing with any physical problem. It is the greatest of them all, the ability to define the user's physics, in the form of PDEs for any kind of problem. The ability to give any arbitrary PDE does not guarantee that its solvers would be able to solve it. Some of the advantages of the software have been briefly stated below:

- Easy, seamless interface between heat transfer and other physics fields.
- Ability to modify governing equations.
- Flexibility in selection in the form of direct, iterative, or mixed solvers, and segregated, or fully coupled solutions.

11.2.2.4 Mesh refinement

COMSOL primarily uses FEM to compute single and 27 Multiphysics simulations. Whenever using FEM, it is important to remember that the accuracy of the solution is linked to the mesh size. As mesh size decreases towards zero (leading to a model of infinite size), one has to move toward the exact solution for the equations they are solving. Although one is limited by finite computational resources and time they will have to rely on an approximation

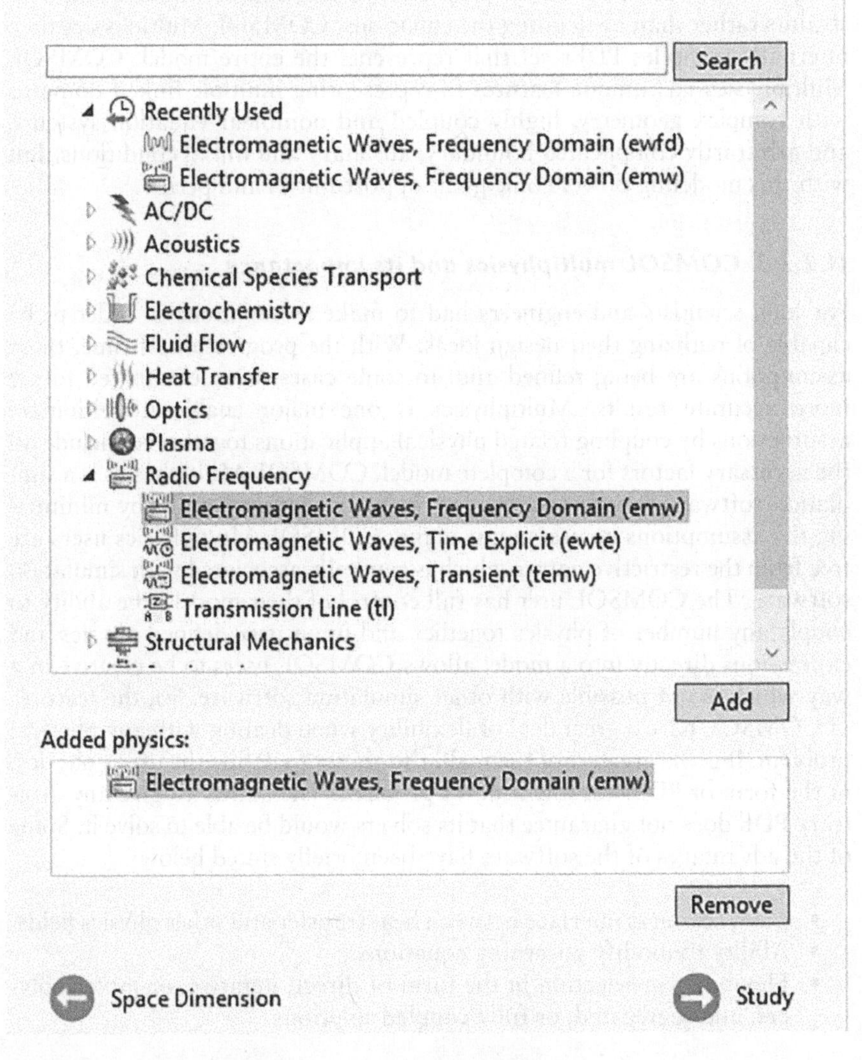

Figure 11.13 Selecting RF module using COMSOL version 4.4.

of the real solution. The objective of the simulation is to minimize the difference ("error") between the exact and the approximated solution and to ensure that the error is below some accepted tolerance level that will vary from project to project based on your design and analysis goals.

We need to track a characteristic output parameter from our simulation as we vary the mesh size and determine at which mesh size the

Table 11.1 Material properties used in the optical fiber

	Core	*Cladding*
Material	Silica glass	Doped silica glass
Refractive index	3.4754	1.444

parameter has converged on the correct value. However, convergence criteria will depend on design and analysis goals. Finer element size is chosen for mesh analysis.

11.2.2.5 Study of a temperature sensor in COMSOL version 4.4

The loss spectra and mode distribution are analyzed by using COMSOL Multiphasic software. The simplicity of the design, high sensitivity, and wide detection range of the proposed sensor make it competitive for the reported temperature sensor.

The proposed design consists of only one ring of two different dimensions of air holes with diameters d_1 and d_2. Λ (pitch) is the center to center distance between the air holes. To ensure the dual core an elliptical air hole with dimension of a (major axis) and b (minor axis) is placed at the center of the core. The circular air holes are placed at a distance of 4 μm from the center. For the establishment of the surface plasmon wave, the two air holes in the horizontal axis are chosen for smaller diameter intentionally (Figure 11.13).

Some material properties were used as default in the study which is presented in Table 11.1. These properties are valid for the free space wavelength of 1.55 μm, which is in the infrared region where most of the fiber optic communication takes place. Modal analysis of the nanowire was performed as a part of the study. Associated sweeps were done to examine the dependency of the nanowire performance on the core radius. The core radius was varied from 0.2 to 1.4 μm. In this setting, the corresponding electric field intensities and effective RIs were determined for various values of core radius. The standard meshing tool was used with the mesh sequence type selected as physics-controlled mesh and element size was tuned to "Extra fine". Figure 11.14 shows the selection of standard mesh analysis from the model builder window, and Figure 11.15 shows the meshed geometry of the silicon nanowire cross-section in 2D.

Effective mode index lies between RIs of the two materials. The fundamental mode has the maximum index. Hence to find the fundamental mode certainly, the mode index is set to search around the core index. Mode analysis frequency can be obtained corresponding to a free space wavelength of 1.55 μm which has been shown in Figure 11.16. Then after computing the analysis and changing the expression from the setting window for the solution we obtained the result.

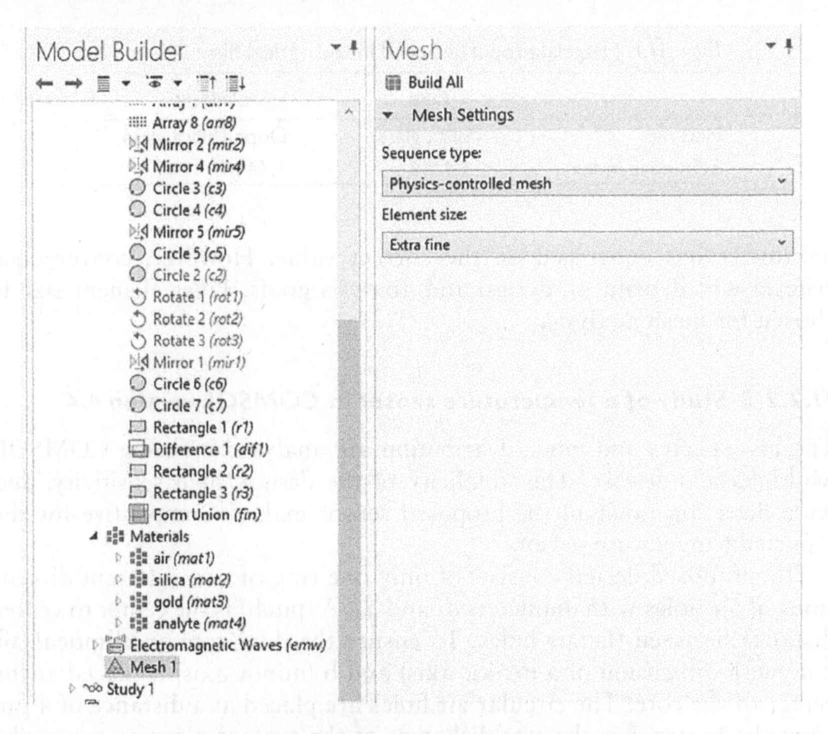

Figure 11.14 Selecting sequence type and element size for mesh analysis.

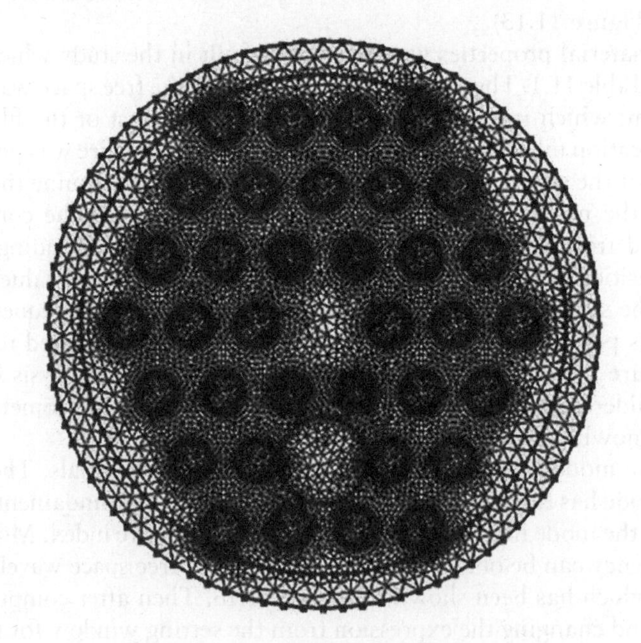

Figure 11.15 Structure of the FEM in COMSOL for a temperature sensor.

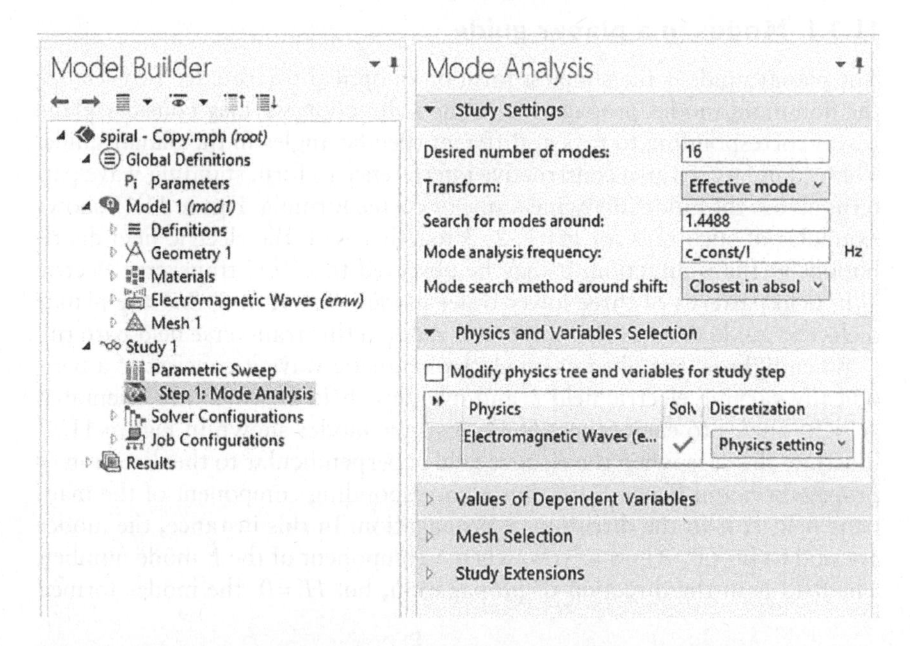

Figure 11.16 Selection of effective mode index and mode analysis frequency.

11.3 PART 3: MODAL ANALYSIS OF NANOFIBERS AND PCFs

The characteristics of the dual-core PCF sensor are studied using FEM, and the structure is improved according to the numerical simulation results. PCF is filled with the analyte that has no influence on the wavelength sensitivity of the sensor which means those holes can be replaced by small air holes. The wavelength sensitivity can be tuned by adjusting the sizes of the other large air holes which are liquid ones. The dynamic detection range of RI is from 1.33 to 1.51. In particular, high linearity is obtained in the range of 1.44–1.51.

In order to obtain an improved model for the propagation of light in a waveguide, electromagnetic wave theory must be considered. The basis for the study of electromagnetic wave propagation is provided by Maxwell's equations. All the field components are assumed to have a functional form of exp[$i(\omega t-\beta z)$] and the governing wave equations, which is deduced from Maxwell's equations in terms of the electric field vector $\vec{\boldsymbol{E}}$ takes the form, where is the dielectric constant of the medium which is complex for metal and lossy materials, is the relative permeability of the medium, is the free space wave number, is the angular frequency and β is the propagation constant.

$$\vec{\nabla}\times\frac{1}{\mu_r}\vec{\nabla}\times\vec{E}-K_0^2\varepsilon_r\vec{E}\quad=0 \tag{11.2}$$

11.3.1 Modes in a planar guide

The planar guide is the simplest form of an optical waveguide. To visualize the dominant modes propagating in the z-direction we may consider plane waves corresponding to rays at different specific angles in the planar guide. These plane waves give constructive interference to form standing wave patterns across the guide following a sine or cosine formula. Figure 11.17 shows examples of such rays for $m=1, 2, 3$ together with the electric field distributions in the x-direction. It may be observed that "m" transverse electric (TE) field patterns of three lower-order models ($m=1, 2, 3$) "n" the planar dielectric guide denotes the number of zeros in this transverse field pattern.

When light is described as an electromagnetic wave it consists of a periodically varying electric field E and magnetic field H which are orientated at right angles to each other. The transverse modes shown in Figure 11.17 illustrate the case when the electric field is perpendicular to the direction of propagation and hence $E_z=0$, but a corresponding component of the magnetic field H is in the direction of propagation. In this instance, the modes are said to be TE. Alternatively, when a component of the E mode number. The field is in the direction of propagation, but $H_z=0$, the modes formed

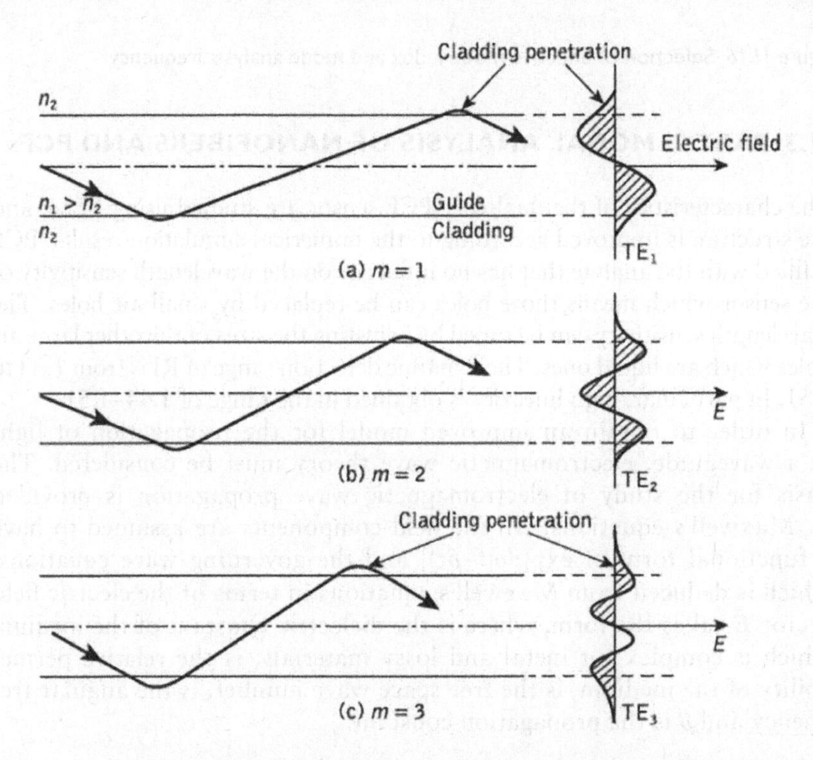

Figure 11.17 Physical model showing the ray propagation and the corresponding transverse electric (TE) field patterns of three lower-order models ($m = 1, 2, 3$) in the planar dielectric guide. Denotes the number of zeros in this transverse field pattern.

are called transverse magnetic *l*. The mode numbers are incorporated into this nomenclature by referring to the TE_m and TM_m modes, as illustrated for TE modes shown in Figure 11.17. When the total field lies in the transverse plane, transverse electromagnetic waves exist where both E_z and H_z are zero. Although transverse electromagnetic waves occur in metallic conductors (e.g., coaxial cables) they are seldom found in optical waveguides.

The connection between the standard HE, EH, TE, and thulium mode designations and also the LP*m* mode designations. The mode subscripts *m* and *n* square measure associated with the electrical field strength profile for a selected LP mode. Their square measure is generally 2*m* field maxima round the circumference of the fiber core and *n* field maxima along a radius vector. Furthermore, it may be observed from Figure 11.18 that the notation for labeling the HE and EH modes has changed from that specified for the exact solution in the cylindrical waveguide mentioned previously. The subscript m in the LP notation now corresponds to HE and EH modes with labels $m+1$ and $m-1$, respectively. The electric field intensity profiles for the lowest three LP modes, together with the electric field distribution of their constituent exact modes, are shown in Figure 11.18. It may

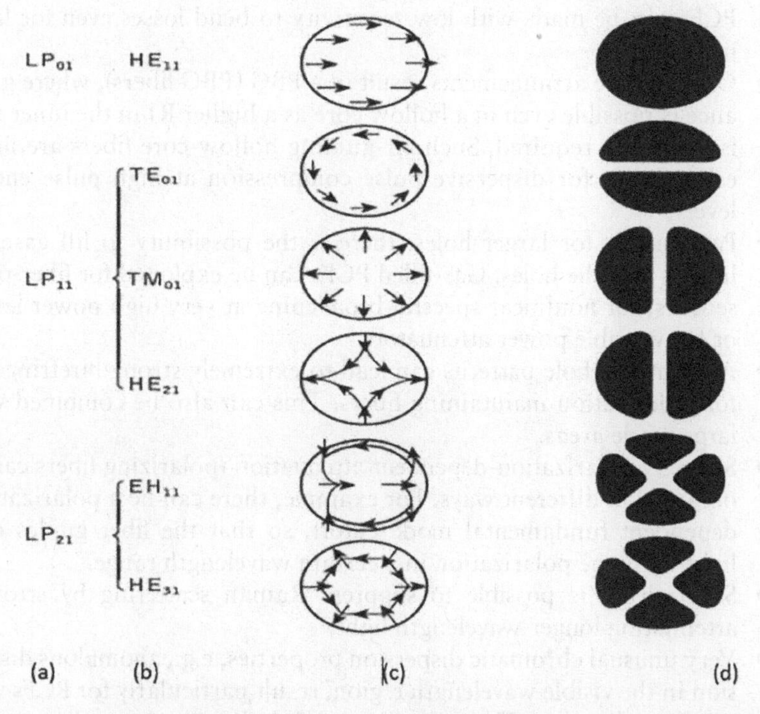

Figure 11.18 The electric field configurations for the three lowest LP modes illustrated in terms of their constituent exact modes: (a) LP mode designations; (b) exact mode designations; (c) electric field distribution of the exact modes; (d) intensity distribution of E_x for the exact modes indicating the electric field intensity profile for the corresponding LP modes.

be observed from the field configurations of the exact modes that the field strength in the transverse direction (E_x or E_y) is identical for the modes which belong to the same LP mode. Hence the origin of the term "linear".

11.3.2 Properties achievable by design

PCFs with different designs of the hole patterns (concerning the basic geometry of the lattice, the relative size of the holes, and possibly small displacements) can have the following remarkable properties, strongly depending on the design details:

- It is possible to obtain a very high numerical aperture of, e.g., 0.6 or 0.7 of multimode fibers (also for the pump cladding of a double-clad fiber).
- Single-mode guidance over very wide wavelength regions (endlessly single-mode fiber) is obtained for small ratios of hole size and hole spacing.
- Extremely small or extremely large mode areas (possibly with a lower numerical aperture than possible with a conventional fiber) are possible. These lead to very strong or very weak optical nonlinearities. PCFs can be made with low sensitivity to bend losses even for large mode areas.
- Certain hole arrangements result in a PBG (PBG fibers), where guidance is possible even in a hollow core as a higher RI in the inner part is no longer required. Such air-guiding hollow-core fibers are interesting, e.g., for dispersive pulse compression at high pulse energy levels.
- Particularly for larger holes, there is the possibility to fill gases or liquids into the holes. Gas-filled PCFs can be exploited for fiber-optic sensors, for nonlinear spectral broadening at very high power levels, or for variable power attenuators.
- Asymmetric hole patterns can lead to extremely strong birefringence for polarization-maintaining fibers. This can also be combined with large mode areas.
- Strongly polarization-dependent attenuation (polarizing fibers can be obtained in different ways. For example, there can be a polarization-dependent fundamental mode cutoff, so that the fiber guides only light with one polarization in a certain wavelength range.
- Similarly, it is possible to suppress Raman scattering by strongly attenuating longer-wavelength light.
- Very unusual chromatic dispersion properties, e.g., anomalous dispersion in the visible wavelength region, result particularly for PCFs with small mode areas. There is substantial design freedom, allowing for different combinations of desirable parameters.

- Coreless end caps can be fabricated simply by fusing the holes near the fiber end with heat treatment. The sealed end facets allow for larger mode areas at the fiber surface and thus a higher damage threshold, e.g., for amplifying intense nanosecond pulses.
- Multicore designs are possible, e.g., with a regular pattern of core structures in a single fiber, where there may or may not be some coupling between the cores.

11.3.3 Effective index

An effective index is a characteristic property of the material. An effective RI for an optical waveguide provides us with a measure of the overall phase delay of a light beam in it. We can define RI n as the quantification of the phase change per unit length in a waveguide to that in a vacuum. Here, the phase delay per length is assumed to have been caused by the medium. The effective index η_{eff} has the analogous meaning for light propagation in a waveguide; the value of phase constant of the waveguide or β (for some wavelength) is the η_{eff} times the vacuum wavenumber. This relation can be expressed with the following equations:

$$\beta = \frac{2\pi}{\lambda} \tag{11.3}$$

$$\text{or } \beta = \eta_{eff} k_0 \tag{11.4}$$

$$\text{or, } \eta_{eff} = \frac{\beta}{k_0} \tag{11.5}$$

Here $k_0 = \left(\frac{2\pi}{\lambda}\right)$ is the wavenumber and λ is the wavelength.

Usually, in a single-mode fiber, the effective RI has a value that falls between RIs of core and cladding. This gave rise to a misconception that RI is a kind of weighted average of RI of core and cladding of the waveguide. The weight factors are determined by the fractions of optical power propagating in core and cladding. Consider a step-index multimode fiber with a high numerical aperture, i.e., with a very large index step. In that case, all fiber modes propagate only in the core except the higher-order ones. This might lead us to believe the effective index of all modes closely matches the core index. But on the contrary, higher-order modes still have significantly lower effective indices. They experience a smaller phase shift per unit length in spite of propagation through the same material.

Essentially it is the fact that higher-order modes contain more pronounced plane wave components (spatial Fourier components) with a larger angular offset from the fiber axis. So, in a sense, it is a matter of different propagation directions and not of different materials. However, both the effects become relevant in the case of fibers having a lower numerical aperture. For multimode waveguides, the effective RI depends not only on the wavelength but also on the mode of light propagation. So, it is also called the modal index. The effective index is not just a material property, it depends on the whole waveguide design too.

11.3.4 Effective area

The effective area can be defined as a quantitative measure of the area which a waveguide effectively covers in the transverse dimensions. All nonlinear effects depend upon the intensity of the electromagnetic field in the medium. However, it is the total optical power entering and leaving the fiber that is usually measured. The measured optical power leaving a fiber is simply the integral of the electromagnetic field over the entire fiber cross-section. The electromagnetic field is given by

$$E = \sqrt{E_x^2 + E_y^2} \tag{11.6}$$

where E_x is the x-component of electric field and E_y is the y-component of an electric field. The effective area is defined as

$$A_{\text{eff}} = \frac{\iint \left(|E|^2 \, dxdy\right)^2}{\iint |E|^4 \, dx\, dy} \tag{11.7}$$

The normalized effective area can be expressed as

$$A = \frac{A_{\text{eff}}}{R^2} \tag{11.8}$$

However, the normalized effective area can often be more significant or useful. The fundamental mode being confined in the core possesses a lower normalized effective area compared to the higher-order modes. The field in a single-mode fiber is not evenly distributed or even fully contained within the core. It is larger at the fiber axis than near the core-cladding interface and extends into the cladding to a degree depending on the actual RI profile. In our work, the effective area has been calculated corresponding to both the core and the cladding area, because the field distribution is not uniform and hence is a non-vanishing part of the propagation in the cladding region of the fiber.

11.3.5 Confinement factor

The confinement factor can be defined as the fraction of the total power residing in the silicon core. In our work, the power has been calculated from the pointing vector using all six components of the electric and magnetic fields [18]. Confinement factor can be expressed by the equation,

$$C_f = \frac{P_c}{P_o} \tag{11.9}$$

11.3.6 Nonlinearity

The terms linear and nonlinear in optics mean intensity-independent and intensity-dependent phenomena, respectively. When the output signal strength does not vary in direct proportion to the input signal strength, then it is called the nonlinearity of the device. For a nonlinear device, the output-to-input amplitude ratio (also called gain) depends on the strength of the input signal (Figure 11.19).

For optical fibers, even moderate optical power leads to high optical intensities because the light is confined to a small transverse region. In addition, light often propagates over considerable distances in fiber. For these reasons, nonlinear effects due to fiber nonlinearities often have a substantial effect. The power dependence of RI is responsible for the Kerr effect. This is the most common nonlinear effect in fiber. Certainly, this means that the phase delay in the fiber gets larger if the optical intensity increases. This can be described via an increase of RI in proportion to the intensity.

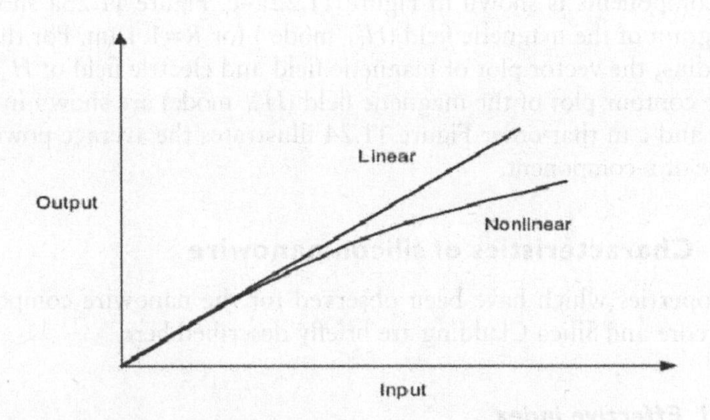

Figure 11.19 Linear and nonlinear interaction.

11.4 PART 4: ANALYSIS OF DIFFERENT OPTICAL FIBER

The results obtained after the simulation of the designed nanowires are represented in this chapter. And later on, the results for Silica nanowire as the optical sensor was shown. Nanowire is designed following the process that has already been explained in Section 11.2.

11.4.1 Silicon nanowire

11.4.1.1 Different optical modes

In this section, different modes of Silicon nanowire with silica cladding have been studied, where η_{core} =3.4754 and η_{clad}=1.444 at 1.55 µm wavelength for core radius r_{core}=1.4 µm. For the nomenclature LP_{mn} was used where "m" and "n" represent mode orders in azimuthal and radial directions, respectively. The modes are dominated by either H_x or H_y field. Modes are classified as x-polarized or y-polarized where x-polarized mode is with dominant H_y and E_x field and y-polarized mode is with dominant H_x and E_y field. The H^x_{mn} and H^y_{mn} notation is used here which is identical to that of the LP_{mn} modes. Dominant H-field component is identified by H_x or H_y. All the studies here have been considered with dominant H_x field if not mentioned otherwise. Only a finite number of guided propagation modes are seen in a silicon nanowire waveguide and the intensity distribution of which have a finite extent around the waveguide core. The waveguide structure and optical frequency in very important in this case. Because of the number of guided modes, their transverse amplitude profiles and their propagation constants depend on the details of waveguide structure and optical frequency. Figures 11.20 and 11.21 show different modes for the magnetic field x component. The electric field of H^x_{01} modes of x, y, and z-components is shown in Figure 11.22a–c. Figure 11.23a shows the 3D diagram of the magnetic field (H^x_{01} mode) for R=1.4 µm. For the same core radius, the vector plot of magnetic field and electric field of H^x_{01} mode and the contour plot of the magnetic field (H^x_{01} mode) are shown in Figure 11.23b and c in that order Figure 11.24 illustrates the average power flow of mode of z-component.

11.4.2 Characteristics of silicon nanowire

The properties which have been observed for the nanowire composed of Silicon core and Silica Cladding are briefly described here.

11.4.2.1 Effective index

In this thesis work, the Silicon core of radius R with a circular geometry has been scrutinized, with a Silica cladding. Figure 11.25 shows the variation of effective indices for a Silica circular optical waveguide as the radius is reduced and enters the nanowire regime. For the fundamental mode H_{01}^{x} variation in the effective index with respect to radius is observed in Figure 11.25 where the effective index η_{eff} satisfies (3.2). Since the quasi-TM and TE fundamental H_{01}^{x} and H_{01}^{y} of the waveguide are degenerate for the circular optical waveguide the effective indices of H_{01}^{y} are not shown here. The higher-order modes H_{11}^{x} (HE_{01}or LP_{11}), H_{21}^{x} (HE_{11}) or LP_{21} and H_{02}^{x} (HE_{21} or LP_{02}) [27] are also shown in Figure 11.25. From Figure 11.25 it is observed that at first effective indices of the modes decrease slowly with decreasing core radius bus as the modes approach their cutoff conditions these reduce rapidly. It can also be noted that when the radius is increased

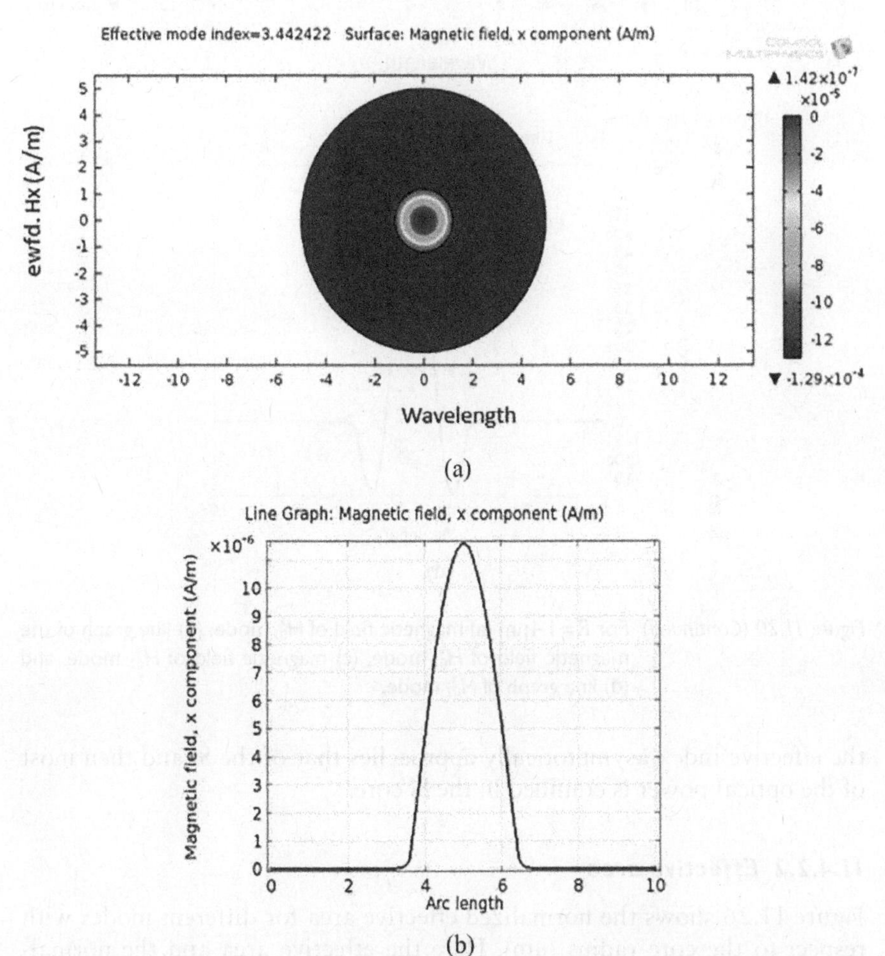

Figure 11.20 For R = 1.4 μm (a) magnetic field of H_{01}^{x} mode, (b) line graph of the magnetic field of H_{01}^{x} mode, (c) magnetic field of H_{02}^{x} mode, and (d) line graph of H_{02}^{x} mode.

(Continued)

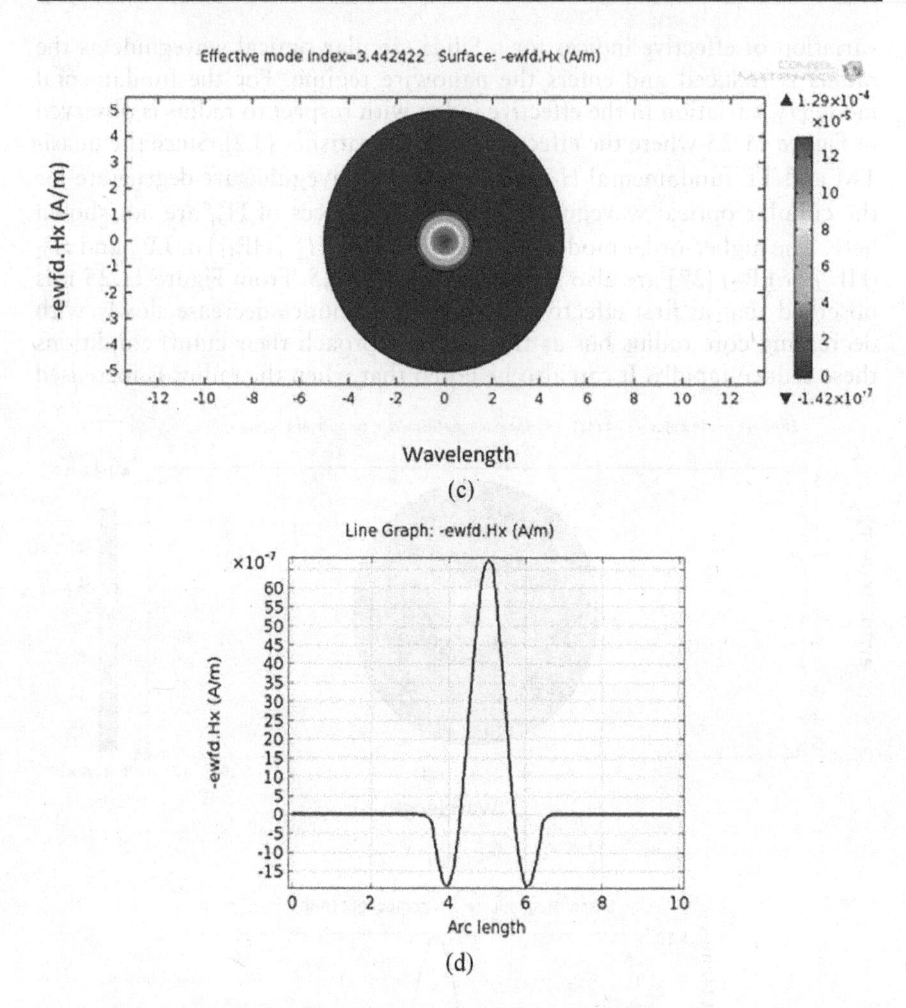

Figure 11.20 (Continued) For R = 1.4 µm (a) magnetic field of H^x_{01} mode, (b) line graph of the magnetic field of H^x_{01} mode, (c) magnetic field of H^x_{02} mode, and (d) line graph of H^x_{02} mode.

the effective index asymptotically approaches that of the Si and then most of the optical power is confined in the Si core.

11.4.2.2 Effective area

Figure 11.26 shows the normalized effective area for different modes with respect to the core radius (µm). Here the effective area and the normalized effective area can be calculated for equations 11.5 and 11.6, respectively. The fundamental mode possesses a lower normalized effective area

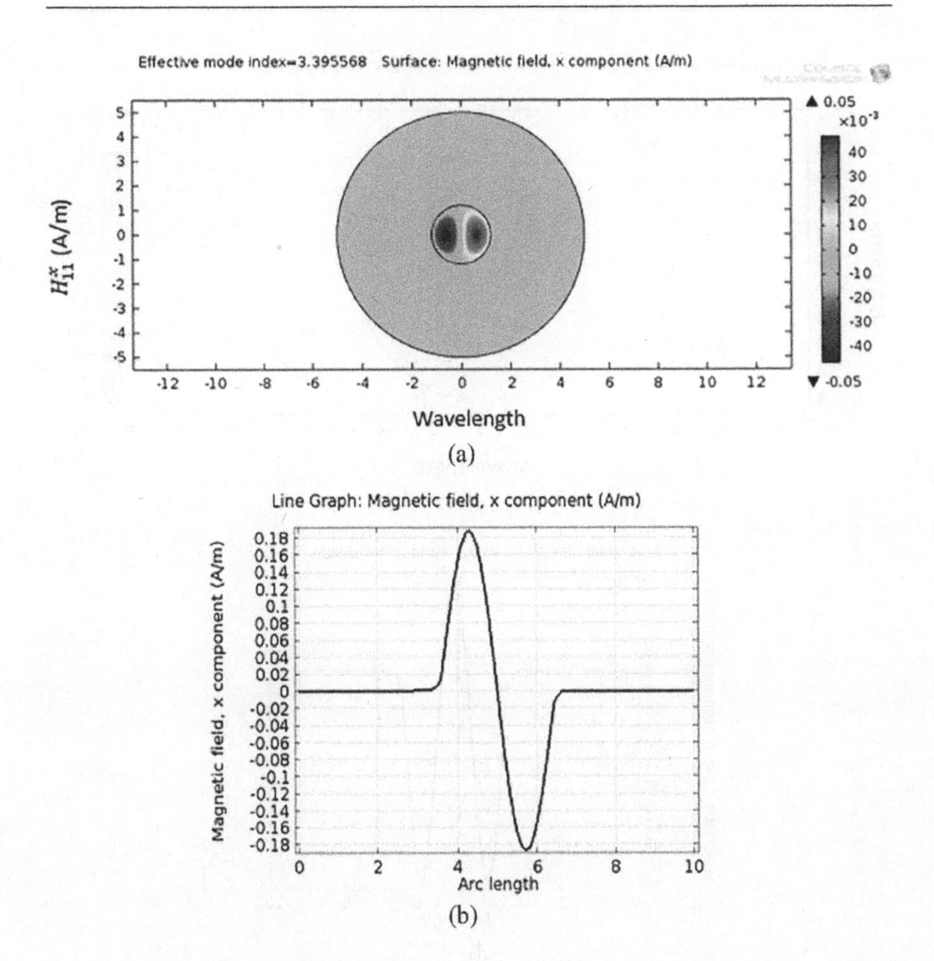

Figure 11.21 For R=1.4 μm (a) magnetic field of H_{11}^{x} mode, (b) line graph of the magnetic field of H_{11}^{x} mode, (c) magnetic field of H_{21}^{x} mode, and (d) line graph H_{21}^{x} mode. *(Continued)*

compared to the higher-order modes, as this mode is more confined to the core. As the core radius is reduced, the corresponding normalized effective area increases slightly until the mode approaches the cutoff region. With the further reduction in R, the value of the effective area increases rapidly, as the mode spreads into the cladding region. The normalized effective areas for the H_{02}^{x}, H_{11}^{x}, H_{21}^{x} and modes increase as they approach their cutoff region. It is observed from Figure 11.26 that the mode has a lower normalized effective area for a higher core radius but it approaches the cutoff condition before H_{11}^{x} and H_{21}^{x} modes when the core radius is reduced.

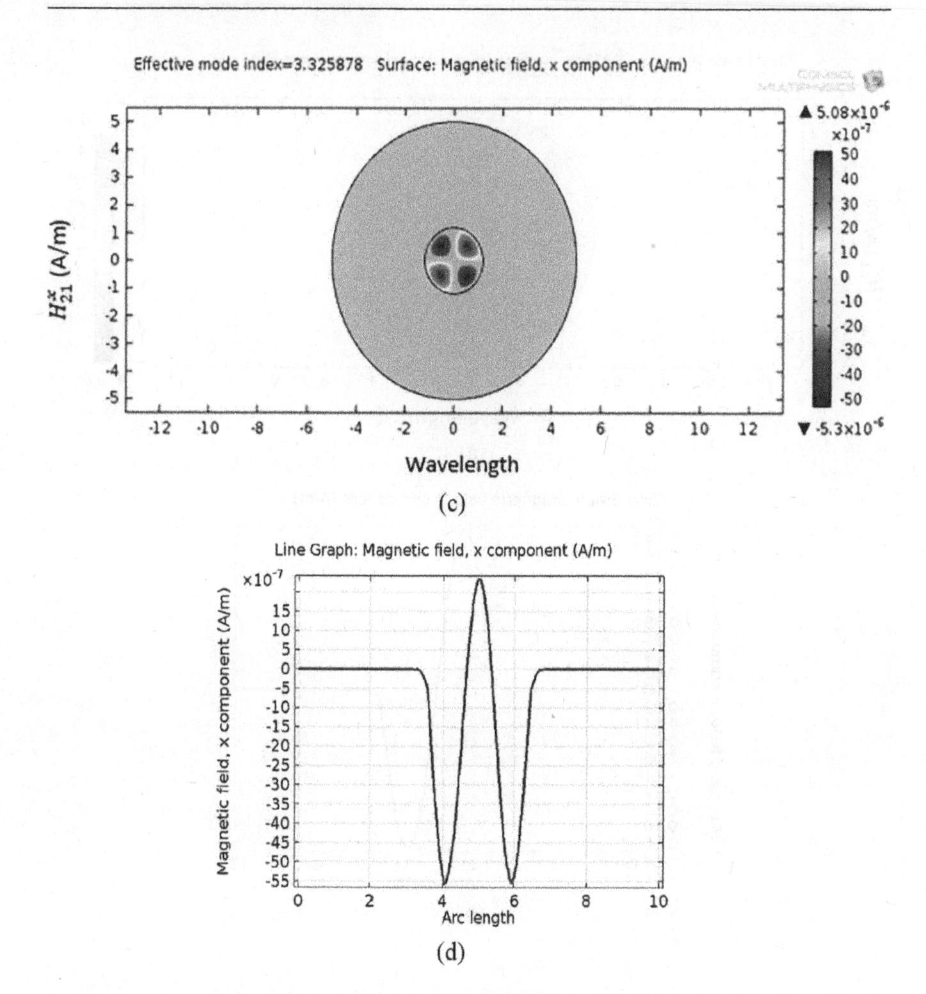

Figure 11.21 (Continued) For R= 1.4 µm (a) magnetic field of H_{11}^{x} mode, (b) line graph of the magnetic field of H_{11}^{x} mode, (c) magnetic field of H_{21}^{x} mode, and (d) line graph H_{21}^{x} mode.

11.4.2.3 Confinement factor

The confinement factor is defined as the fraction of the total power residing in the Silicon core. The comparison of the confinement factor with the core radius for fundamental mode and higher-order modes is shown in Figure 11.27. The confinement factor is derived from equation 11.8. From the graph, it can be observed that the fundamental mode is near the cutoff region by the rapid increase in normalized effective area and rapid decrease in confinement factor beneath $R=200$ nm. From Figure 11.27 it can also be observed that the cutoff radii for the H_{01}^{x}, H_{11}^{x}, H_{21}^{x} and H_{02}^{x} modes are approximately 160, 210, 340, and 410 nm, respectively, and these values

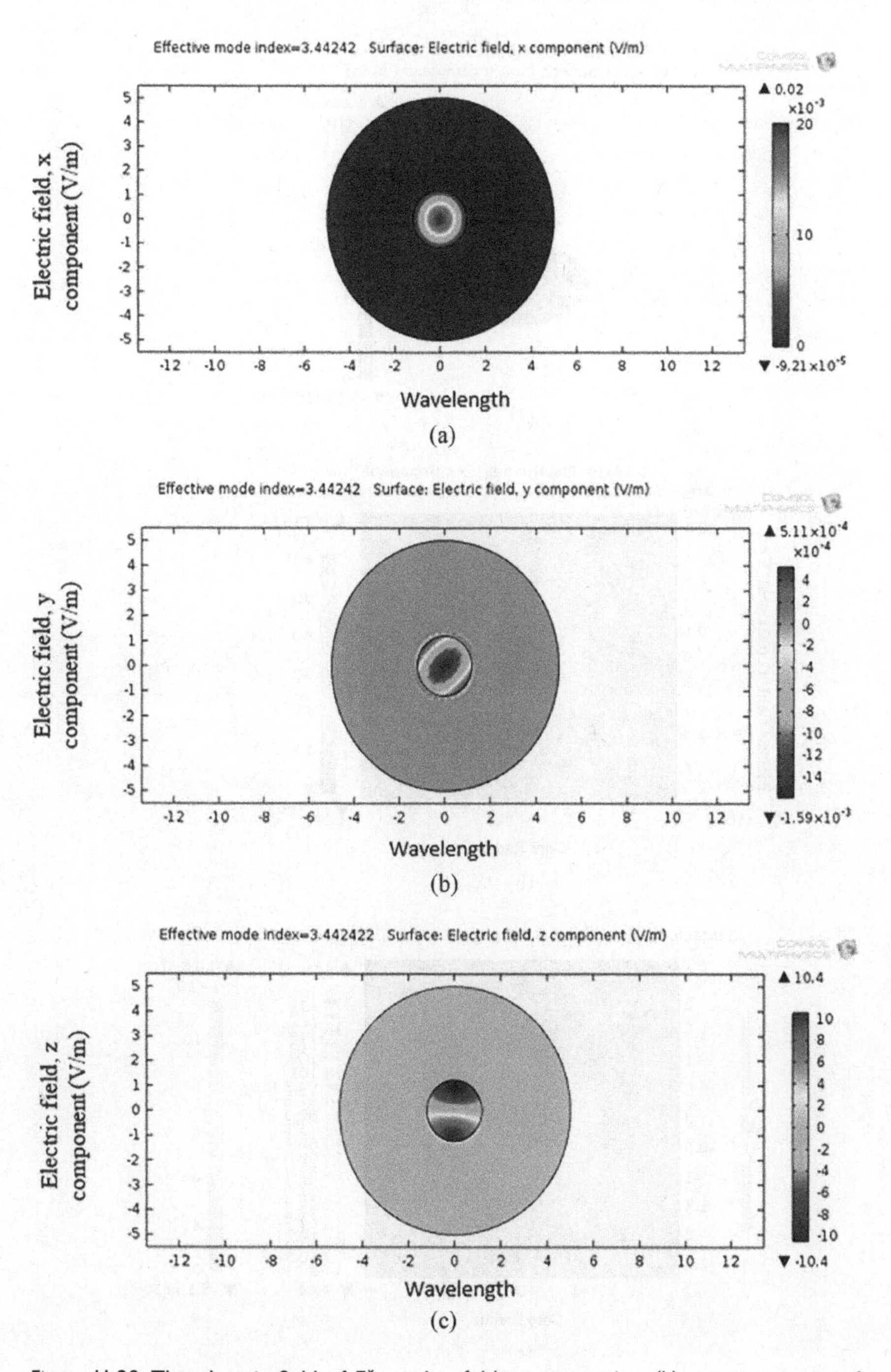

Figure 11.22 The electric field of E_{01}^{x} mode of (a) *x*-component, (b) *y*-component, and (c) *z*-component.

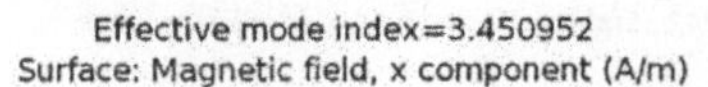

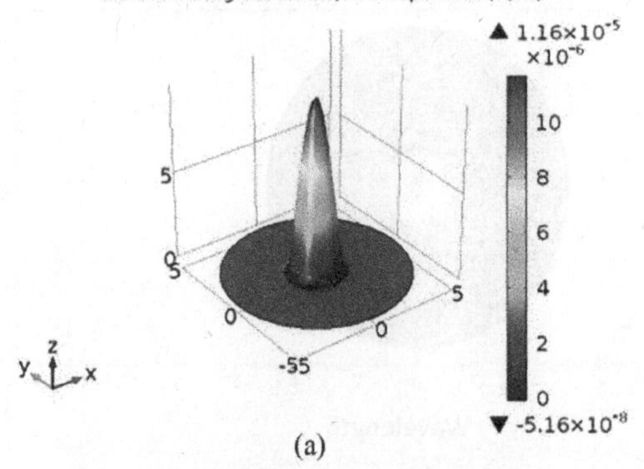

(a)

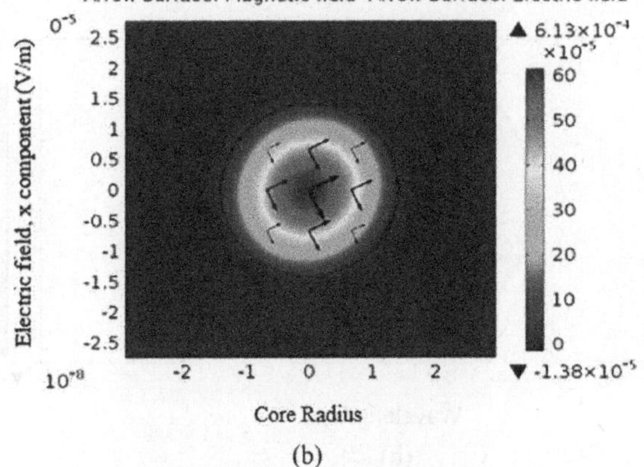

(b)

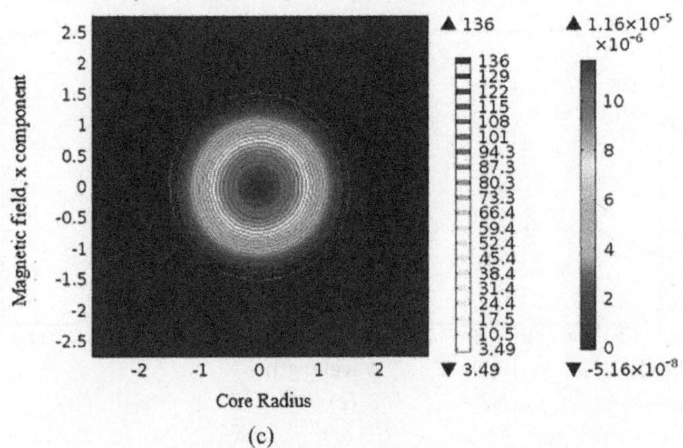

(c)

Figure 11.23 For $R = 1.4\,\mu m$ of mode H_{01}^{x} (a) 3D diagram of the magnetic field, (b) vector plot of magnetic field and electric, and (c) contour plot of the magnetic field.

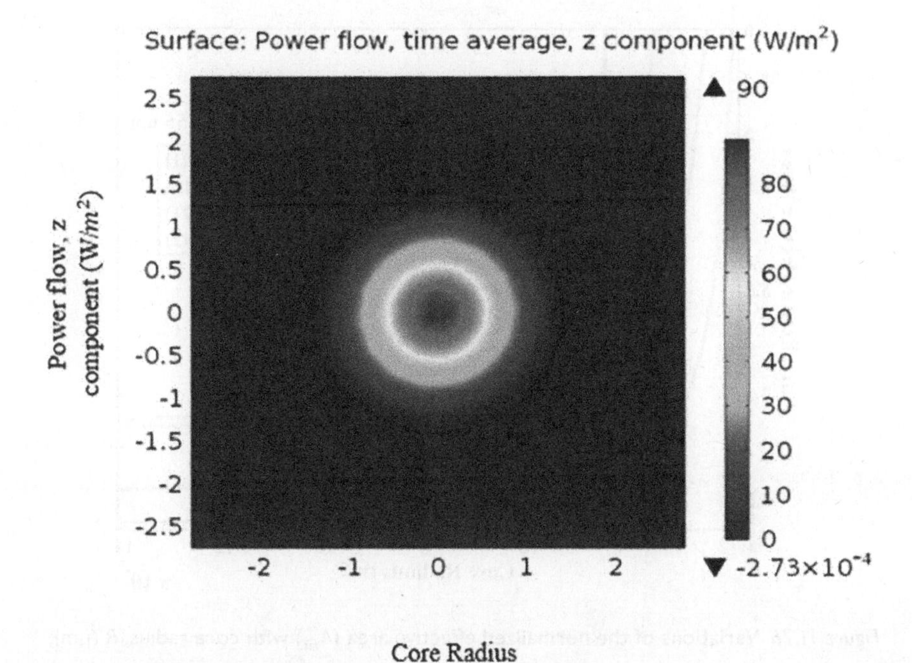

Figure 11.24 Average power flow of H_{01}^{x} mode of z-component.

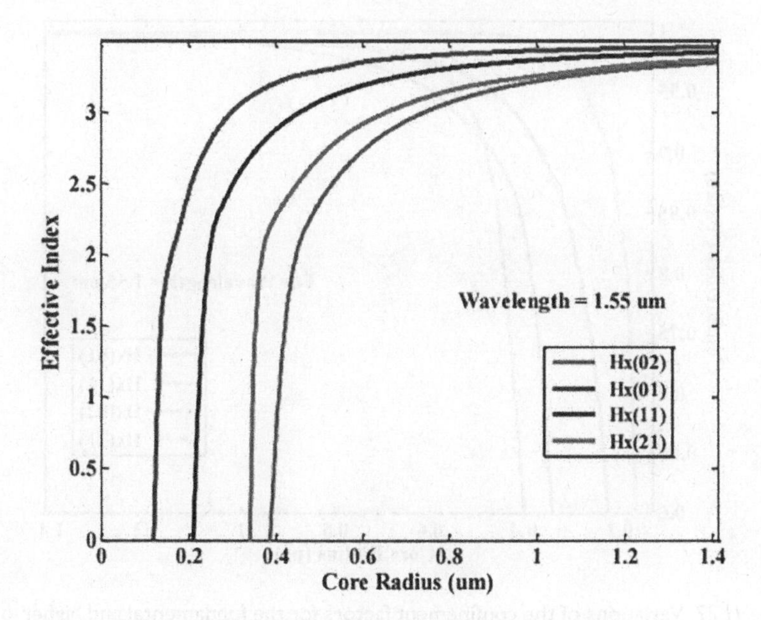

Figure 11.25 Variations in the effective indices with the core radius.

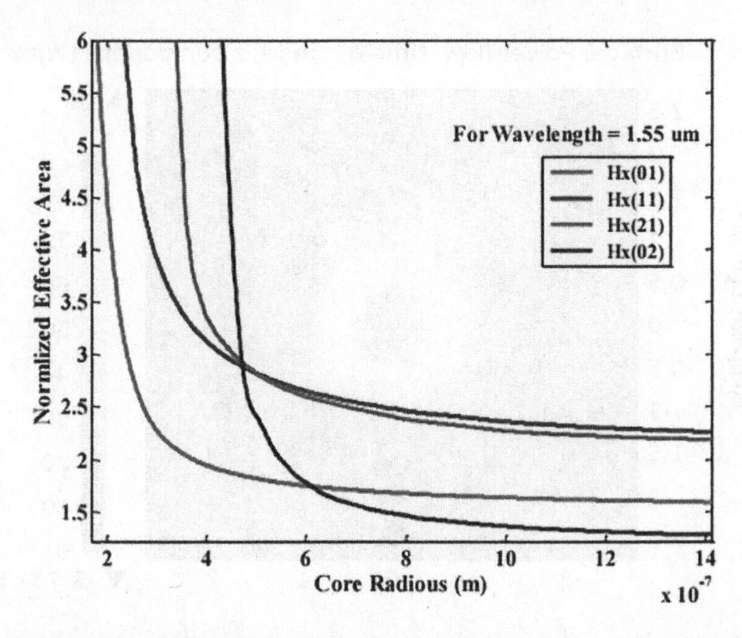

Figure 11.26 Variations of the normalized effective area (A_{eff}) with core radius, R (μm).

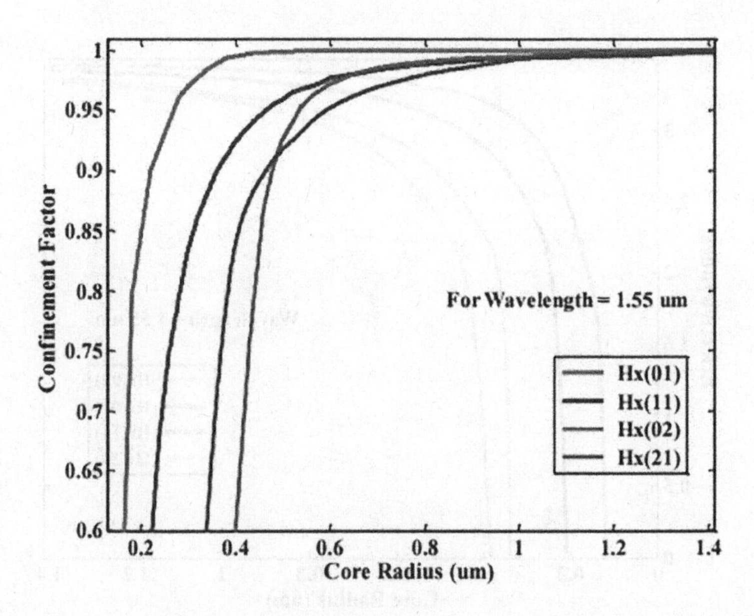

Figure 11.27 Variations of the confinement factors for the fundamental and higher-order modes with core radius, R (μm).

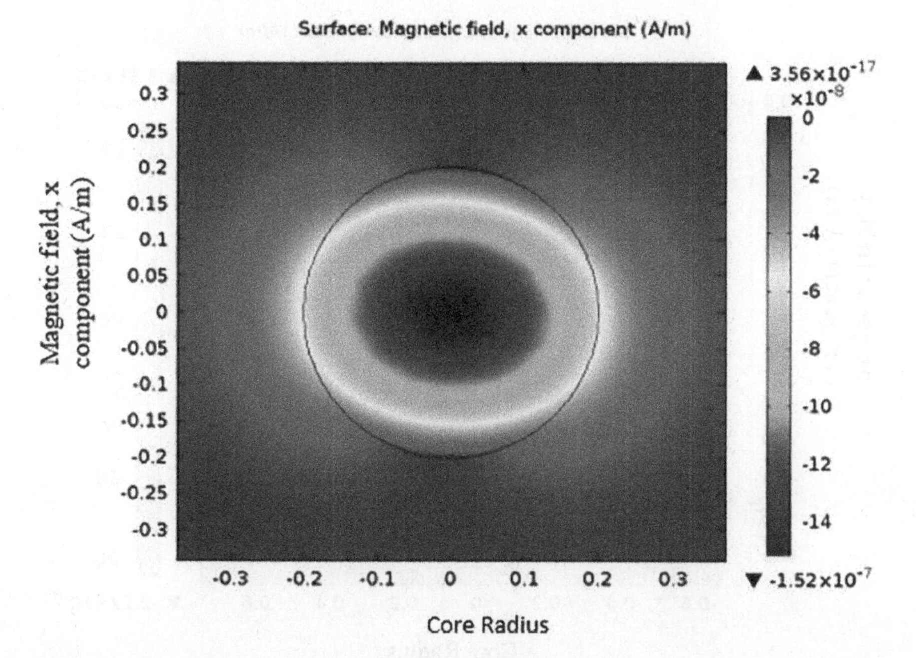

Figure 11.28 H_x field for the quasi H_{01}^{x} mode with core radius, R=200 nm.

also comply with the values obtained from Figure 11.25. It is also observed from Figure 11.27 that the H_{02}^{x} mode has a higher confinement factor (and lower normalized effective area) for higher core radius but it reduces faster than the H_{11}^{x} and H_{21}^{x} modes when the core radius is reduced.

11.4.2.4 H-field mode profile

It has been observed that when the radius of the core R, is much smaller than the wavelength size, the fundamental mode may not be circular, even though the nanowire structure is rotationally symmetric. The x component of the H field is dominant for the H_{01}^{x} mode and the H_x field is shown in Figure 11.28 where core radius, R=200 nm. It can be observed that the maximum field intensity is at the center of the core and decreases along the radial direction. Figure 11.28 shows that the mode shape is not circular rather it is dumbbell-shaped due to the spreading of H_x field in the horizontal direction. Through the H_x field is dominant for the H_{01}^{x} mode the other two components are also present. Figure 11.29 shows the H_y field which has four peaks with alternate positive and negative signs. Though its magnitude is smaller than the dominant H_x field, it is not trivial. Figure 11.30 shows the H_z field from which it can be noted that it has two peaks along the X-axis and its value along the Y-axis is zero.

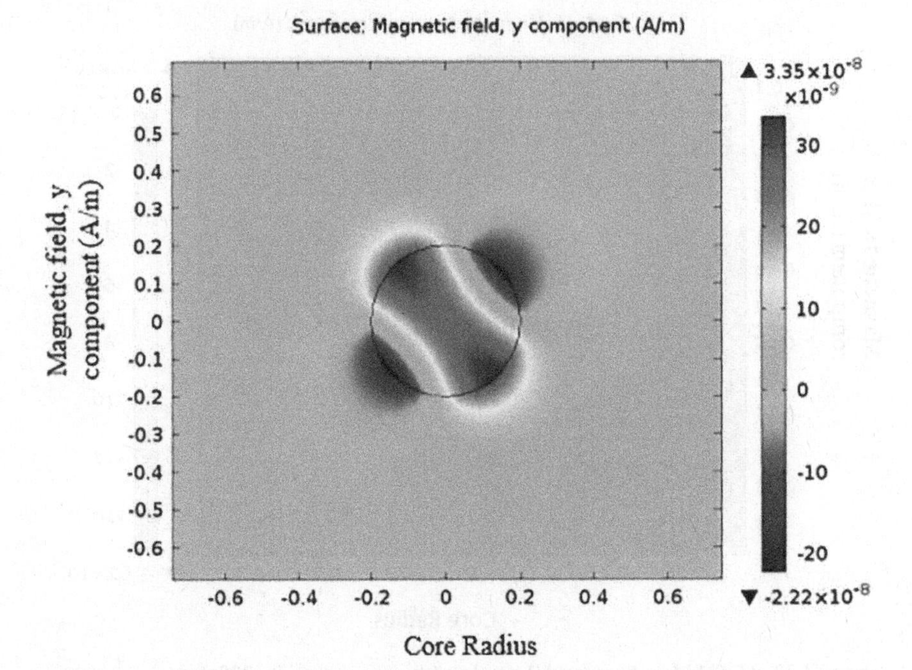

Figure 11.29 H_y field for the quasi H_{01}^x mode with core radius, $R = 200$ nm.

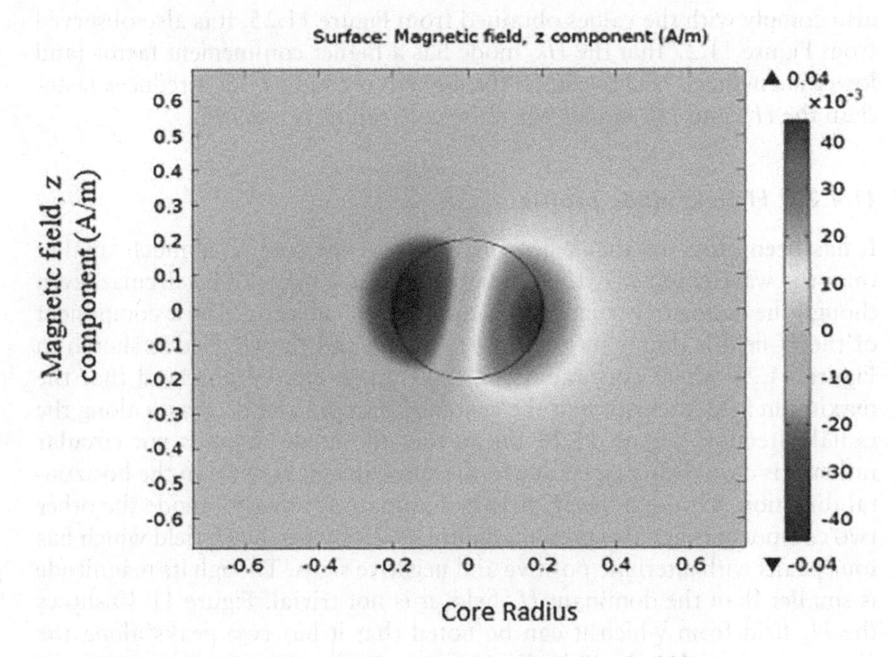

Figure 11.30 H_z field for the quasi H_{01}^x mode with core radius, $R = 200$ nm.

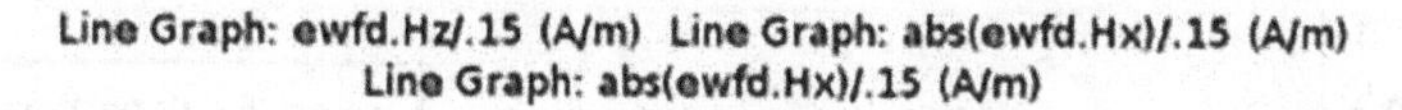

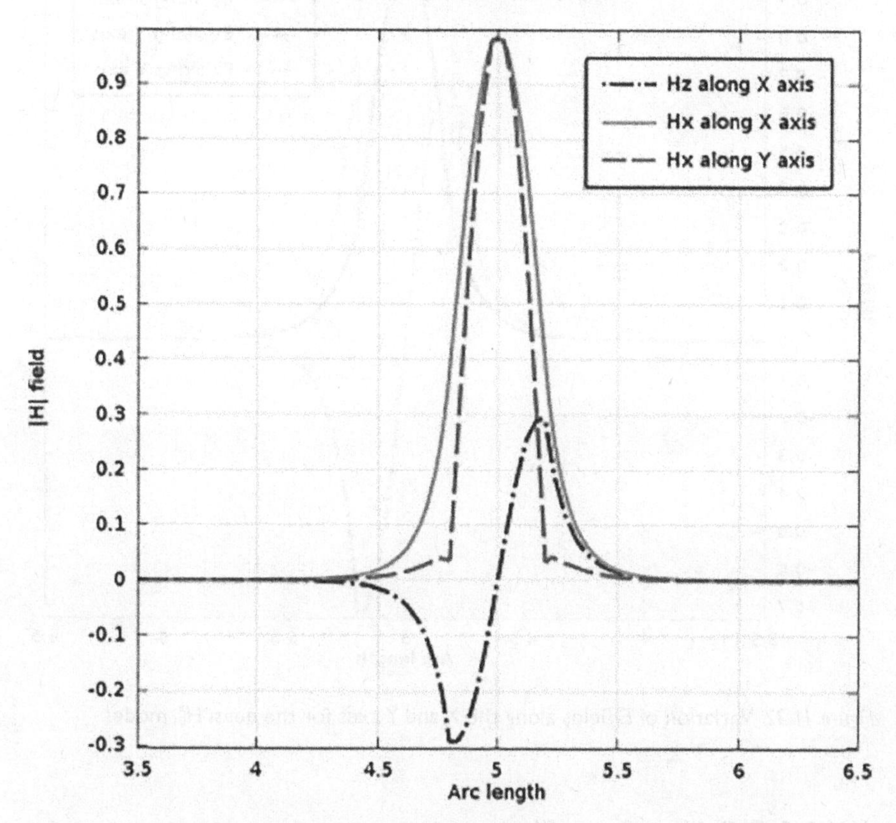

Figure 11.31 Variation of the *H* fields along the *X* and *Y* axes for the quasi H_{01}^{x} mode.

Figure 11.31 shows the variation of H_x and H_z field along the axes when $R = 200$ nm. For fundamental H_{01}^{x} mode, the boundary condition for magnetic and electric fields exists along the *Y* and *X* axes, respectively. For this mode field exists but its value is zero along the *X* and *Y* axes, which is not shown here. Similarly, the value of H_z is zero along the *Y*-axis but nonzero along the *X*-axis. From Figure 11.32 it can be observed that the H_x field varies more slowly along the X-axis and the variation of the H_x field along the *Y*-axis is almost identical for all the values of *R*. As a result, the spot-size diameter along the *X*-axis is significantly larger compared to that along the *Y*-axis for a smaller core radius. It is also observed that the maximum value of H_z field is about 30% of the maximum H_x value. The variation of the dominant H_x field profile is not identical in the direction of *X* and *Y* axes because of boundary conditions for the H_x field component which leads to an elliptical profile of the H_x field for the fundamental H_{01}^{x} mode.

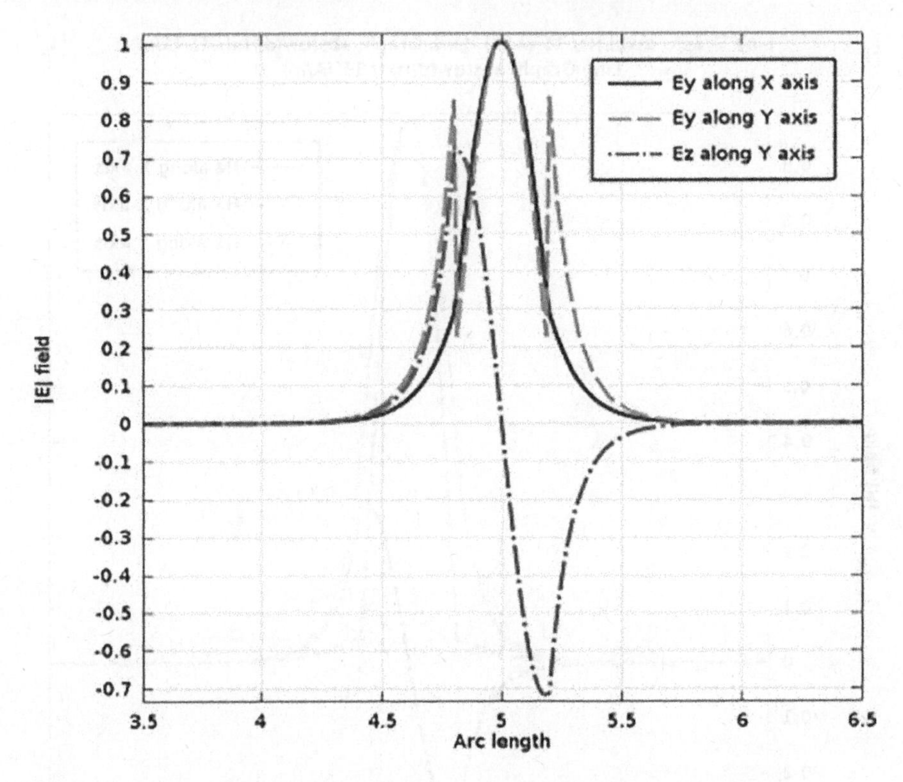

Figure 11.32 Variation of E-fields along the X and Y axes for the quasi H_{01}^{x} mode.

11.4.2.5 E-field mode profile

Once the magnetic field, H, is obtained from this H-field formulation, the corresponding E-field can be calculated from $\nabla \times H$ Maxwell's equations. For the mode of H_{mn}^{x}, the E_y field is dominant and directly related to the H_x field with an additional contribution from $\frac{\delta H_z}{\delta x}$. The variation of the electric field E_y component along X and Y is shown in Figure 11.32. From Figure 11.32 it is observed that E_y field is continuous along the X-axis as E_y is tangent to the core at $x=R$ but discontinuous along Y-axis because E_y field in normal to the interface at $y=R$. Here it is also shown that variation of E-field along X and Y-axis direction is not identical, which rise the asymmetry of the optical beam profile. From Figure 11.32 H_y field is zero along X and Y axes so that E_x field is zero along the X and Y axes and is not shown in Figure 11.33. E_z field variation along Y-axis is shown here when $R=200\,\text{nm}$. Along X-axis the E_z field is zero because of the boundary condition of an electric wall for mode H_{01}^{x}.

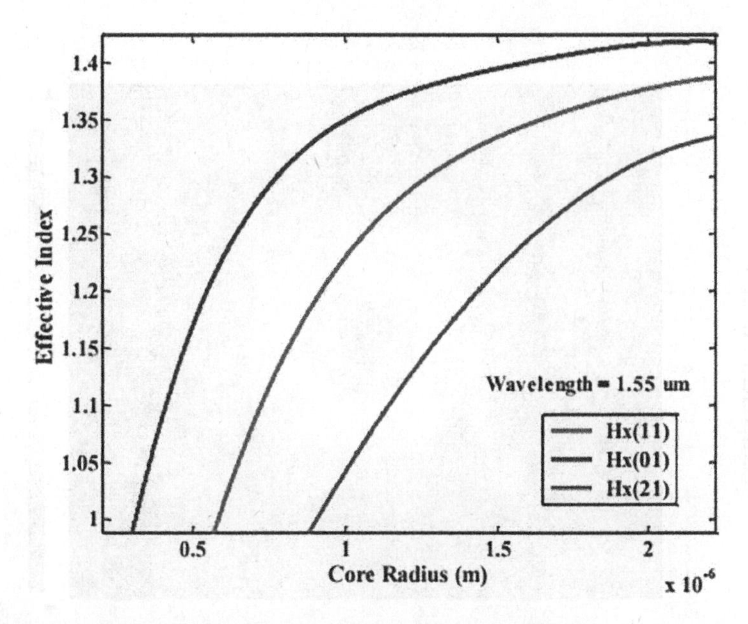

Figure 11.33 Variation of the effective indices with the core radius.

11.4.3 Silica nanowire

The properties which have been observed for the nanowire composed of Silica core and air cladding are briefly described here silica circular optical waveguide, the variation of effective indices is observed as the radius of the waveguide is reduced and enters the nanowire region. For the fundamental mode, variation in effective index with respect to core radius is shown in Figure 11.33. The fundamental effective index η_{eff} satisfies equation 11.4. Since the quasi-TM and TE fundamental H_{01}^{x} and H_{01}^{y} of the waveguide are degenerate for the circular optical waveguide the effective indices of H_{01}^{y} have not been presented here (Figure 11.34).

11.4.4 Sensitivity

The phase shift ($\Delta\varphi$) of the sensing arm can be obtained as,

$$\Delta\varphi = (\beta - \beta_0)\cdot L = \Delta\beta \cdot L \tag{11.10}$$

where L is the effective length of the sensitive area, β_0 and β are the initial and instant propagation constants of the light in the sensitive area, respectively. At a given temperature and pressure, when a certain amount of the

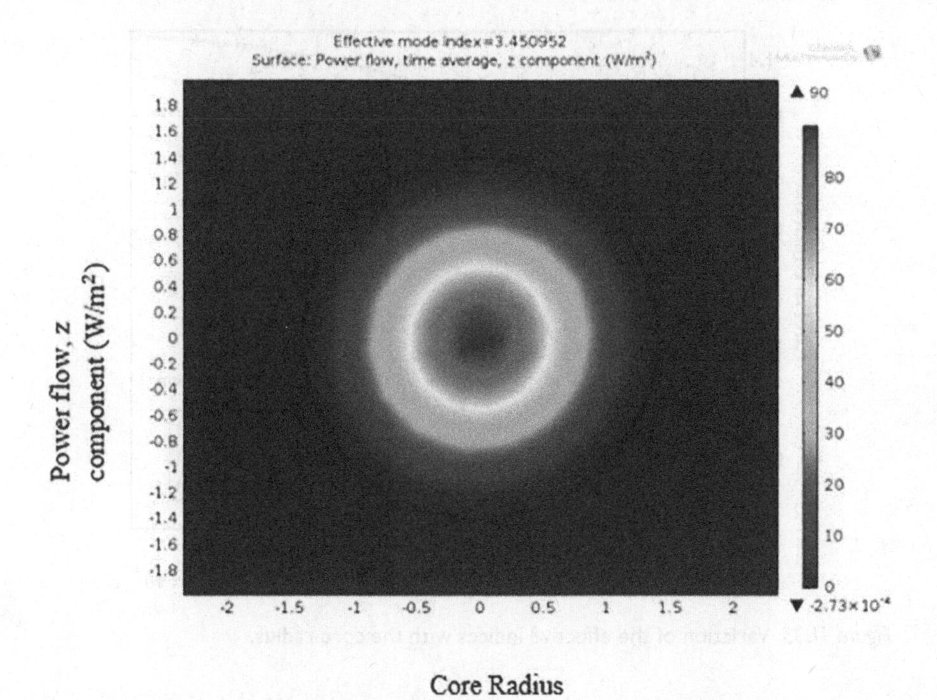

Figure 11.34 Average power flow of H^x_{01} mode of z-component for Silica.

specimen with initial RI n_{so} =1.4 is added into a solution (water) an RI of n_c=1.355 the overall index n_s as can be approximately obtained as,

$$n_s = n_c \cdot (1 - C) + n_{so} \cdot C \tag{11.11}$$

C the molar concentration of the specimen in the solution.

The sensitivity of sensors can be normalized [27],

$$S_N = \frac{1}{L} \cdot \frac{d(\Delta\varphi)}{dn_s} \tag{11.12}$$

which can be written as,

$$S_N = \frac{\dfrac{d\Delta\beta}{dC}}{\dfrac{dn_s}{dC}} \tag{11.13}$$

11.5 PART 5: SPR-BASED TEMPERATURE SENSOR

11.5.1 Geometry of SPR-based temperature sensor

This section discusses the schematic geometry of the SPR-based PCF temperature sensor (see Figure 11.35). To enhance the coupling between a plasmonic mode and a core guided mode and to reduce the plasmonic to plasmonic mode coupling is the purpose of the design. Equating the propagation constants of the two modes, implying that the effective RIs of the two modes have to be shut is required by phase-matching theoretically. The effective index of its core guided mode is about that of a core material which for most sensible materials is above 1.45 in the case of a single-mode fiber. Here we assume that the PCF is made of silica, ethanal, gold, air. In order to further improve the temperature sensitivity, Zhao Yong [19] et al. have reported an optical fiber temperature sensor based on the SPR principle. The optical fiber probe coated with a silver film was encapsulated into a capillary filled with alcohol to achieve enhanced sensitivity which was a combination of the alcohol characteristic (with a high thermal coefficient) and SPR technology. Silica and sensing medium their effective RIs are close. At a certain wavelength regime, the energy of a core-guided mode is transferred to a significant increase in loss will be observed at this wavelength regime, as the plasmonic mode is highly lossy.

In Figure 11.35 a dual-core PCF temperature sensor based on SPR is demonstrated using FEM. The dual-core structure makes the fabrication process easier by depositing metal and temperature coefficient material outside the PCF. The proposed design consists of only one ring of two different dimensions of air holes with diameters d_1 and d_2. Λ (pitch) is the center to

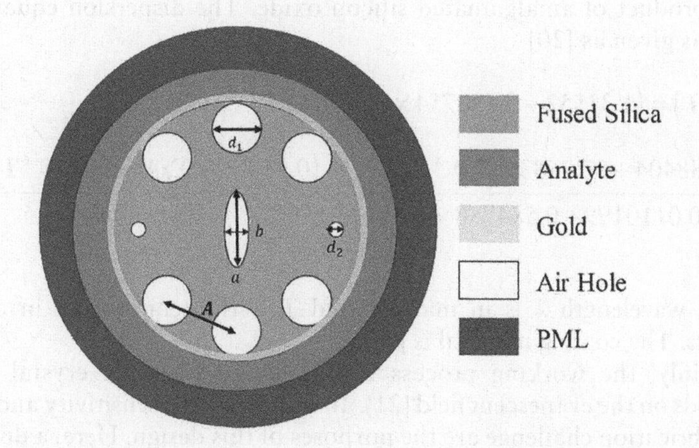

Figure 11.35 Cross-sectional view of the proposed dual-core-designed PCF temperature sensor.

Table 11.2 Optimum parameters of the proposed sensor

$A(\mu m)$	d_1 (μm)	d_2 (μm)	$a(\mu m)$	$b(\mu m)$	d (nm)
4	1	0.6	0.5	1.5	40

center distance between the air holes. To ensure the dual core an elliptical air hole with dimensions of a (major axis) and b (minor axis) is placed at the center of the core. The circular air holes are placed at a distance of 4 μm from the center. For the establishment of the surface plasmon wave, the two air holes in the horizontal axis are chosen for smaller diameter intentionally. Therefore, the proposed design confirms the formation of a dual core and the optimum parameters for the computational analysis of the temperature sensor are tabulated in Table 11.2.

The loss spectra and mode distribution are analyzed by using COMSOL Multiphasic software. The simplicity of the design, high sensitivity, and wide detection range of the proposed sensor make it competitive for the reported temperature sensor.

11.5.2 Temperature dependence of dielectric constant

When RIs of the constituents of sensors vary with temperature, the phase-matching wavelength between the core guided mode and plasmonic mode changes. Therefore, the temperature changes of the silica, sensing liquid, and also aluminous coating can shift the absorption peak of the spectrum recorded at the tip of the fiber. We tend to discuss the properties of the constituents of the sensing element (mainly temperature and wavelength-dependent properties), that are primarily based on wide applied theories. For the current analysis, we've thought of the PCF to be a product of amalgamated silicon oxide. The dispersion equation of silica is given as [20]

$$n^2(\lambda,T) = \left(1.31552 + 0.690754\times 10^{-5}T\right) + \frac{\left(0.788404+0.295835\times10^{-4}T\right)\lambda^2}{\lambda^2-\left(0.0110199+0.584758\times10^{-6}T\right)} + \frac{\left(0.91316+0.548368\times10^{-6}T\right)\lambda^2}{\lambda^2-100} \quad (11.14)$$

where wavelength λ is in microns and T is the temperature in degrees Celsius. The coating material is gold.

Mainly, the working process of SPR-based photonic crystal sensor depends on the evanescent field [21]. To maximize the sensitivity and lessen the fabrication challenge are the purposes of this design. Here, a dual core is placed at the two opposite sides of the elliptical air hole. The advantage of a dual-core sensor over a single-core sensor is that the evanescent field can easily reach the metal surface and excite the electron effectively. So, to reduce the distance between the metal surface and evanescent field dual core is introduced. It is also noted that the small diameter of the air hole in the horizontal direction confirms the generation of SPW. The background material of the sensor is fused silica that is characterized by the Sllmeier equation [22]. The fabrication of the designed PCF sensor is done by the stack and draw technique [23]. Then, gold (plasmonic material) is deposited outside of the PCF by the chemical vapor deposition technique [24]. The plasmonic material gold is characterized by the Drude–Lorentz model [25].

$$\varepsilon(\omega) = \varepsilon_1 + i\varepsilon_2 = \varepsilon_\infty \frac{\omega_p^2}{\omega(\omega + i\omega_c)} \tag{11.15}$$

where ω_c is the collision frequency, ω_p is the plasma frequency, and ε_∞ is associated with the absorption peaks at high frequency ($\omega >> \omega_c$). For gold, we take ε_∞ =9.75, ω_0 = 1.3659 × 10^{16}, and ω_c = 1.45 × 10^{14}, which fits well into the experimental data in the literature [26].With temperature due to thermal expansion the plasma frequency varies and can be written as [27]

$$\omega_p = \omega_{p0} \times \exp\left(-\frac{T - T_0}{2} \times \alpha_v (T_0) \right) \tag{11.16}$$

where T_0 is the room temperature (273.15 K, 20°C), ω_{p0}, is the plasma frequency at T_0, And α_v is the thermal volume expansion coefficient of metal.

The collision frequency ω_c of bulk metal depends on two factors: one corresponding to the phonon-electron scattering (ω_{cp}) and the other to the electron–electron scattering (ω_{ce}). It is given by [28,29].

$$\omega_c = \omega_{cp} + \omega_{ce} \tag{11.17}$$

By the Lawrence model the electron–electron scattering frequency is given as [8]

$$\omega_{ce}(T) = \frac{1}{6}\pi^4 \frac{\Gamma\Delta}{hE_F}\left[(k_B T)^2 + \left(\frac{h\omega}{4\pi^2} \right) \right] \tag{11.18}$$

where E_F is the Fermi energy of metal electrons, h is the Planck's constant, and k_B is the Boltzmann constant. Γ and Δ are defined in [30].

Then ω_c can be modeled using the phonon–electron scattering model of Holstein [31] as

$$\omega_{cp}(T)=\omega_0[\frac{2}{5}+4\left(\frac{T}{T_D}\right)^5\int_0^{\frac{T_D}{T}}\frac{z^4dz}{e^z-1} \tag{11.19}$$

where T_D is Debye temperature, and T is the temperature in degrees Kelvin. Here ω_0 is a constant that can be evaluated from DC conductivity data. However, the values thus obtained do not fit the experimental data [32]. We use an approximate treatment similar to that in [33] in the following simulations. The value of ω_c is known in equation 11.15. Together with equations 11.17 and 11.18, we can calculate ω_{cp} at 298.15 K. Then ω_0 in equation 11.19 is determined, and ω_{cp} at other temperatures can be calculated.

Thus, the frequency and temperature dependence of RI of gold can be calculated by using equations 11.15–11.19. It should be noted that, for the calculation of thermal expansion of film, one should not use the linear thermal expansion coefficient ($\alpha_L=\alpha_V/3$) of the bulk material. Since the film may only be expanded into the normal direction, one has to employ a correct thermal expansion coefficient (α_L') for the expansion of the film thickness. The corresponding expression is

$$\alpha_L'=\alpha_L\frac{1+\mu}{1-\mu} \tag{11.20}$$

where μ is the Poisson number of the metal. The values of the metal parameters used for numerical simulation are given in Table 11.2.

To enhance the sensitivity for temperature-sensing applications, a dielectric material with a high-value thermo-optic coefficient (dn/dT) is needed in the second-layer holes. The sensing medium is considered to be a liquid and its RI is evaluated by

$$\eta=\eta_{\text{liquid}}+\left(\frac{dn}{dT}\right)(T-T_0) \tag{11.21}$$

where $dn/dt=-4\times10^{-4}(°C^{-1})$, η_{liquid} is RI of the liquid at the room temperature T_0 = 25°C By neglecting the material dispersion of the liquid and let us assume $\eta_{\text{liquid}}=1.35$ from the spectral wavelength from 500 to 1000 nm at 25°C.

11.5.3 Silica-bounded plasmonic mode and sensing medium bounded plasmonic mode

There exist two plasmonic modes for each mode order when a metallic layer is used to coat the dielectric. These two modes are known as silica-bounded plasmonic mode and sensing medium bounded plasmonic mode. The fundamental optical field distribution of the proposed temperature sensor is shown in Figure 11.36 for realizing the core guided mode and plasmonic mode. Figure 11.36a and b shows the core guided mode for x and y-polarization whereas Figure 11.36c and d shows the plasmonic mode for x and y-polarization. In this paper, the performance of the proposed sensor is analyzed both for x-polarization and y-polarization light. The phase matching occurs at a certain wavelength known as resonance wavelength. In this resonance wavelength, the energy of the core-guided mode is transferred to a plasmonic mode. A significant loss will be observed at this resonance wavelength as a plasmonic mode is highly lossy. When RIs of the sensors vary with temperature, the resonance wavelength between the core guided mode and plasmonic mode changes. Therefore, with the temperature change, the absorption peak can be shifted.

11.6 PART 6: PERFORMANCE ANALYSIS OF TEMPERATURE SENSOR

In this section, FEM is used to analyze the performance of the temperature sensor in the sensing ranges from 0°C to 80°C.

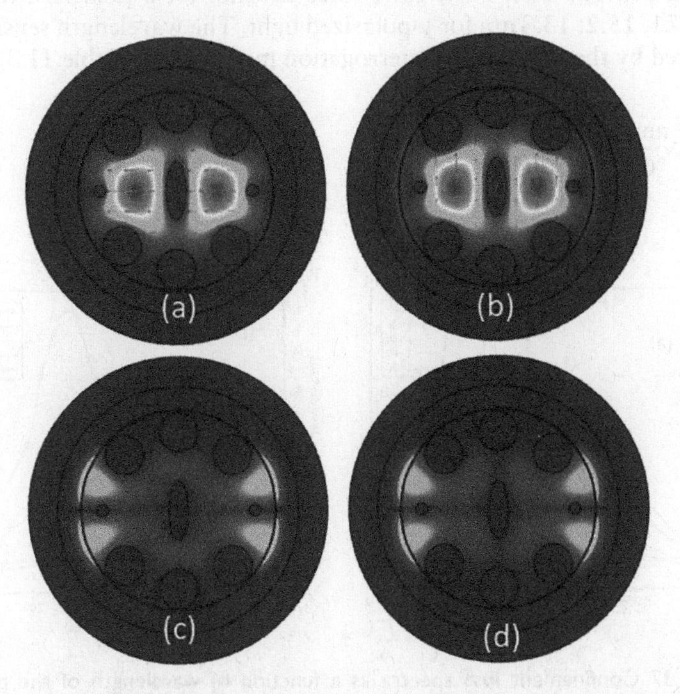

Figure 11.36 The electric field distribution of the temperature sensor for fundamental mode (a) x-polarization, (b) y-polarization and SPR modes, (c) x-polarization, and (d) y-polarization.

11.6.1 Confinement loss

The equation of confinement loss curve is depicted in [17].

$$A = 40\pi.\frac{Im(n_{eff})}{\ln(10)\lambda}$$

$$A \approx 8.686 \times k_o \cdot Im(n_{eff}) \times 10^4 \text{ dB/cm} \tag{11.22}$$

where n_{eff} is the imaginary part of effective RI and k_o is the free space wavenumber. The computer simulation results are shown in Figure 11.37 where confinement loss curves as a function of wavelength for a range of temperature are presented for the optimum parameters.

One can observe from Figure 11.37a and b that with the increasing temperature the loss depth is decreasing both for x-polarized and y-polarized light. The sharp loss peak is obtained at 669.2, 649.8, 634.1, 620.2 and 607.4.2 nm for x-polarized light and at 678.7.3, 657.2, 640.1, 624.9 and 611.6 nm for y-polarized light for the temperature of 0°C, 20°C, 40°C, 60°C, and 80°C, respectively. The wavelength difference between the two adjacent loss curves is 19.4, 15.7, 13.9, 12.8 nm for x-polarized light and 21.5, 17.1, 15.2, 13.3 nm for y-polarized light. The wavelength sensitivity is measured by the wavelength interrogation method [18] (Table 11.3),

$$S_\lambda\left[\frac{\text{nm}}{°\text{C}}\right] = \frac{d\lambda_{peak}(T)}{dT} \tag{11.23}$$

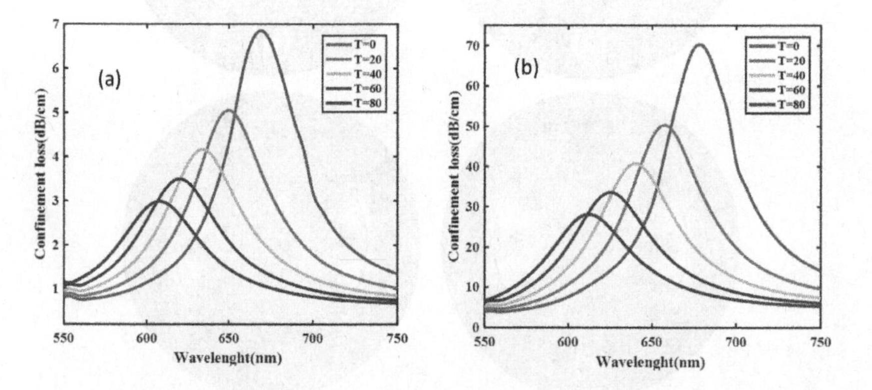

Figure 11.37 Confinement loss spectra as a function of wavelength of the proposed temperature sensor for a range of temperature (a) x-polarization and (b) y-polarization with optimum parameters.

Table 11.3 Performance analysis for different temperatures with the optimum parameters

Temperature T (°C)	*Polarization mode*	*Temperature difference (°C)*	*Max. resonance wavelength (nm)*	*Resonance peak shift (nm)*	*Wavelength sensitivity pm/°C*
0	y pol	20	678.7	21.5	1075
	x pol		669.2	19.4	970
20	y pol	20	657.2	17.1	855
	x pol		649.8	15.7	785
40	y pol	20	640.1	15.2	760
	x pol		634.1	13.9	695
60	y pol	20	624.9	13.3	665
	x pol		620.2	12.8	640
80	y pol	20	611.6	-	-
	x pol		607.4	-	-

11.6.2 Effect of variation of the designed parameters on sensor performance

Now if the structural parameter fluctuates from its optimum value during the fabrication process the effects of variation of parameters on the sensor performance will be observed in this section.

11.6.2.1 Variation of gold layer thickness

First, the effect of variation of thickness of a gold layer is carefully investigated by considering Figures 11.38 and 11.39 in terms of wavelength sensitivity for gold thickness 35 and 45 nm, respectively, while other parameters are kept constant at the optimum value. Figure 11.38a and b shows that the sharp loss peak is obtained at 645.9, 628.7, 614.3, 601.3, and 589.4 nm for x-polarized light and at 654.4, 635.9, 620.3, 607, and 594.9 nm for y-polarized light for the temperature of 0°C, 20°C, 40°C, 60°C, and 80°C, respectively. The wavelength difference between two adjacent loss curves is 17.9, 14.4, 13, 11.9 nm for x-polarized light and 18.5, 15.6, 13.3, 12.1 nm for y-polarized light. Using equation 11.23 the maximum wavelength sensitivity is 895 pm/°C for x-polarized light and 925 pm/°C for y-polarized light.

Figure 11.39a and b expresses similar information for x-polarized light and y-polarized light but 45 nm thickness. It is easily shown from Figure 11.39a and b that the resonance wavelength is obtained at 686.4, 666.2, 649.5, 634, and 621 nm for x-polarized light and at 694.9, 673.3, 655.2, 638.7, and 625.3 nm for y-polarized light for the temperature of 0°C, 20°C, 40°C, 60°C, and 80°C, respectively.

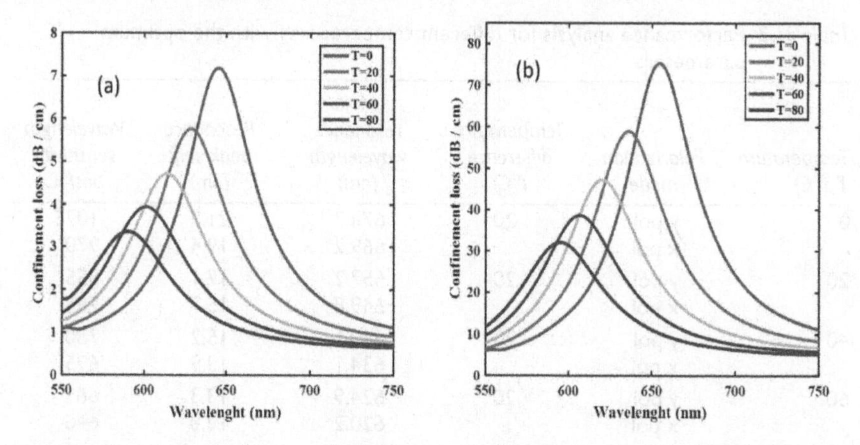

Figure 11.38 Confinement loss spectra as a function of wavelength of the proposed temperature sensor for a range of temperature (a) x-polarization and (b) y-polarization with gold layer thickness 35 nm.

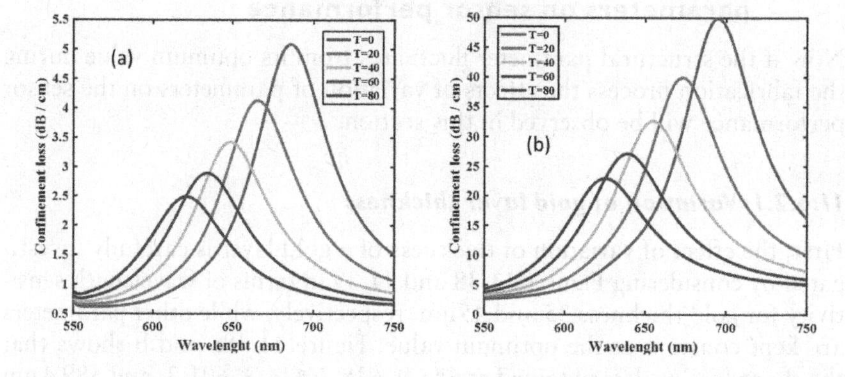

Figure 11.39 Confinement loss spectra as a function of wavelength of the proposed temperature sensor for a range of temperature (a) x-polarization and (b) y-polarization with gold layer thickness 45 nm.

When the wavelength difference between two adjacent loss curves is compared then the results of the comparison are 20.2, 16.7, 15.5, 13 nm for x-polarized light and 21.7, 18.1, 16.5, 13.4 nm for y-polarized light. By means of sensitivity equation 11.23, the obtained maximum wavelength sensitivity is 1,010 pm/°C for x-polarized light and 1,085 pm/°C for y-polarized light. Therefore, from the above simulation result, it is concluded that with the increasing gold thickness the sensitivity also increases. However, surface plasmonic waves are very sensitive to the gold layer. When the gold thickness is kept at 36, 40, and 45 nm, then the variation of the resonance peak of the proposed sensor is shown in Figure 11.40 for the

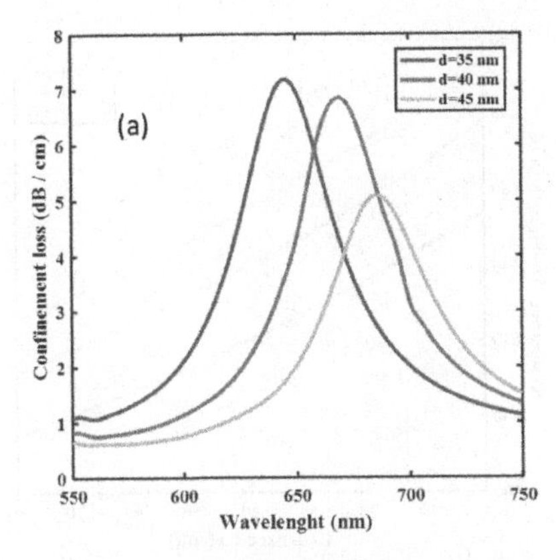

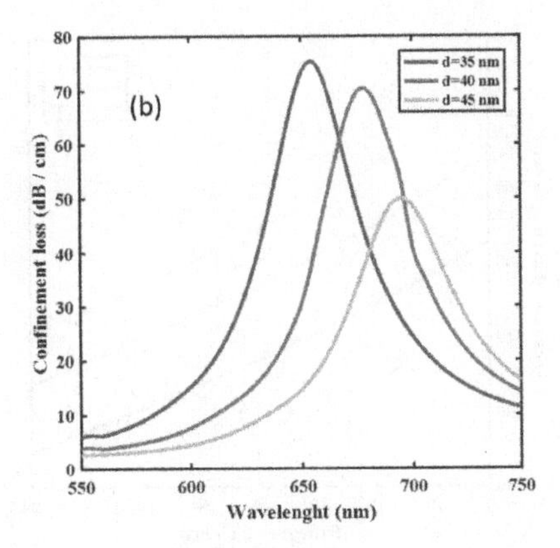

Figure 11.40 Effect of gold layer thickness on the sensitivity of the proposed sensor (a) *x*-polarization and (b) *y*-polarization.

temperature of 0°C. The resonance peak shifts to a larger wavelength with the increase of the thickness of the gold. When the gold thickness is thicker than 40 nm then a lower resonance peak is observed with an increasing curve width. On the other hand, if the thickness is thinner than 40 nm then a higher resonance peak is obtained with narrower curve width as shown in Figure 11.40. There is a problem of penetration for an electric field if the gold thickness is too much thicker. Moreover, the loss of the sensor is very high if the thickness is too low. Therefore, the optimal thickness of gold for better performance of the sensor is chosen at 40 nm.

However, the variation of the plasmonic peak due to the variation of temperature for the sensor is relatively stable even with other gold layer thicknesses, as depicted in Figure 11.41a and b for *x*-polarized light and *y*-polarized light, respectively. The figures show that temperature resonance wavelength curves are nearly linear.

11.6.2.2 Variation of diameter of the air hole

In this section, the effect of variation of small diameter d_1 is considered as in Figure 11.42a and b for *x*-polarization and *y*-polarization light, respectively, when the diameter d_1 is set at 0.6 μm and 1 μm while other parameters are kept constant. One can observe from Figure 11.42a that for *x*-polarization light the maximum sensitivity is 945 pm/°C and for *y*-polarization light this value is increased to 1,010 pm/°C as seen from Figure 11.42b. Therefore, it is decided that with the increasing diameter d_1 of the air hole of the PCF the maximum sensitivity is decreased linearly. Moreover, the larger air hole reduces the chance of light confinement in the core region. Therefore, there be an optimal value. The optimum value for diameter d_1 is kept at 0.6 μm for this particular design.

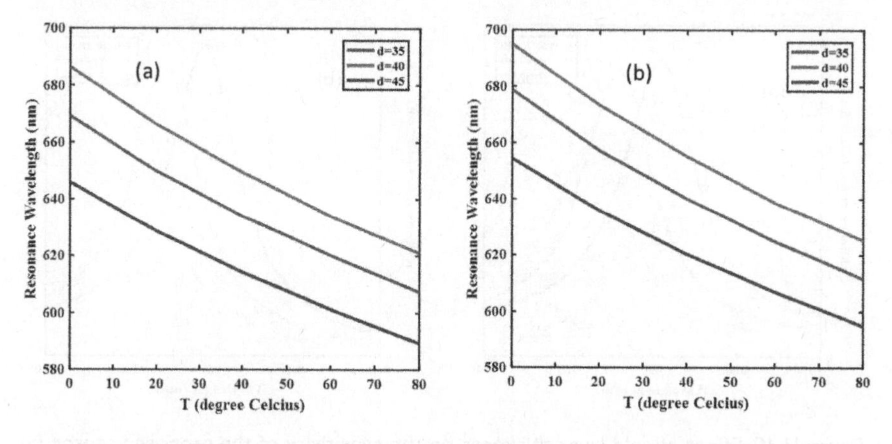

Figure 11.41 Stability test for various gold thicknesses on sensor performance (a) *x*-polarization and (b) *y*-polarization.

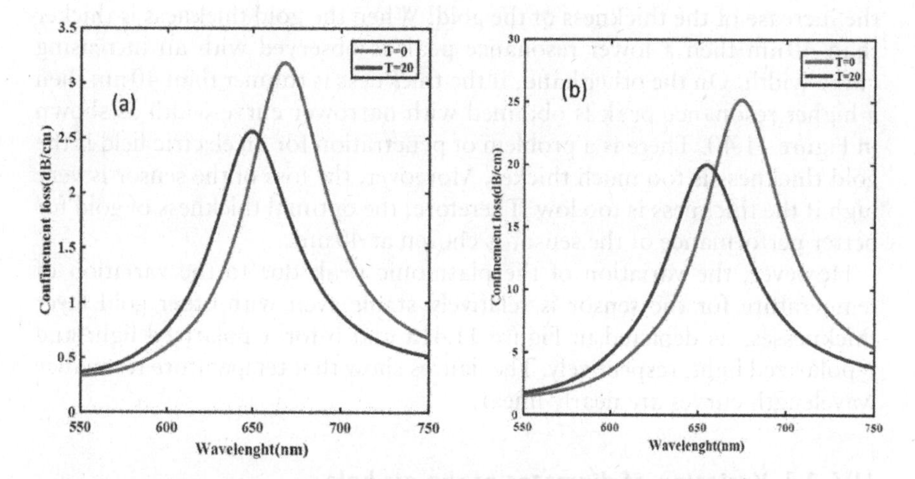

Figure 11.42 Effect of air hole diameter on the sensitivity of the proposed sensor (a) *x*-polarization and (b) *y*-polarization.

11.7 PART 7: CONCLUSION AND FUTURE WORK

11.7.1 Conclusion of the work

In this work, the properties of silicon nanowire and silica nanowire have been presented using software based on FEM. Due to the presence of all six components, the nanowires having small radii are fully vectorial. Hence, the modal solutions of the optical waveguides were approached with a fully vectorial formulation. Studies of characteristics of both silicon nanowire and silica nanowire are presented in the work. These characteristics include variations of effective indices, effective area, dispersion, power confinement, etc. The study shows while all three components of *H*-field are continuous across the dielectric interfaces, the electric field components are not. It has further been shown in this work that the modes in such nanowires with strong index contrast have hybridness or show the presence of the nondominant orthogonal polarization. It has been noted in the case of waveguides with small dimensions that the dispersion properties can be greatly controlled for various linear and nonlinear applications. The nonlinear properties of silicon nanowires have also been studied in this work. Afterward, a detailed study of optical sensing of the nanowires has been presented. A Mach–Zehnder-based optical sensor has been studied for this part. A phase shift has been noted due to the presence of a sensing arm and a reference arm in the sensor. Assuming the specimen to be polystreptavidin, characteristics such as propagation constant, propagation constant difference, and power fraction have been studied and plotted against variations of core diameters. The sensitivity of the optical sensor has been studied against variations of the molar concentration for two different radii of the nanowire range. The concept of "the thinner the wire, the higher the sensitivity" has been presented in this work. Finally, the study has suggested that optical sensing with nanowires may set forth a new approach to miniaturized optical sensors with high sensitivity.

In brief, a dual core structured with gold-coated outside of the PCF temperature sensor using SPR has been proposed and evaluated by FEM. The computational results indicate that the obtained wavelength sensitivity of the proposed sensor is 970 pm/°C for *x*-polarization and 1,075 pm/°C for *y*-polarization mode. The detection range of the temperature sensor is high about 0°C–80°C. Furthermore, plasmonic material and analytes are used in the outmost of the PCF that reduces the operational complexity of the proposed structure. Considering high sensitivity and wide detection range of temperature, this dual-core PCF temperature sensor can be a potential candidate for monitoring temperature in the medical, manufacturing industry, environmental monitoring, and transformer oil.

11.7.2 Future scope of work

Nanofibers have a huge scope in the coming generations, and there is always room for improvement in every nanowire-related work. Many discoveries are expected in the short run based on the uniqueness of nanowire building blocks. The nanowire technology can be used for applications in the three different effects such as the self-phase modulation, cross-phase modulation, and four-wave mixing. At a high power level, nanowire technology can be further developed for stimulating effects such as the stimulated Brillouin scattering and stimulated Raman scattering. The sensitivity parameter of nanowires desires many opportunities to be enhanced. The reduced footprint of the nanowire sensor may allow sensing in an environment of a smaller scale, support integration of sensor array with higher density, and require fewer samples. But in practice, to realize a nanowire sensor modeled in this work, some challenges such as instability of the nanowire-assembled couplers and oversensitivity of the whole system have to be overcome, which require further experimental investigation in the future. The optical sensing part of the work can be made further refined in the future by studying with sensors made up of different specimens to designate the best among them. Other properties of nanowire as biosensors can also be explored in future works. Studies relating to the improvement of performance and stability of nanowires while making them more cost-effective will go a long way in the growth of nanowire technology.

REFERENCES

1. M. Abbasi, M. Soroosh, E. Namjoo, "Polarization-insensitive temperature sensor based on liquid filled photonic crystal fiber," *Optik* vol. 168 (2018) pp. 342–347.
2. J. I. A. Rashid, J. Abdullah, N. A. Yusof, and R. Hajian, "The development of silicon nanowire as sensing material and its applications," *Journal of Nanomaterials*, vol. 213 (2003) Article ID 328093,p. 16.
3. G. Tian, K. Pan, Y. Chen et al., "Vertically aligned anatase TiO_2 nanowire bundle arrays: Use as Pt support for counter electrodes in dye-sensitized solar cells," *Journal of Power Sources*, vol. 238 (2013) pp. 350–355.
4. F. Shahdost-fard, A. Salimi, E. Sharifi, and A. Korani, "Fabrication of a highly sensitive adenosine aptasensor based on covalent attachment of aptamer on to chitosan-carbon nanotubes-ionic liquid nanocomposite," *Biosensors and Bioelectronics*, vol. 48 (2013) pp. 100–107.
5. L. Qian, J. Mao, X. Tian, H. Yuan, and D. Xiao, "In situ synthesis of CuS nanotubes on Cu electrode for sensitive nonenzymatic glucose sensor," *Sensors and Actuators B*, vol. 176 (2013) pp. 952–959.
6. Y. Ding, Y. Liu, J. Parisi, L. Zhang, and Y. Lei, "A novel NiO-Au hybrid nanobelts based sensor for sensitive and selective glucose detection," *Biosensors and Bioelectronics*, vol. 28, no. 1, p. 393
7. "Based sensor for sensitive and selective glucose detection," *Biosensors and Bioelectronics*, vol. 28, no. 1, p. 393
8. Y. Sun, S. H. Yang, L. P. Lv et al., "A composite film of reduced graphene oxide modified vanadium oxide nanoribbons as a free standing cathode material for rechargeable lithium batteries," *Journal of Power Sources*, vol. 241, p. 168.
9. J. Wu, Sh Li, M. Shi, X. Feng, "Photonic crystal fiber temperature sensor with high sensitivity based on surface plasmon resonance," *Optical Fiber Technology*, vol. 43, p. 109.

10. Q. Liu, S. Li, H. Chen, Z. Fan, J. Li, "Photonic crystal fiber temperature sensor based on coupling between liquid-core mode and defect mode," *IEEE Photonics Journal*, vol. 7, p. 1.
11. E. R. Vera, C. M. B. Cordeiro, P. Torres, "Highly sensitive temperature sensor using a Sagnac loop interfereometer based on a side-hole photonic crystal fiber filled with metal," *Applied Optics*, vol. 56 (2017) pp. 156–162.
12. X. Li, Y. Zhao, X. Zhou, L. Cai, "High sensitivity all-fiber Sagnac Interferometer temperature sensor using a selective ethanol-filled photonic crystal fiber," *Instrumentation Science & Technology*, 2017.
13. Y. Zhao, Z. Q. Deng, and H. F. Hu, "Fiber-optic SPR sensor for temperature measurement," *IEEE Transactions on Instrumentation and Measurement*, vol. 64 (2015) pp. 3099–3104.
14. S. J. Qiu, Y. Chen, F. Xu, and Y. Q. Lu, "Temperature sensor based on an isopropanol-sealed photonic crystal fiber in-line interferometer with enhanced refractive index sensitivity," *Optics Letter*, vol. 37 (2012) p. 863.
15. G.P. Nikishkov, Introduction to the Finite Element Method [Online].
16. Available: http://citeseerx.ist.psu.edu/viewdoc/download?doi=10.1.1.521.9293 & rep=rep 1 & type=pdf.
17. Available: http://web.mit.edu/16.810/www/16.810_L4_CAE.pdf.
18. Available: https://en.wikipedia.org/wiki/COMSOL_Multiphysics.
19. Available: https://www.rpphotonics.com/effective_refractive_index.html.
20. N. Kejalakshmy, A. Agrawal, Y. Aden, D. M. H. Leung, B. M. A. Rahman, K. T. V. Grattan, "Characterization of silicon nanowire by use of full-vectorial finite element method," *Applied Optics*, vol. 49, no. 16 (2010) pp. 3173–3181.
21. Y. Zhao, Z. Q. Deng, and H. F. Hu, "Fiber-optic SPR sensor for temperature measurement," *IEEE Transactions on Instrumentation*, vol. 64, no. 11 (2015) p. 3099.
22. G. Ghosh, M. Endo, and T. Iwasaki, "Temperature-dependent Sellmeier coefficients and chromatic dispersions for some optical fiber glasses," *Journal of Lightwave Technology*, vol. 12 (1994) pp. 1338–1342.
23. A. Rifat et al., "Photonic crystal fiber based plasmonic sensors," *Sensors and Actuators B: Chemical*, vol. 243 (May 2017) pp. 311–325.
24. Q. Liu, Sh. Li, H. Chen, Zh. Fan, J. Li, "Photonic crystal fiber temperature sensor based on coupling between liquid-core mode and defect mode," *IEEE Photonics Journal*, vol.7 (2015) pp. 450–509.
25. E. F. Chillcce, C. M. B. Cordeiro, L. C. Barbosa, and C. H. B. Cruz, "Telluritephotonic crystal fiber made by a stack-and-draw technique," *Journal of Non-Crystalline Solids*, vol. 352, no. 32 (2006) pp. 3423–3428.
26. A. Rifat, R. Ahmed, G. A. Mahdiraji, and F. M. Adikan, "Highly sensitive d- shaped photonic crystal fiber-based plasmonic biosensor in visible to near-IR," *IEEE Sensors Journal*, vol. 17, no. 9 (Jun. 2017) pp. 2776–2783.
27. Q. Liu, S. G. Li, H. L. Chen, J. S. Li, and Z. K. Fan, "High-sensitivity Plasmonic temperature sensor based on photonic crystal fiber coated with nanoscale gold film," *Applied Physics Express*, vol. 8, no. 4 (Mar. 2015).
28. P. B. Johnson and R. W. Christy, "Optical constants of the noble metals," *Physical Review B* vol. 6 (1972) pp. 4370–437.
29. K. Lin, Y. Lu, Z. Luo, R. Zheng, P. Wang, and H. Ming, "Numerical and experimental investigation of temperature effects on the surface plasmon resonance sensor," *Chinese Optics Letters*, vol.7 (2009) pp. 428–431.

30. K. Sharma and B. D. Gupta, "Theoretical model of a fiber optic remote sensor based on surface plasmon resonance for temperature detection," *Optical Fiber Technology*, vol. 12 (2006) pp. 87–100.
31. R. Beach and R. Christy, "Electron-electron scattering in the intra band optical conductivity of Cu, Ag, and Au," *Physical Review*, vol. 16 (1977) pp. 52–77.
32. W. E. Lawrence, "Electron-electron scattering in the low temperature resistivity of the nobel metals," *Physical Review B*, vol. 13 (1976) pp. 5316–5319.
33. T. Holstein, "Optical and infrared volume absorptivity of metals," *Physical Review B*, vol. 96 (1972) p. 535.
34. K. Ujihara, "Reflectivity of metals at high temperatures," *Journal of Applied Physics*, vol. 43 (1972) pp. 2376–2383.
35. S. Herminghaus and P. Leiderer, "Surface plasmon enhanced transient thermo reflectance," *Applied Physics*, vol. 51 (1990) pp. 350–353.

Index

Taylor & Francis eBooks

www.taylorfrancis.com

A single destination for eBooks from Taylor & Francis with increased functionality and an improved user experience to meet the needs of our customers.

90,000+ eBooks of award-winning academic content in Humanities, Social Science, Science, Technology, Engineering, and Medical written by a global network of editors and authors.

TAYLOR & FRANCIS EBOOKS OFFERS:

A streamlined experience for our library customers

A single point of discovery for all of our eBook content

Improved search and discovery of content at both book and chapter level

REQUEST A FREE TRIAL

support@taylorfrancis.com